Carbon, Nitrogen and Phosphorus Cycling in Forest Soils

Carbon, Nitrogen and Phosphorus Cycling in Forest Soils

Special Issue Editor

Robert G. Qualls

MDPI • Basel • Beijing • Wuhan • Barcelona • Belgrade

MDPI

Special Issue Editor
Robert G. Qualls
University of Nevada
USA

Editorial Office
MDPI
St. Alban-Anlage 66
4052 Basel, Switzerland

This is a reprint of articles from the Special Issue published online in the open access journal *Forests* (ISSN 1999-4907) from 2017 to 2018 (available at: https://www.mdpi.com/journal/forests/special_issues/CNP_soil)

For citation purposes, cite each article independently as indicated on the article page online and as indicated below:

LastName, A.A.; LastName, B.B.; LastName, C.C. Article Title. *Journal Name* **Year**, *Article Number*, Page Range.

ISBN 978-3-03897-682-0 (Pbk)
ISBN 978-3-03897-683-7 (PDF)

Cover image courtesy of Ann E. Russell.

Contents

About the Special Issue Editor . . . . **vii**

Preface to "Carbon, Nitrogen and Phosphorus Cycling in Forest Soils" . . . . **ix**

Ann E. Russell, Stephanie N. Kivlin and Christine V. Hawkes
Tropical Tree Species Effects on Soil pH and Biotic Factors and the Consequences for
Macroaggregate Dynamics
Reprinted from: *Forests* **2018**, *9*, 184, doi:10.3390/f9040184 . . . . **1**

Zhan Xiaoyun, Guo Minghang and Zhang Tibin
Joint Control of Net Primary Productivity by Climate and Soil Nitrogen in the Forests of
Eastern China
Reprinted from: *Forests* **2018**, *9*, 322, doi:10.3390/f9060322 . . . . **15**

Zachary W. Carter, Benjamin W. Sullivan, Robert G. Qualls, Robert R. Blank,
Casey A. Schmidt and Paul S.J. Verburg
Charcoal Increases Microbial Activity in Eastern Sierra Nevada Forest Soils
Reprinted from: *Forests* **2018**, *9*, 93, doi:10.3390/f9020093 . . . . **28**

Ewa Błońska and Jarosław Lasota
Soil Organic Matter Accumulation and Carbon Fractions along a Moisture Gradient of
Forest Soils
Reprinted from: *Forests* **2017**, *8*, 448, doi:10.3390/f8110448 . . . . **44**

Daniel H. Howard, John T. Van Stan, Ansley Whitetree, Lixin Zhu and Aron Stubbins
Interstorm Variability in the Biolability of Tree-Derived Dissolved Organic Matter (Tree-DOM)
in Throughfall and Stemflow
Reprinted from: *Forests* **2018**, *9*, 236, doi:10.3390/f9050236 . . . . **57**

Bharat M. Shrestha, Scott X. Chang, Edward W. Bork and Cameron N. Carlyle
Enrichment Planting and Soil Amendments Enhance Carbon Sequestration and Reduce
Greenhouse Gas Emissions in Agroforestry Systems: A Review
Reprinted from: *Forests* **2018**, *9*, 369, doi:10.3390/f9060369 . . . . **67**

Angang Ming, Yujing Yang, Shirong Liu, Hui Wang, Yuanfa Li, Hua Li, You Nong, Daoxiong
Cai, Hongyan Jia, Yi Tao and Dongjing Sun
Effects of Near Natural Forest Management on Soil Greenhouse Gas Flux in *Pinus massoniana*
(Lamb.) and *Cunninghamia lanceolata* (Lamb.) Hook. Plantations
Reprinted from: *Forests* **2018**, *9*, 229, doi:10.3390/f9050229 . . . . **85**

Tomohiro Yokobe, Fujio Hyodo and Naoko Tokuchi
Seasonal Effects on Microbial Community Structure and Nitrogen Dynamics in Temperate
Forest Soil
Reprinted from: *Forests* **2018**, *9*, 153, doi:10.3390/f9030153 . . . . **99**

Antonietta Fioretto, Michele Innangi, Anna De Marco, Cristina Menta, Stefania Papa,
Antonella Pellegrino and Amalia Virzo De Santo
Discriminating between Seasonal and Chemical Variation in Extracellular Enzyme Activities
within Two Italian Beech Forests by Means of Multilevel Models
Reprinted from: *Forests* **2018**, *9*, 219, doi:10.3390/f9040219 . . . . **116**

Qingshui Ren, Hong Song, Zhongxun Yuan, Xilu Ni and Changxiao Li
Changes in Soil Enzyme Activities and Microbial Biomass after Revegetation in the Three
Gorges Reservoir, China
Reprinted from: *Forests* **2018**, *9*, 249, doi:10.3390/f9050249 . . . . **134**

Fanpeng Zeng, Xin Chen, Bin Huang and Guangyu Chi
Distribution Changes of Phosphorus in Soil–Plant Systems of Larch Plantations across
the Chronosequence
Reprinted from: *Forests* **2018**, *9*, 563, doi:10.3390/f9090563 . . . . **147**

**Li Zhang, Ao Wang, Fuzhong Wu, Zhenfeng Xu, Bo Tan, Yang Liu, Yulian Yang,
Lianghua Chen and Wanqin Yang**
Soil Nitrogen Responses to Soil Core Transplanting Along an Altitudinal Gradient in an Eastern
Tibetan Forest
Reprinted from: *Forests* **2018**, *9*, 239, doi:10.3390/f9050239 . . . . **158**

**Yo-Jin Shiau, Chung-Wen Pai, Jeng-Wei Tsai, Wen-Cheng Liu, Rita S. W. Yam,
Shih-Chieh Chang, Sen-Lin Tang and Chih-Yu Chiu**
Characterization of Phosphorus in a Toposequence of Subtropical Perhumid Forest Soils Facing
a Subalpine Lake
Reprinted from: *Forests* **2018**, *9*, 294, doi:10.3390/f9060294 . . . . **170**

Shusheng Yuan, Tongtong Tang, Minchao Wang, Hao Chen, Aihua Zhang and Jinghua Yu
Regional Scale Determinants of Nutrient Content of Soil in a Cold-Temperate Forest
Reprinted from: *Forests* **2018**, *9*, 177, doi:10.3390/f9040177 . . . . **184**

Xiaofeng Zheng, Jie Yuan, Tong Zhang, Fan Hao, Shibu Jose and Shuoxin Zhang
Soil Degradation and the Decline of Available Nitrogen and Phosphorus in Soils of the Main
Forest Types in the Qinling Mountains of China
Reprinted from: *Forests* **2017**, *8*, 460, doi:10.3390/f8110460 . . . . **196**

**Natasha M. I. Godoi, Sabrina N. dos S. Araújo, Salatiér Buzetti, Rodolfo de N. Gazola,
Thiago de S. Celestrino, Alexandre C. da Silva, Thiago A. R. Nogueira and
Marcelo C. M. Teixeira Filho**
Soil Chemical Attributes, Biometric Characteristics, and Concentrations of N and P in Leaves
and Litter Affected by Fertilization and the Number of Sprouts per the *Eucalyptus* L'Hér. Strain
in the Brazilian Cerrado
Reprinted from: *Forests* **2018**, *9*, 290, doi:10.3390/f9060290 . . . . **205**

About the Special Issue Editor

Robert G. Qualls is Professor Emeritus at the Department of Natural Resources and Environmental Science at the University of Nevada, USA. He received a B.Sc. in Biology, with Honors in Biology and Honors in Creative Writing at the University of North Carolina. He received an M.S.P.H. in Environmental Science and Engineering from the University of North Carolina and a Ph.D. in Ecology from the University of Georgia, doing research at the Coweeta Hydrologic Laboratory. He was later an Assistant Research Professor at Duke University, working on biogeochemical cycles in the Everglades of Florida. He has taught courses in Microbial Ecology, Wetland Ecology and Management, Forest and Range Soils, Soil Genesis and Classification, Natural Resource Ecology, Biodiversity, Conservation and Humans, and Ecology of Flowing Waters. After some early research work in ultraviolet light disinfection of bacteria and viruses, and kinetics of reactions of oxidizing chlorine with humic substances, his research work has centered on biogeochemistry of forests, wetlands and streams. This work has included dissolved organic matter dynamics, primary succession and soil organic matter formation, root production during succession, formation and decomposition of humic substances, protein content in dissolved organic matter, bioavailability of N and P in rivers and lakes, and ecophysiology of invasive plants. Dr. Qualls has also been a Visiting Professor at Yokohama National University in Japan. He received the "Pioneer of Disinfection Award" from the International Water Association and the Water Environment Association. He currently serves as Editor for the journal *Forests*. Outside the academic arena, he was also a member of the USA team at the World Masters Track and Field Championship in 2010 and 2016 in the 5000 m, 1000 m runs and the 4x800m relay.

Preface to "Carbon, Nitrogen and Phosphorus Cycling in Forest Soils"

The majority of carbon stored in the soils of the world is stored in forests. The refractory nature of some portions of forest soil organic matter also provides the slow, gradual release of organic nitrogen and phosphorus to sustain long term forest productivity. Contemporary and future disturbances, such as climatic warming, deforestation, short rotation sylviculture, the invasion of exotic species, and fire, all place strains on the integrity of this homeostatic system of C, N, and P cycling. On the other hand, the CO_2 fertilization effect may partially offset losses of soil organic matter, but many have questioned the ability of N and P stocks to sustain the CO_2 fertilization effect.

Despite many advances in the understanding of C, N, and P cycling in forest soils, many questions remain. For example, no complete inventory of the myriad structural formulae of soil organic N and P has ever been made. The factors that cause the resistance of soil organic matter to mineralization are still hotly debated. Is it possible to "engineer" forest soil organic matter so that it sequesters even more C? The role of microbial species diversity in forest C, N, and P cycling is poorly understood. The difficulty in measuring the contribution of roots to soil organic C, N, and P makes its contribution uncertain. Finally, global differences in climate, soils, and species make the extrapolation of any one important study difficult to extrapolate to forest soils worldwide.

In the emerging literature, topics of special current interest in the study of forest soil C, N, and P cycling include subjects such as:

- forest soil C stocks and climate change,
- ability of soil N and P mineralization to sustain increased productivity due to CO_2 fertilization,
- causes of recalcitrance in soil organic matter mineralization,
- contribution of roots to soil C and N,
- methane production and oxidation in forest soils,
- soil C, N, and P during forest succession,
- effects of invasive species, forest management practices, or fire on C, N, and P cycling,
- the effect of biochar (charcoal) in forest soils
- roles of microbes and soil fauna on C, N, and P cycling, e.g. mycorrhizal fungi
- stable isotope studies of C and N cycling,
- new methods for the study of C, N, and P cycling.

The 16 papers in this book cover a geographically diverse range of forest ecosystems on four continents. The studies also cover a range of forest types: tropical rainforest, tropical savannah, subtropical mixed deciduous and needleleaf forest, temperate broadleaf and needleleaf forests, subalpine and alpine needleleaf forests, boreal needleleaf forest, as well as tropical and temperate forest plantations. The chapters are arranged by subject in the following order, those concerned largely with carbon cycling (including net primary productivity, soil organic matter, and greenhouse gasses), microbial ecology (community composition and enzyme activity), and N or P cycling.

Robert G. Qualls

Special Issue Editor

forests

MDPI

Article

Tropical Tree Species Effects on Soil pH and Biotic Factors and the Consequences for Macroaggregate Dynamics

Ann E. Russell [1,*], Stephanie N. Kivlin [2] and Christine V. Hawkes [3]

[1] Department of Natural Resource Ecology & Management, Iowa State University, Ames, IA 50011, USA
[2] Department of Ecology and Evolutionary Biology, University of Tennessee, Knoxville, TN 37996, USA; skivlin@utk.edu
[3] Department of Integrative Biology, University of Texas at Austin, Austin, TX 78712, USA; chawkes@austin.utexas.edu
*   Correspondence: arussell@iastate.edu; Tel.: +1-515-294-5612

Received: 9 March 2018; Accepted: 2 April 2018; Published: 4 April 2018

Abstract: Physicochemical and biotic factors influence the binding and dispersivity of soil particles, and thus control soil macroaggregate formation and stability. Although soil pH influences dispersivity, it is usually relatively constant within a site, and thus not considered a driver of aggregation dynamics. However, land-use change that results in shifts in tree-species composition can result in alteration of soil pH, owing to species-specific traits, e.g., support of nitrogen fixation and Al accumulation. In a long-term, randomized complete block experiment in which climate, soil type, and previous land-use history were similar, we evaluated effects of individual native tropical tree species on water-stable macroaggregate size distributions in an Oxisol. We conducted this study at La Selva Biological Station in Costa Rica, in six vegetation types: 25-year-old plantations of four tree species grown in monodominant stands; an unplanted Control; and an adjacent mature forest. Tree species significantly influenced aggregate proportions in smaller size classes (0.25–1.0 mm), which were correlated with fine-root growth and litterfall. Tree species altered soil pH differentially. Across all vegetation types, the proportion of smaller macroaggregates declined significantly as soil pH increased ($p \leq 0.0184$). This suggests that alteration of pH influences dispersivity, and thus macroaggregate dynamics, thereby playing a role in soil C, N, and P cycling.

Keywords: soil structure; soil pH; Oxisol; variable-charge soils; aluminum accumulator

1. Introduction

Soil structure is a 'master integrating variable' that is often linked to nutrient cycling because it is associated with so many soil properties, including water-holding capacity and stabilization of soil organic C (SOC) [1]. Soil structure can be characterized by the relationships among soil aggregates, which are defined as relatively discrete clusters of particles. Small aggregates may themselves be clustered into larger aggregates [2,3]. Aggregates in soils range in size from microns to millimeters in diameter, with macroaggregates generally defined as 0.25 to <8 mm. Aggregation dynamics are controlled by factors that affect binding/cementation, flocculation/dispersion, and arrangement of soil particles [4].

Cementing and binding agents can be of mineralogical or biotic origin. Highly weathered soils that are rich in Fe and Al oxides, e.g., Oxisols, allophanic soils, and Ultisols have very stable microaggregates [5–7]. Most interpret this as evidence that oxides and hydroxides of Al and Fe, as well as amorphous aluminosilicates, are the dominant stabilizing agents in these soils. In soils derived from volcanic mudflows, solid silt- to sand-size mineral grains appeared to provide the nuclei for aggregate development [8]. Biotic binding agents include fine roots, fungi, bacteria, and soil fauna. Microbial polysaccharides were important in stabilizing microaggregates (1–20 μm), while plant detritus served as nuclei for development of larger aggregates (20–300 μm), according to studies of a Mollisol and an Alfisol [9]. Compared with temperate-zone soils, aggregation in highly weathered tropical soils is expected to be controlled relatively more by mineralogical cementing [10].

Factors that disperse or flocculate soil clays, notably soil pH, also influence aggregate formation and stabilization [11]. Most conceptual frameworks of aggregate dynamics do not explicitly include this mechanism, however [1,10,12,13]. The inherent assumption is that site-level soil pH is relatively constant, and thus alteration of soil pH is not included as a driver of aggregation within a site that does not receive applications such as lime. This assumption of a relatively constant soil pH within a site may not apply under land-use change, especially where shifts in species composition occur. Given that plant species, especially trees, can alter soil pH [14–16], changes in species composition could influence aggregation via alteration of soil pH. Thus, including the effect of plant species on soil pH would improve our conceptual framework of aggregation dynamics.

The objectives of this study were to determine the extent to which vegetation type influenced macroaggregate structure in an Oxisol, and to evaluate the influence of binding agents, both mineralogical and biological, and alteration of pH on soil aggregation. In a unique tropical experimental field setting in which soil type, climate, forest age and previous land-use history were similar across plots [17], we assayed macroaggregate structure in six vegetation types. These included: four tree species, each grown in 25-year-old monodominant plantations; an unplanted Control that regenerated naturally; and a mature forest. All four of the tree species are broad-leaved, evergreen, and native, but differ in multiple traits that influence detrital inputs and decomposition [18–20]. One species, *Pentaclethra macroloba* (Willd.) Kuntze, is a nodulated legume, known to acidify soil at this site. Another species, *Vochysia guatemalensis* Donn. Smith, is an Al-accumulator, known to increase the soil pH [16].

We evaluated three hypotheses:

1. Macroaggregate size distribution and chemistry differ among the six vegetation types. The four species in plantations and the Control plots would have started with similar soil structure at the beginning of this experiment, 25 years prior to our study. Thus, observed differences among vegetation types in the chemistry and quantity of organic matter (OM) inputs [18,20] would influence soil aggregation, depending on the relative importance of their binding material in this soil. A lack of difference in macroaggregate structure would indicate over-riding control by mineralogical interactions.

2. Macroaggregate structure is correlated with fine-root growth, litterfall, fungal, and microbial effects on aggregation. Fine-root and litterfall additions are expected to bind aggregates and thus contribute to their stabilization via multiple chemical and physical mechanisms [11]. Fungal hyphae can reorient clay particles, bind particles with extracellular polysaccharides, and enmesh particles, thereby influencing soil aggregation [21] (Ternan et al., 1996). Mycorrhizal fungi can act at multiple scales, from the plant community to the soil mycelium, to influence soil aggregation [22]. In addition, microbial biomass C, rather than fungal hyphae alone, is often associated with factors that stabilize soil organic C (SOC), and could thus affect soil structure [23]. Microbial activity, however, differs temporally, and with aggregate size, soil type, cropping system and management [24,25].

3. Macroaggregate structure is correlated with soil pH. The tree species in this experiment have altered soil pH [16,26] and this could alter aggregation processes. Previously, it was hypothesized that in these variable-charge soils, changes in soil pH above or below the point of zero charge (PZC) would disperse colloids [16]. This effect on colloidal stability could thus influence macroaggregate stability.

2. Materials and Methods

2.1. Study Site and Experimental Design

This study was conducted in a tropical rainforest at La Selva Biological Station (10°26' N, 83°59' W) in Costa Rica. The mean annual temperature is 25.8 °C, and the mean annual rainfall is 4000 mm, with rainfall averaging >100 mm in any month [27]. The parent material is considered to be weathered andesitic/basaltic Pleistocene lava flows [28]. These soils may have received more recent inputs, ash or lahars, from nearby volcanoes [29]. The soil was classified as Mixed Haplic Haploperox [30]. Mean soil C and N concentrations in the surface (0–15 cm) layer range from 44 to 55 and 3.4 to 4.2 g/kg, respectively [26]. The point of zero charge (PZC) in these variable-charge soils has not been measured, but a reasonable assumption is that the PZC is ≤4, as it is in many comparable Oxisols [31,32].

The study site was deforested in ~1955 and pastured until abandonment in 1987 [20]. In 1988, a randomized complete block experiment containing four blocks was initiated. The experiment includes four species: *Hieronyma alchorneoides* L., *Pentaclethra macroloba*, a nodulated legume, *Virola koschnyi* Warb., and *Vochysia guatemalensis*, an Al-accumulator (see Russell et al. [20] for a more complete description). The site is hilly, with elevation ranging from 44 to 89 m. To ensure that topographic effects did not create a bias, each block was centered on a hilltop, and randomization of plot assignment to a species was stratified such that across the four blocks, each species was represented in each topographic position (hilltop, slope and slope bottom). In Block 3 of *Vochysia*, a stand-level lightning event killed nearly 75% of the trees in 2011, so that plot was not used for this study. Each plot was 50 × 50 m (0.25 ha) and divided into four quadrants. Trees were planted at a spacing of 3 × 3 m. For the first three years, understory vegetation was cleared manually. Trees were thinned by 50% at age four in *Hieronyma* and *Vochysia*, the fast-growing species. There was no other management of the plots, except for trail maintenance. For more information about the original design, see Russell et al. [20].

Two types of reference vegetation were also included to provide a basis of comparison for the four planted tree species. (1) An unplanted Control was contained within the original experimental design; (2) In the mature forest, we established a fifth block (150 × 200 m in size). Situated <150 m from the experimental plots, soil, climate, and vegetation prior to 1955 were similar to the adjacent experiment. We sampled four 50 m × 50 m plots randomly selected from within this block. *Pentaclethra* is the dominant species in La Selva mature forest, accounting for 36–38% of the estimated aboveground biomass of trees [33]. Thus, the complete study design included 23 plots (6 vegetation types x 4 blocks, less 1 plot for *Vochysia*).

2.2. Field Methods

Soil was sampled for macroaggregate analysis once in late January 2013, as previous studies indicated a lack of seasonal variation (data unpub.). Twenty subsamples were taken in the 0–15 cm layer, using a 3.2-cm diameter push-tube soil sampler at randomly selected locations within three of the four quadrants of each plot. The fourth quadrant was not sampled because soil compaction had occurred during recent erection of a moveable 40-m tower for canopy sampling. The 60 subsamples from each plot were composited into a single sample per plot and mixed well. A subsample of ~800 g of field-moist soil was gently passed through an 8-mm sieve. Of that sample, 150 g was stored at field moisture in the refrigerator at 3 °C for the macroaggregate analyses, while the remaining soil was air-dried for other measurements. Both field-moist and air-dried soils were transported to Iowa State University for subsequent analysis. Soil pH was determined on samples collected from the same depth interval in 2011 when all four quadrants of each plot were sampled.

For assessments of fungal abundance and microbial biomass C, soils were collected at four dates over two years: February 2012, September 2012, February 2013, and September 2013. At each date, we collected five soil cores (10 cm deep × 2.5 cm diameter) in each plot, which were located as one core randomly positioned in each of the four plot quadrants, plus a fifth core from a random position in the full plot. The cores were combined to create a single plot-level sample and homogenized by sieving to 2 mm. There were 23 soil samples per date for a total of 76 samples over two years. The homogenized soils were subsampled for quantification of fungal abundance and microbial biomass C, described below. See Kivlin and Hawkes [34,35] for additional details.

Fine litter production (litterfall of leaves, flowers, fruits, branches ≤1 cm diameter, and frass) was measured every two weeks for two years, 2011–2013, using four 1.3 × 0.4 m litter traps per plot as in Raich et al. [18]. The samples were dried at 65 °C soon after collection, weighed and then ground as described in Russell et al. [16].

Fine-root growth, as an assay for fine-root detrital inputs to soil, was measured in 2012–2013 (0–15 cm depth) using 2-mm mesh ingrowth cores as described by Valverde et al. [36] and Russell et al. [20]. Root ingrowth on a length basis was determined by elutriating the sample, separating roots from detritus, and quantifying root length with a WinRHIZO image analysis system (www.regentinstruments.com/products/rhizo/Rhizo.html).

2.3. Laboratory Analyses

Macroaggregate size distributions in the different vegetation types were evaluated by means of a wet-slaking method that is based on the quantification of the size (diameter) class of macroaggregates according to their ability to resist slaking in water [3]. Air-dried Oxisol samples are usually resistant to wet-sieving [10], so we modified the method of Elliott [3] by using field-moist samples. Before wet-sieving, the initial gravimetric soil moisture content of samples was determined to ensure that the variability in soil moisture did not exceed 10%. The soil moisture content was determined by drying 10-g subsamples at 105 °C for 48 h. The mean soil moisture content was 43% and ranged from 38 to 47% among the samples. Thus, we concluded that initial differences were small enough not to require moisture adjustments prior to wet-slaking. For the procedure, we weighed approximately 100 g of field-moist soil and recorded the mass of each sample to two decimal places. No additional water was added to the samples before wet-sieving. The weighed field-moist sample was then transferred to a Petri dish, 150 mm in diameter and 25 mm in height, that was lined with Whatman Number 1 filter paper and covered while awaiting analysis. We transferred one sample at a time to the top of a set of nested sieves with mesh sizes of 4, 2, 1, 0.5, and 0.25 mm. The nested sieves were held in place by a frame consisting of a round base, two rods as long as the nested sieves, and a clamping device at the top of the sieves that secured the rods. The nested sieves were attached to a lever that raised and lowered the sieves into a 38-L cylinder that was 2/3 filled with water. The set of sieves was then lowered into water just until all soil on all sieves was submerged at all points of lever movement. The sample was wet-sieved by gentle mechanical action that moved the set of sieves 11 cm at a constant rate of 118 cycles per minute in the water for 5 min.

The sample remaining on each sieve was then transferred to a separate aluminum pan. Soil materials that were lodged in the sieve were not included in any of the aggregate size distributions. Thus, the sum of the sample recovered in the 0.25–4.0 mm size distributions is a conservative measure of the total within that size range. We dried samples in a forced-air oven for 48 h. A drying temperature of 60 °C is often used in macroaggregate studies [3], but 55 °C was recommended to avoid N loss, given that samples would be subsequently analyzed for N (C. Cambardella, pers. comm.). The dried subsample within each size class was then weighed and finely ground to <0.25 mm in a mortar and pestle for analysis of C and N.

Total C and N in macroaggregates were measured by dry combustion, using a ThermoFinnigan Flash EA 1112 elemental analyzer (Thermo Fisher Scientific, Waltham, MA, USA). Total P was extracted in aqua regia ($HCl:HNO_3$ = 4:1 by volume) following the method of Crosland et al. [37]. Phosphorus in the digests was determined by inductively coupled plasma atomic emission spectrometry (ICP-AES). Total organic P was determined by the ignition method [38], in which organic P in the soil is converted to inorganic P by oxidation at 550 °C. Phosphorus in ignited and unignited subsamples is then extracted with 0.5 N H_2SO_4 for 16 h. Phosphorus in the extracts was determined via ICP-AES. Organic P in the soil was calculated by subtracting P concentration in the unignited sample from P concentration in the ignited sample. Litterfall Al was measured by microwave-assisted acid digestion and analyzed using inductively coupled plasma optical emission spectroscopy [39]. Soil pH was measured using a stirred slurry of 10 mL sieved, air-dried soil in 25 mL deionized water [40,41].

To quantify fungal hyphae, soils were first air-dried for 24 h and ground with a mortar and pestle to disrupt clay aggregates. Hyphae were extracted from soil using 3% sodium hexametaphosphate, collected via vacuum filtration, stained with acid fuchsin, and quantified by microscopy at 160× with at least 100 fields of view per sample [42,43]. Both septate (decomposers and pathogens) and aseptate (arbuscular mycorrhizal fungi) hyphae, as well as all hyphae inside of soil aggregates, were recorded and summed for total abundance, which is reported as length per g dry soil. Microbial biomass C (MBC) was quantified as the difference between chloroform fumigated and unfumigated soils extracted in a 1:5 ratio of 0.5M K_2SO_4 [44,45]. Microbial biomass data are reported as µg C per g dry soil. Carbon was quantified by combustion (Apollo 9000 TOC Analyzer, Teledyne Tekmar, Mason, OH, USA) [34,35].

2.4. Data Analysis

The experimental unit was the plot. For variables with multiple measurements per plot, the plot mean was used in the analysis. Inclusion of the mature forest in analyses resulted in a randomized incomplete block design. Blocks were treated as a random effect. We tested for homogeneity of variances and normality of distributions. For Al in litterfall, the variances were heterogeneous, so tests were performed on natural log-transformed data. Computations were done using a generalized linear model with both random and fixed effects, the Statistical Analysis System (SAS) mixed model [46]; inclusion of the mature forest required use of the Satterthwaite adjustment for degrees of freedom. Pairwise comparisons for significant overall F-tests were performed using *p*-values adjusted by Tukey's Honestly Significant Difference method.

To test for differences among the vegetation types in macroaggregate size distributions and element (C, N, P) content (H1), we analyzed response variables for the five macroaggregate size classes in the six vegetation types. This Analysis of Variance (ANOVA) model included terms for size class (S), vegetation type (V) and S × V interaction. The response variables regarding macroaggregates were: fraction of total sample; C, N, inorganic P and organic P content; and C:N. The S × V interactions were significant; therefore, separate ANOVAs were conducted for each macroaggregate size class.

To evaluate the strength of the four hypothesized biotic effects on macroaggregate fractions (H2), we tested the four factors simultaneously by conducting partial regression residual analysis. This analysis quantifies the effect of a particular explanatory variable after the effects of all the other explanatory variables in the model have been taken into consideration [47]. This multiple regression model included one response variable, the fraction of macroaggregates within a size class. The four predictor (=explanatory) variables tested were: (a) fungal hyphae in aggregates (mm g^{-1} dry soil); (b) microbial biomass C (µg C g^{-1} dry soil); (c) fine-root growth (length basis, 0–15 cm depth, cm m^{-2} yr^{-1}); and (d) Al flux in fine litterfall (kg ha^{-1} yr^{-1}). With $n = 23$ plots for these analyses, it was not appropriate to include more explanatory variables in the model. For all explanatory variables, we used the mean over all sample times in this analysis. Exploratory analyses of microbial variables indicated that results were relatively invariable, regardless of whether we used the values from one sample time that occurred <1 month after macroaggregate sampling or mean values over the four sample times. Results were also similar between the variables of total hyphae and hyphae within aggregates. Of these two variables, we report only the latter, with values based on the mean of four sample times. Pearson correlational analyses were conducted to test for relationships between the fraction of macroaggregates within the smallest size class and soil pH.

3. Results

3.1. Macroaggregate Size Distribution and Element Concentrations under Different Vegetation (H1)

In the surface horizon, the largest three size classes of water-stable macroaggregates (4–8, 2–4, and 1–2 mm) accounted for 72% of the total dry mass of aggregates, with no significant differences among the vegetation types in any of these three size classes (Figure 1a). In the smallest size classes, however, vegetation (V) had a significant impact on the fraction within the size class (S), with the effect of vegetation differing by size class ($p = 0.0012$, S × V) (Table 1, Figure 1a). In the mature forest, the fraction of macroaggregates in the 0.5–1.0-mm size class was significantly higher than in the other vegetation types, with the exception of *Pentaclethra*, which is the dominant species in the mature forest. In the 0.25–0.5-mm size, the vegetation types differed with: Mature forest > *Pentaclethra* > *Vochysia* > *Hieronyma*, *Virola* and Control (Figure 1b).

Concentrations of N, organic P and inorganic P in aggregates differed significantly among vegetation types (Table 1; Table A1). Nitrogen concentrations were higher in *Vochysia* than in *Hieronyma*, *Pentaclethra* and *Virola* (Figure 2a). *Hieronyma* had the highest C:N values, whereas *Pentaclethra*, the mature forest and Control had the lowest (Figure 2a). Both organic and inorganic P concentrations were highest in *Virola* and lowest in the mature forest and Control (Figure 2b).

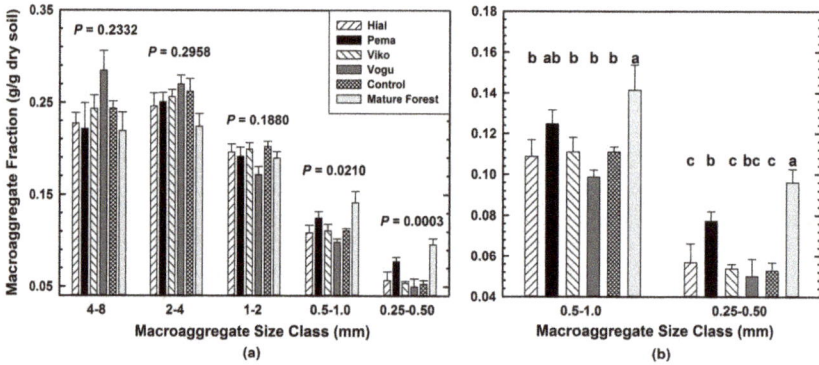

Figure 1. Effect of vegetation type on macroaggregate size distributions. (**a**) All size classes; (**b**) Detail of smaller size classes with significant differences denoted by letters.

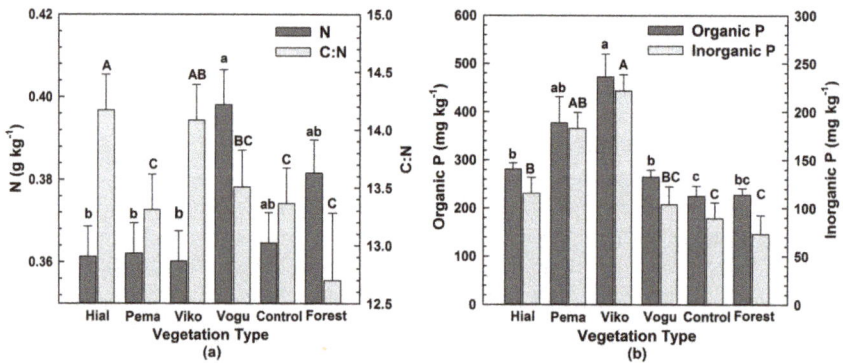

Figure 2. Effect of vegetation type on: (**a**) N concentrations (left *y*-axis) and C:N (right *y*-axis) and (**b**) Organic P (left *y*-axis) and Inorganic P (right *y*-axis).

Table 1. Statistical results (*p* values) for analysis of macroaggregates by size class.

Response Variable	Explanatory Variable		
	Size (S)	Vegetation (V)	S × V
Fraction (g g^{-1} dry soil) [1]	<0.0001	0.8952	0.0012
C (g kg^{-1})	0.3155	0.0781	0.9935
N (g kg^{-1})	0.7791	0.0478	0.9984
C:N	0.0420	0.0024	0.9367
Inorganic P (mg kg^{-1}) [2]	0.7682	0.0008	0.9750
Organic P (mg kg^{-1})	0.9019	0.0006	0.9998
C Fraction (gC g^{-1} dry soil)	<0.0001	0.5782	0.2863
N Fraction (gN g^{-1} dry soil)	<0.0001	0.2100	0.3230
Inorganic P Fraction (gP kg^{-1} dry soil)	<0.0001	0.0039	0.8511
Organic P Fraction (gP kg^{-1} dry soil)	<0.0001	0.0003	0.7911

[1] 'Fraction' refers to the proportion of the total dry soil contained with a given size class of macroaggregate. [2] For P, only the three largest size categories (4–8, 2–4, and 1–2 mm) were included in the analysis due to insufficient material in the two smallest size classes.

All values represent the mean across all size classes of macroaggregates, given no significant differences among size classes for these variables. Small letters denote differences among vegetation types in N and Organic P concentrations; capital letters signify differences for C:N and inorganic P.

Concentrations of C, N, organic P and inorganic P did not differ significantly among the macroaggregate size classes (Table 1; Table A1). Thus, larger macroaggregates (>1 mm), which accounted for the majority of the water-stable aggregates (72%), contained the largest fraction of macroaggregate C, N and P, (Figures 3 and 4, calculated from Table A1, as the product of the element concentration within a size class and the fraction of total soil within that size class, respectively). These data are expressed as the fraction of total soil, with means across all vegetation types, given no significant differences among vegetation types. The differences in elemental concentrations translated into significant effects of the vegetation on the N and P (organic and inorganic) fractions, but not on C fractions contained in macroaggregates (Table 1).

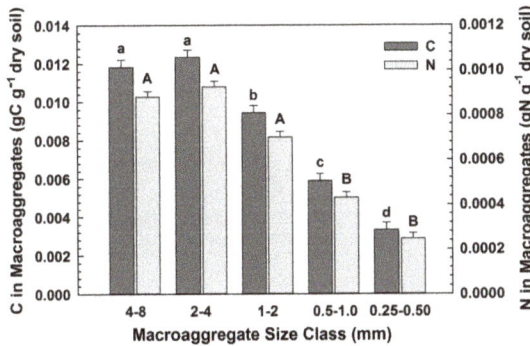

Figure 3. Fractions of C and N contained in macroaggregate size classes. Significant differences among size classes for C (left) are denoted by small letters, for N (right) by capital letters.

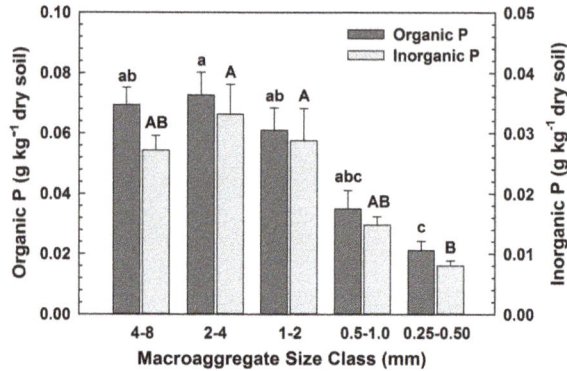

Figure 4. Fractions of organic and inorganic P contained in macroaggregate size classes. Significant differences among size classes for organic P (left) are denoted by small letters, for inorganic P (right) by capital letters.

3.2. Biotic Effects on Macroaggregate Size Distributions (H2, H3)

We evaluated several biotic factors that could potentially differ among the vegetation types and thus influence macroaggregate formation differentially: fungal hyphae within aggregates; microbial biomass C; fine-root growth; and flux of Al in litterfall (Table A2). These explanatory variables were included in the multiple regression model for which the response variable was the fraction of total macroaggregate dry mass contained within a size class (Table A1). Fine-root growth was significantly correlated with the dry mass macroaggregate fraction in two size classes (4–8 and 1–2 mm) (Table 2). The flux of Al in litterfall was significantly correlated with the dry mass macroaggregate fraction in the smallest size classes (0.25–0.5 mm) (Table 2).

Table 2. Relationships between macroaggregate size fractions (g g^{-1} dry soil) and four biological factors.

Explanatory Variables [1]	Macroaggregate Size Class (mm)				
	4–8	2–4	1–2	0.5–1.0	0.25–0.50
	p-value [2]				
Hyphae in Aggregates	0.2185	0.9067	0.4062	0.6491	0.9383
Microbial Biomass C	0.7711	0.7080	0.2663	0.7559	0.2316
Fine Root Growth	**0.0013**	0.8827	**0.0077**	0.7732	0.5268
Al in Litterfall	0.7450	0.1382	0.4991	0.1079	**0.0201**

[1] The variables included: fungal hyphae within macroaggregates (mm g^{-1} dry soil); microbial biomass C (µg C g^{-1} dry soil); fine root growth (cm cm^{-2} yr^{-1}); and Al flux in litterfall (kg Al ha^{-1} yr^{-1}). All belowground measurements are for the 0–15 cm depth interval. [2] The *p* values are results of partial regression analysis (*n* = 23 plots).

The effect of vegetation on soil pH influenced macroaggregate fractions in the smallest size classes (Table 3; Table A2). Across all plots, macroaggregates <1 mm in diameter declined with increasing soil pH (Figure 5a,b). These two macroaggregate size classes were the smallest measured in this study. All data are for the 0–15 cm mineral soil layer. Results are based on Pearson's analyses.

Table 3. Relationships between macroaggregate fractions (g g^{-1} dry soil) and soil pH.

Test Statistic	Macroaggregate Size Class (mm)				
	4–8	2–4	1–2	0.5–1.0	0.25–0.50
p	0.5240	0.4122	0.5489	0.0563	**0.0055**
r	0.1400	0.1794	0.1319	0.4032	0.5595

8

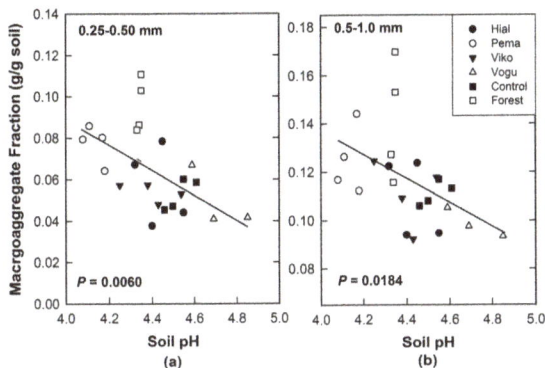

Figure 5. Correlations between soil pH and C contained in macroaggregates. (**a**) Data are for the smallest size fractions, 0.25–0.50 mm and (**b**) Size fractions 0.5–1.0 mm.

4. Discussion

4.1. Cementing Effect of Soil Mineralogy

The similarly high proportion of larger macroaggregates (1–8 mm) across all vegetation types indicated that mineral interactions were relatively more important than biotic factors in binding soil particles in this Oxisol. This was expected under the current conceptual framework [10,13], and given the presence of Fe and Al oxides in this soil. However, total Fe (hydr)oxide content does not always correlate with aggregate stability. Duiker et al. [48] found that the crystallinity of the Fe (hydr)oxides was important; poorly crystalline (hydr)oxides, with their larger and more reactive surface area, had a higher degree of aggregation than crystalline Fe (hydr)oxides, even when present in lower concentrations.

The similarities among the macroaggregate size classes in elemental concentrations and C:N in this study indicated a lack of aggregate hierarchy resulting from the cementing effect of the mineral interactions. These results were consistent with those from another Oxisol in which oxides were the stabilizing agent [10]. Macroaggregate size in our study thus did not appear to represent stages of formation or decomposition of aggregates as in conceptual models for temperate soils (e.g., [1]).

4.2. Binding Effects of the Biota

Despite the strong potential for oxides to affect aggregation in this Oxisol, biotic factors exerted an influence over the smaller macroaggregates (0.25–1 mm) in this experiment, as indicated by significant differences among tree species in these size classes. These aggregates comprised the smallest proportion of the total soil in the *Vochysia* treatment, and the largest fraction in the *Pentaclethra* treatment (Figure 1a). Of the potential explanatory biotic factors that we evaluated, only fine-root growth and Al flux in litterfall were significantly correlated with macroaggregate structure. The quantity and chemistry of plant detritus have been shown to influence aggregation in numerous studies cited by Bronick and Lal [11]. In addition, plant roots can perform multiple tasks that would bind soil particles, including realigning and fastening them together and exuding chemical cementing agents [11]. Thus, it is reasonable that differences among the tree species in fine-root growth [16] were correlated with macroaggregate structure. Similarly, Rillig et al. [22] found that aggregation increased with root length density.

Microbial factors (fungal abundance and microbial biomass C) were not correlated with macroaggregate structure (Table 2). While soil microbial processes and their interactions with plants can also influence binding, and thus aggregation, differences in microbial community composition and function were not strongly correlated with tree species at our study site [34,35]. Studies at other sites have had similar results [49,50]. Thus, it is not surprising that these variables were not correlated with macroaggregate structure. For fungal taxonomic richness and phylogenetic diversity, the relationships with tree species composition were also not significant [34,35]. Similarly, microbial community composition was poorly predicted by plant species composition and functional traits in a study at Barro Colorado Island, Panama, [51].

4.3. Dispersivity (pH) Effects of Tree Species

Of the many factors that influence soil structure, soil pH has been noted, but is often overlooked [11], and is not generally included in conceptual frameworks of controls over aggregation (e.g., [1,13]. In previous studies at our site, we found that the planted tree species significantly altered soil pH. At the start of this experiment in 1988, mean ($\pm$S.E.) soil pH across this site that had been in pasture for >30 years was 4.52 ($\pm$0.02) [17]. After 25 years under the mono-dominant plantation treatment of *Pentaclethra*, the nodulated legume, soil pH had declined to as low as 4.08 in one plot (treatment mean $\pm$ S.E.: 4.14 $\pm$ 0.02), and under *Vochysia*, the Al-accumulating species, soil pH had increased to as 4.85 in one plot (treatment mean: 4.71 $\pm$ 0.08) (Table A2; [16]. These pH changes over time occurred without additions of lime or other chemicals. Other studies of nodulated legumes that support microbial N_2 fixation have demonstrated that ammonification and subsequent nitrification are increased by these species, therefore releasing H^+ and acidifying soil [52]. At the other end of the spectrum of species effects on soil pH, species that accumulate Al could take up H^+ as a consequence of changing the equilibrium between Al minerals and soluble Al [53]:

$$Al(OH)_3 + 3H^+ \rightleftharpoons Al^{3+} + 3H_2O \tag{1}$$

In a previous study, we hypothesized that in these variable-charge soils, the observed range of pH values under the different vegetation types—nearly one pH unit—would be sufficient to create observable differences in aggregation and dispersion of soil colloids across vegetation types. We predicted that aggregation would be greatest at the hypothesized PZC of $\leq$4 and would decrease as pH increased [16]. In this study, the *Pentaclethra* treatment had the lowest soil pH and the highest fraction of macroaggregates <1 mm, whereas *Vochysia* had the highest pH and lowest fraction of macroaggregates in this smaller size range (Figure 1). The result that across all treatments the smallest macroaggregate fractions declined with increasing pH (Figure 5) is consistent with the hypothesis that aggregation would be negatively correlated with soil pH in this variable-charge soil.

5. Conclusions

This long-term tropical experimental setting allowed us to test whether biotic effects exerted control over macroaggregate structure in an Oxisol in which Fe and Al oxides were expected to be the major cementing agents. We found no significant differences among vegetation types in water-stable macroaggregate fractions for aggregates >1 mm, but for aggregates $\leq$1 mm, differences were significant. Biotic factors that could mediate these effects, e.g., fine-root growth and Al in litterfall, were significantly correlated with several size classes of macroaggregate fractions, but microbial and fungal abundance within macroaggregates had no significant correlation. After 25 years of growth, the planted tree species had altered soil pH by a range across all plots of nearly one pH unit. These differences in pH were correlated with smaller aggregate fractions ($p = 0.0563$, $r = 0.40$ for 0.5–1.0 mm size; $p = 0.0055$, $r = 0.56$ for 0.25–050 mm size), suggesting an important linkage between these two master soil variables, pH and macroaggregate structure, such that both were influenced by tree species. Increases or decreases in pH mediated by plant traits could release elements that are physically protected within macroaggregates and thereby influence soil C, N and P cycling.

Acknowledgments: This material is based upon work supported by the National Science Foundation under Grant No. 1119223, 1119169, and 1120015. We are grateful to Ricardo Bedoya, Flor Cascante, Marlon Hernández,

Melissa Sánchez, and Eduardo Paniagua for assistance in the field and laboratory. Amy Morrow and David Denhaan conducted the C analyses. Cindy Cambardella advised on and Jody Ohmacht assisted with the wet-slaking methods. John Kovar and Jay Berkey were responsible for the soil P analyses, for which we are grateful. Katherine Taylor assisted in many aspects of the laboratory studies. We thank Philip Dixon, Emily Casleton, and Daniel Ries for advice on statistical analyses.

Author Contributions: All three authors conceived and designed the experiments, performed the experiments, and analyzed the data for their respective components; S.N.K. and C.V.H. conducted the fungal and microbial studies, and A.E.R. conducted studies for the other components. A.E.R. wrote the paper.

Conflicts of Interest: The authors declare no conflict of interest.

Appendix A

Table A1. Macroaggregate dry mass and elemental concentration by diameter size class and vegetation type in experimental site in Costa Rica. 'Fraction' refers to the proportion of the total dry soil mass recovered after wet sieving. Means ($\pm$S.E.) are for $n = 4$ samples (3 in *Vochysia*) from the 0–15 cm depth.

Vegetation	Size Class	Fraction of Dry Mass	C	N	C:N	Organic P	Inorganic P
	mm	g/g	g/kg	g/kg		mg/kg	mg/kg
Hieronyma	4–8	0.227 ± 0.012	48.85 ± 1.87	3.57 ± 0.09	13.71 ± 0.45	247.81 ± 36.39	110.97 ± 13.19
	2–4	0.246 ± 0.014	48.85 ± 2.22	3.53 ± 0.13	13.85 ± 0.43	283.93 ± 19.23	113.16 ± 15.27
	1–2	0.196 ± 0.009	50.78 ± 2.37	3.58 ± 0.13	14.17 ± 0.45	292.23 ± 10.48	119.34 ± 12.73
	0.5–1	0.109 ± 0.008	52.51 ± 2.57	3.65 ± 0.16	14.44 ± 0.68	323.97	125.28
	0.25–0.5	0.057 ± 0.009	54.93 ± 2.36	3.75 ± 0.12	14.68 ± 0.53	318.39	108.47
Pentaclethra	4–8	0.221 ± 0.029	48.20 ± 4.95	3.61 ± 0.37	13.35 ± 0.09	400.50 ± 103.38	175.37 ± 66.12
	2–4	0.251 ± 0.010	46.08 ± 3.53	3.57 ± 0.24	12.87 ± 0.24	390.63 ± 101.75	160.37 ± 55.96
	1–2	0.191 ± 0.010	46.94 ± 3.33	3.57 ± 0.27	13.17 ± 0.21	304.86 ± 142.77	258.00 ± 101.07
	0.5–1	0.125 ± 0.007	46.94 ± 3.64	3.60 ± 0.26	13.64 ± 0.20	488.09	156.27
	0.25–0.5	0.077 ± 0.005	49.16 ± 3.74	3.75 ± 0.24	13.51 ± 0.36	408.80	164.50
Virola	4–8	0.243 ± 0.015	49.76 ± 1.81	3.56 ± 0.09	13.98 ± 0.36	428.60 ± 12.71	139.40 ± 10.64
	2–4	0.256 ± 0.008	50.14 ± 2.29	3.60 ± 0.08	13.90 ± 0.35	412.34 ± 162.27	240.51 ± 99.42
	1–2	0.199 ± 0.007	50.31 ± 3.02	3.55 ± 0.14	14.16 ± 0.41	524.92 ± 96.56	232.16 ± 92.47
	0.5–1	0.111 ± 0.007	51.79 ± 3.14	3.63 ± 0.15	14.24 ± 0.51	535.67	234.02
	0.25–0.5	0.054 ± 0.002	52.00 ± 2.43	3.67 ± 0.08	14.15 ± 0.41	509.14	240.11
Vochysia	4–8	0.285 ± 0.021	52.50 ± 2.61	3.92 ± 0.22	13.41 ± 0.31	233.30 ± 8.44	100.00 ± 5.52
	2–4	0.270 ± 0.010	53.10 ± 2.02	3.82 ± 0.18	13.91 ± 0.22	232.07 ± 15.32	102.48 ± 5.44
	1–2	0.171 ± 0.009	53.19 ± 2.16	3.87 ± 0.18	13.77 ± 0.27	290.10 ± 24.19	99.88 ± 6.75
	0.5–1	0.099 ± 0.003	57.76 ± 2.01	4.17 ± 0.19	13.85 ± 0.15	293.99	102.09
	0.25–0.5	0.050 ±0.009	57.10 ± 1.98	4.10 ± 0.26	13.98 ± 0.41	356.02	100.18
Control	4–8	0.244 ± 0.008	47.45 ± 2.51	3.58 ± 0.13	13.23 ± 0.33	242.59 ± 48.55	103.52 ± 18.15
	2–4	0.262 ± 0.014	47.22 ± 1.69	3.62 ± 0.14	13.05 ± 0.21	228.57 ± 47.77	104.40 ± 23.58
	1–2	0.202 ± 0.006	48.63 ± 2.60	3.65 ± 0.16	13.30 ± 0.22	207.83 ± 43.05	95.42 ± 22.29
	0.5–1	0.111 ± 0.002	49.56 ± 2.95	3.66 ± 0.17	13.53 ± 0.34	166.38	81.69
	0.25–0.5	0.053 ± 0.004	50.96 ± 2.60	3.72 ± 0.14	13.70 ± 0.28	260.81	60.83
Mature Forest	4–8	0.219 ± 0.021	50.85 ± 2.97	3.94 ± 0.17	12.88 ± 0.29	221.43 ± 26.57	72.51 ± 5.47
	2–4	0.224 ± 0.014	50.43 ± 2.76	4.00 ± 0.13	12.59 ± 0.31	214.55 ± 34.30	76.75 ± 9.96
	1–2	0.190 ± 0.007	47.54 ± 2.79	3.74 ± 0.16	12.71 ±	257.37 ± 18.74	69.18 ± 8.14
	0.5–1	0.142 ± 0.012	46.87 ± 3.18	3.67 ± 0.14	12.74 ±	158.26	68.30
	0.25–0.5	0.096 ± 0.007	47.04 ± 2.76	3.74 ±0.12	12.57 ±	251.20	78.68

Notes. Sample mass in the two smallest size classes was insufficient for analyses of P in all blocks. Values represent the composited samples across blocks of a given species and size class.

Table A2. Four biological factors and soil pH data used in evaluating relationships with macroaggregate structure.

Vegetation	Hyphae in Aggregates	Microbial Biomass C	Fine-Root Growth	Al in Litterfall	Soil pH
	mm g^{-1} Dry Soil	µg C g^{-1} Dry Soil	cm cm^{-2} yr^{-1}	kg Al ha^{-1} yr^{-1}	
Hieronyma	16,588 ± 1150	1022 ± 108	99 ± 8	10 ± 2	4.43 ± 0.05
Pentaclethra	19,201 ± 818	933 ± 49	70 ± 9	8 ± 1	4.14 ± 0.02
Virola	17,438 ± 1853	781 ± 180	91 ± 9	14 ± 1	4.40 ± 0.06
Vochysia	12,021 ± 1818	1171 ± 33	181 ± 15	180 ± 17	4.71 ± 0.08
Control	15,420 ± 1543	949 ± 139	106 ± 10	83 ± 14	4.53 ± 0.03
Mature Forest	16,548 ± 721	1033 ± 89	94 ± 11	2 ± 1	4.34 ± 0.01

Notes. Values represent means ($\pm$S.E.) over all sample times. Data were published previously [16,34]. Data are for the 0–15-cm interval.

References

1. Jastrow, J.D.; Amonette, J.E.; Bailey, V.L. Mechanisms controlling soil carbon turnover and their potential application for enhancing carbon sequestration. *Clim. Chang.* **2007**, *80*, 5–23. [CrossRef]
2. Tisdall, J.M.; Oades, J.M. Organic matter and water-stable aggregates in soils. *Eur. J. Soil Sci.* **1982**, *33*, 141–163. [CrossRef]
3. Elliott, E. Aggregate structure and carbon, nitrogen, and phosphorus in native and cultivated soils. *Soil Sci. Soc. Am. J.* **1986**, *50*, 627–633. [CrossRef]
4. Payne, D. Soil structure, tilth and mechanical behaviour. *Russells Soil Cond. Plant Growth* **1988**, *11*, 378–411.
5. El-Swaify, S. Physical and mechanical properties of oxisols. In *Soils with Variable Charge*; Theng, B.K.G., Ed.; Offset Publ.: Palmerston North, New Zealand, 1980; pp. 303–324.
6. Warkentin, B.; Maeda, T. Physical and mechanical characteristics of Andisols. *N. Z. Soc. Soil Sci.* **1980**, *12*, 281–302.
7. Igwe, C.A.; Zarei, M.; Stahr, K. Fe and Al oxides distribution in some ultisols and inceptisols of southeastern Nigeria in relation to soil total phosphorus. *Environ. Earth Sci.* **2010**, *60*, 1103–1111. [CrossRef]
8. Spycher, G.; Rose, S.L.; Sollins, P.; Norgren, J.; Young, J.L.; Kermit Cromack, J. Evolution of Structure in a chronosequence of andesitic forest soils. *Soil Sci.* **1986**, *142*, 164–178. [CrossRef]
9. Tiessen, H.; Stewart, J.W. Light and electron microscopy of stained microaggregates: The role of organic matter and microbes in soil aggregation. *Biogeochemistry* **1988**, *5*, 312–322. [CrossRef]
10. Oades, J.; Waters, A. Aggregate hierarchy in soils. *Soil Res.* **1991**, *29*, 815–828. [CrossRef]
11. Bronick, C.J.; Lal, R. Soil structure and management: A review. *Geoderma* **2005**, *124*, 3–22. [CrossRef]
12. Sollins, P.; Homann, P.; Caldwell, B.A. Stabilization and destabilization of soil organic matter: Mechanisms and controls. *Geoderma* **1996**, *74*, 65–105. [CrossRef]
13. Six, J.; Feller, C.; Denef, K.; Ogle, S.; de Moraes Sa, J.C.; Albrecht, A. Soil organic matter, biota and aggregation in temperate and tropical soils-Effects of no-tillage. *Agronomie* **2002**, *22*, 755–775. [CrossRef]
14. Finzi, A.C.; Canham, C.D.; Van Breemen, N. Canopy tree–soil interactions within temperate forests: Species effects on pH and cations. *Ecol. Appl.* **1998**, *8*, 447–454.
15. Reich, P.B.; Oleksyn, J.; Modrzynski, J.; Mrozinski, P.; Hobbie, S.E.; Eissenstat, D.M.; Chorover, J.; Chadwick, O.A.; Hale, C.M.; Tjoelker, M.G. Linking litter calcium, earthworms and soil properties: An common garden test with 14 tree species. *Ecol. Lett.* **2005**, *8*, 811–818. [CrossRef]
16. Russell, A.E.; Hall, S.J.; Raich, J.W. Tropical tree species traits drive soil cation dynamics via effects on pH: A proposed conceptual framework. *Ecol. Monogr.* **2017**, *87*, 685–701. [CrossRef]
17. Fisher, R.F. Amelioration of degraded rain forest soils by plantations of native trees. *Soil Sci. Soc. Am. J.* **1995**, *59*, 544–549. [CrossRef]
18. Raich, J.W.; Russell, A.E.; Bedoya-Arrieta, R. Lignin and enhanced litter turnover in tree plantations of lowland Costa Rica. *For. Ecol. Manag.* **2007**, *239*, 128–135. [CrossRef]
19. Raich, J.W.; Russell, A.E.; Valverde-Barrantes, O. Fine root decay rates vary widely among lowland tropical tree species. *Oecologia* **2009**, *161*, 325–330. [CrossRef] [PubMed]
20. Russell, A.E.; Raich, J.W.; Arrieta, R.B.; Valverde-Barrantes, O.; González, E. Impacts of individual tree species on carbon dynamics in a moist tropical forest environment. *Ecol. Appl.* **2010**, *20*, 1087–1100. [CrossRef] [PubMed]
21. Ternan, J.; Elmes, A.; Williams, A.; Hartley, R. Aggregate stability of soils in central Spain and the role of land management. *Earth Surf. Process. Landf.* **1996**, *21*, 181–193. [CrossRef]
22. Rillig, M.C.; Mummey, D.L. Mycorrhizas and soil structure. *New Phytol.* **2006**, *171*, 41–53. [CrossRef] [PubMed]
23. Cotrufo, M.F.; Wallenstein, M.D.; Boot, C.M.; Denef, K.; Paul, E. The Microbial Efficiency-Matrix Stabilization (MEMS) framework integrates plant litter decomposition with soil organic matter stabilization: Do labile plant inputs form stable soil organic matter? *Glob. Chang. Biol.* **2013**, *19*, 988–995. [CrossRef] [PubMed]
24. Schutter, M.E.; Dick, R.P. Microbial community profiles and activities among aggregates of winter fallow and cover-cropped soil. *Soil Sci. Soc. Am. J.* **2002**, *66*, 142–153. [CrossRef]
25. Mendes, I.; Bandick, A.; Dick, R.; Bottomley, P. Microbial biomass and activities in soil aggregates affected by winter cover crops. *Soil Sci. Soc. Am. J.* **1999**, *63*, 873–881. [CrossRef]
26. Russell, A.E.; Raich, J.W.; Valverde-Barrantes, O.J.; Fisher, R.F. Tree species effects on soil properties in experimental plantations in tropical moist forest. *Soil Sci. Soc. Am. J.* **2007**, *71*, 1389–1397. [CrossRef]

27. Sanford, R.L., Jr.; Paaby, P.; Luvall, J.C.; Phillips, E. Climate, geomorphology, and aquatic systems. In *La Selva: Ecology and Natural History of a Neotropical Rain Forest*; University of Chicago Press: Chicago, IL, USA, 1994; pp. 19–33.

28. Sollins, P.; Sancho, F.; Ch, R.M.; Sanford, R.L., Jr. So//s and Soil Process Research. In *La Selva: Ecology and Natural History of a Neotropical Rain Forest*; University of Chicago Press: Chicago, IL, USA, 1994; p. 34.

29. Porder, S.; Clark, D.A.; Vitousek, P.M. Persistence of rock-derived nutrients in the wet tropical forests of La Selva, Costa Rica. *Ecology* **2006**, *87*, 594–602. [CrossRef] [PubMed]

30. Kleber, M.; Schwendenmann, L.; Veldkamp, E.; Rößner, J.; Jahn, R. Halloysite versus gibbsite: Silicon cycling as a pedogenetic process in two lowland neotropical rain forest soils of La Selva, Costa Rica. *Geoderma* **2007**, *138*, 1–11. [CrossRef]

31. Chorover, J.; Sposito, G. Surface charge characteristics of kaolinitic tropical soils. *Geochim. Cosmochim. Acta* **1995**, *59*, 875–884. [CrossRef]

32. Anda, M.; Shamshuddin, J.; Fauziah, C.; Omar, S.S. Mineralogy and factors controlling charge development of three Oxisols developed from different parent materials. *Geoderma* **2008**, *143*, 153–167. [CrossRef]

33. Clark, D.B.; Clark, D.A. Landscape-scale variation in forest structure and biomass in a tropical rain forest. *For. Ecol. Manag.* **2000**, *137*, 185–198. [CrossRef]

34. Kivlin, S.N.; Hawkes, C.V. Temporal and spatial variation of soil bacteria richness, composition, and function in a neotropical rainforest. *PLoS ONE* **2016**, *11*, e0159131. [CrossRef] [PubMed]

35. Kivlin, S.N.; Hawkes, C.V. Tree species, spatial heterogeneity, and seasonality drive soil fungal abundance, richness, and composition in Neotropical rainforests. *Environ. Microbiol.* **2016**, *18*, 4662–4673. [CrossRef] [PubMed]

36. Valverde-Barrantes, O.J.; Raich, J.W.; Russell, A.E. Fine-root mass, growth and nitrogen content for six tropical tree species. *Plant Soil* **2007**, *290*, 357–370. [CrossRef]

37. Crosland, A.; Zhao, F.; McGrath, S.; Lane, P. Comparison of aqua regia digestion with sodium carbonate fusion for the determination of total phosphorus in soils by inductively coupled plasma atomic emission spectroscopy (ICP). *Commun. Soil Sci. Plant Anal.* **1995**, *26*, 1357–1368. [CrossRef]

38. Kuo, S. Phosphorus. In *Methods of Soil Analysis Part 3—Chemical Methods*; Sparks, D.L., Page, A.L., Helmke, P.A., Loeppert, R.H., Eds.; Soil Science Society of America, American Society of Agronomy: Madison, WI, USA, 1996; pp. 869–919.

39. Kingston, H.M.; Haswell, S.J. *Microwave-Enhanced Chemistry*; American Chemical Society: Washington, DC, USA, 1997.

40. Díaz-Romeu, R.; Hunter, A. *Metodologías de Muestreo de Suelos, Análisis Químico de Suelos y Tejido Vegetal y de Investigaciones en Invernadero*; Centro Agronómico Tropical de Investigación y Enseñanza (CATIE): Cartago, Costa Rica, 1978.

41. Thomas, G. Soil pH and soil acidity. In *Methods of Soil Analysis Part 3—Chemical Methods*; Soil Science Society of America, American Society of Agronomy: Madison, WI, USA, 1996; pp. 475–490.

42. Brundrett, M.; Melville, L.; Peterson, L. *Practical Methods in Mycorrhiza Research: Based on a Workshop Organized in Conjuction with the Ninth North American Conference on Mycorrhizae, University of Guelph, Guelph, Ontario, Canada*; Mycologue Publications: Devon, UK, 1994.

43. Sylvia, D.M. 3 Quantification of External Hyphae of Vesicular-arbuscular Mycorrhizal Fungi. In *Methods in Microbiology*; Elsevier: Amsterdam, The Netherlands, 1992; Volume 24, pp. 53–65.

44. Scott-Denton, L.E.; Rosenstiel, T.N.; Monson, R.K. Differential controls by climate and substrate over the heterotrophic and rhizospheric components of soil respiration. *Glob. Chang. Biol.* **2006**, *12*, 205–216. [CrossRef]

45. Brookes, P.; Landman, A.; Pruden, G.; Jenkinson, D. Chloroform fumigation and the release of soil nitrogen: A rapid direct extraction method to measure microbial biomass nitrogen in soil. *Soil Biol. Biochem.* **1985**, *17*, 837–842. [CrossRef]

46. Littell, R.; Milliken, G.; Stroup, W.; Wolfinger, R. *SAS Systems for Mixed Models*; SAS Inst. Inc.: Cary, NC, USA, 1996.

47. Neter, J.; Kutner, M.H.; Nachtsheim, C.J.; Wasserman, W. *Applied Linear Statistical Models*; Irwin Chicago: Chicago, IL, USA, 1996; Volume 4.

48. Duiker, S.W.; Rhoton, F.E.; Torrent, J.; Smeck, N.E.; Lal, R. Iron (hydr) oxide crystallinity effects on soil aggregation. *Soil Sci. Soc. Am. J.* **2003**, *67*, 606–611. [CrossRef]

49. Fierer, N.; Jackson, R.B. The diversity and biogeography of soil bacterial communities. *Proc. Nat. Acad. Sci. USA* **2006**, *103*, 626–631. [CrossRef] [PubMed]
50. Prober, S.M.; Leff, J.W.; Bates, S.T.; Borer, E.T.; Firn, J.; Harpole, W.S.; Lind, E.M.; Seabloom, E.W.; Adler, P.B.; Bakker, J.D. Plant diversity predicts beta but not alpha diversity of soil microbes across grasslands worldwide. *Ecol. Lett.* **2015**, *18*, 85–95. [CrossRef] [PubMed]
51. Barberán, A.; McGuire, K.L.; Wolf, J.A.; Jones, F.A.; Wright, S.J.; Turner, B.L.; Essene, A.; Hubbell, S.P.; Faircloth, B.C.; Fierer, N. Relating belowground microbial composition to the taxonomic, phylogenetic, and functional trait distributions of trees in a tropical forest. *Ecol. Lett.* **2015**, *18*, 1397–1405. [CrossRef] [PubMed]
52. Van Miegroet, H.; Cole, D. The Impact of Nitrification on Soil Acidification and Cation Leaching in a Red Alder Ecosystem 1. *J. Environ. Qual.* **1984**, *13*, 586–590. [CrossRef]
53. Essington, M. *Soil and Water Chemistry: An Integrative Approach*; CRC Press: Boca Raton, FL, USA, 2004.

forests

MDPI

Article

Joint Control of Net Primary Productivity by Climate and Soil Nitrogen in the Forests of Eastern China

Zhan Xiaoyun [1], Guo Minghang [1] and Zhang Tibin [1,2,*]

[1] State Key Laboratory of Soil Erosion and Dryland Farming on Loess Plateau, Institute of Soil and Water Conservation, Northwest A&F University, Yangling 712100, China; zhanxiaoyun2005@163.com (Z.X.); mhguo@ms.iswc.ac.cn (G.M.)

[2] Institute of Soil and Water Conservation, Chinese Academy of Sciences & Ministry of Water Resources, Yangling 712100, China

* Correspondence: zhangtibin@163.com; Tel.: +86-29-87012465

Received: 30 March 2018; Accepted: 31 May 2018; Published: 4 June 2018

Abstract: The nature and extent of climate and soil nutrient controls in Chinese forests remain poorly resolved. Here, we synthesized the data on carbon–climate–soil in eastern China, and litter N was firstly taken into consideration, to examine the variation of net primary productivity (NPP) and its driving forces. Results showed that NPP had significant latitude pattern and varied substantially across climate zones. Bivariate analyses indicated that mean annual temperature (MAT), mean annual precipitation (MAP), soil N content (N_{soil}), and annual litter N (N_{re}) were the main controlling factors in spatial pattern of forest NPP. Notably, partial general linear model analysis revealed that MAT, MAP, and N_{re} jointly explained 84.8% of the spatial variation of NPP. Among the three major factors, N_{re} explained more variation of forest NPP than the other two factors, and MAT and MAP affected NPP mainly through the change of litter N rather than via themselves, highlighting the importance of litter N in estimating forest NPP. However, to accurately describe the pattern of forest NPP in China, more detailed field measurements and methodologies on NPP and relevant confounding factors should be addressed in future studies.

Keywords: net primary productivity; climate zone; climate; soil N; litter N

1. Introduction

Net primary productivity (NPP) is a key ecosystem variable and a critical component of the regional and global carbon cycle [1,2]. However, the spatial distribution of forest NPP is still uncertain, due to complicated impacts from various environmental and biological factors, e.g., vegetation distribution, climatic variables, and land use change [3–5]. Therefore, accurate estimation of NPP and its driving forces is essential to understanding terrestrial carbon pools and responses of forest functions to future climate change [6,7].

At a regional scale, NPP is strongly correlated with climatic factors [8–11]. Based on global NPP data, Lieth (1975) [8] developed the climate-driven theory, and described the relationship between climatic factors (annual mean temperature, annual precipitation, and annual evapotranspiration) and NPP in logistic functions. However, it is still unclear whether regional NPP across biomes follows the same pattern [12,13], since the driving factors vary among regions [14–16]. Further, an international coordination for compilation of global NPP data for model validation and development, the Global Primary Productivity Data Initiative (GPPDI), has worked successfully since 1995. However, inadequate observational NPP data seriously inhibit the estimation and modeling of the global carbon cycle and the validation and evaluation of the global carbon models [17].

Globally, forests represent 80% of plant biomass, and 50–60% of annual NPP in terrestrial ecosystems [5,18]. Chinese forests, which cover about half the total land area of China, contain perhaps the widest range of forest types in the world, ranging from boreal forest and mixed coniferous broad-leaved forest in the north, to subtropical evergreen broad-leaved forest, warm temperate coniferous forest, tropical rainforest, and seasonal forest in the south [19,20]. It is regarded that these forests have a significant influence on carbon budget both regionally and globally [21]. A lot of field measurements of forest biomass and NPP estimations are available from multiple sites for the past two decades. However, these data were mostly published in Chinese journals and reports and not accessible to Western scientists. Also, site-based data of forest NPP in China has not yet been synthesized in a consistent manner.

Located in heavily forested area, the North–South Transect of Eastern China (NSTEC) features a high variation in plant composition, climate, and soil substrate materials [22]. It thus provides wide biome heterogeneity to examine spatial pattern of forest NPP. We synthesized the data in the primary literature on NPP in forests within the NSTEC to produce a consistent dataset on NPP, and amassed the data on mean annual temperature (MAT), mean annual precipitation (MAP), soil N content (N_{soil}), and annual litter N (N_{re}). Hobbie (2015) [23] proposed the plant litter feedback paradigm that changes in plant litter traits reinforce patterns of soil fertility and NPP. Thus, we first introduced litter N as available N for plants, which was an improvement over the past studies. Based on these data, we aim to investigate whether temperature and precipitation characterize the pattern of NPP within the transect, to explore whether soil N is a limiting factor for large-scale distribution of NPP, and to test the applicability of the climate-driven theory in eastern China. Besides understanding the causes of variability in ecosystem productivity, the findings are crucial for assessing the potential responses to global climatic change, and are thus incorporated into statistical and simulation models.

2. Materials and Methods

2.1. Study Area

NSTEC provides an ideal platform for exploring the growth of forest in East Asia's monsoon region [22]. NSTEC has a spatial distance of more than 3700.0 km in length, ranging from 108.0° E to 118.0° E for latitude below 40.0° N and from 118.0° E to 128.0° E for latitude above 40.0° N (Figure 1). From north to south, MAP increases from 500.0 mm to about 1800.0 mm, and MAT changes from 1.0 to 22.0 °C, correspondingly. Due to the obvious latitudinal gradients for climate, zonal forest ecosystems occur within the NSTEC, which include cold-temperate coniferous forest, temperate mixed forest, warm-temperate deciduous broadleaved forest, subtropical evergreen broad-leaved forest, and tropical monsoon rainforest. Within NSTEC, 87 observations in 34 plots were included in this study, and geographical distribution of the sites was mapped in Figure 1.

2.2. Datasets

Data on the forest ecosystems within the transect were obtained from published literature (Figure 1, Table S1). The sampling years were not mentioned in most of the original literature, and only the publication years of the literature could be obtained, which varied from 1983 to 2010. The information about the publication time could be seen in Table S1. The times were classified into three groups according to the publication year, and they were 1980s (1983–1989), 1990s (1991–1999), and 2000s (2001–2010). Based on the classification, the temporal variations were analyzed, and there were no significances among climatic factor, soil N, and also NPP for different times (Table S2). Therefore, we take no account of temporal variations of forest NPP, and only spatial pattern of forest NPP was involved. To be clear, "major forest biome" was used to represent a higher-level classification of the "ecological zone", which included boreal, temperate, subtropical, and tropical forests [24]. Each site included site name, latitude, longitude, tree species, MAT, MAP, N_{soil}, N_{re}, and NPP estimations (Table S1). We retrieved missing latitude or longitude information for sites without such data from Google Earth according to site names. Meanwhile, missing MAT and MAP were extracted from the Chinese climate data based on site locations. NPP was expressed, herein, in per unit of oven-dry matter (t ha^{-1} a^{-1}). In order to compare with other studies, NPP may be expressed in g C m^{-2} a^{-1}, where 1.0 g carbon is equivalent to 2.2 g oven-dry matter.

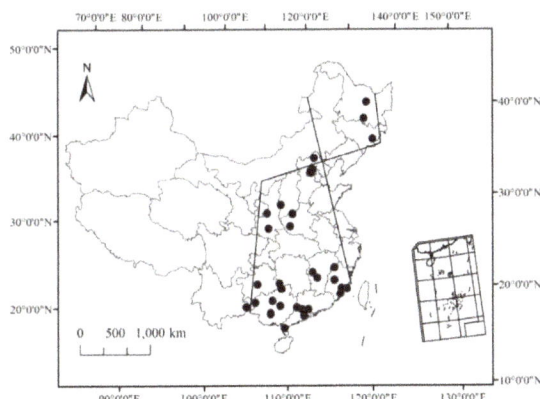

Figure 1. Locations of research plots in this synthesis in eastern China (black dots). The region between the two lines represents the area of the North–South Transect of Eastern China (NSTEC).

2.2.1. Estimation of NPP

Forest NPP was estimated as the sum of increase in the standing crop of vegetation based on the data for biomass, which included net increments of trees, shrubs, and herbs. Biomass data in our study were all from field survey, and data from modeling and regional average were excluded. The sampling methods for measuring forest biomass used by Chinese scientists, however, were quite different. General methods of biomass estimates were given below. Firstly, diameter at breast height (DBH) and the height (H) of each tree in the plot were measured, and then one of the three methods was used: (1) mean tree biomass was measured, and then multiplied by tree density of each plot; (2) allometric equations were used to calculate the biomass of each tree; and (3) allometric equations were used to calculate the mean tree biomass, and multiplied by tree density of each plot [25,26]. Secondly, dead biomass of each tree was the sum of standing dead stem and coarse woody debris [27]. Thirdly, biomass of shrubs and herbs was measured by harvest method within large tree samples. Aboveground tissues were clipped, and the weight of these tissues and the aboveground biomass were calculated based on the plot area. Finally, the biomass of litterfall was determined by monthly collection in three or more 1.0 m × 1.0 m plots, laid out inside a single tree plot. Litterfall components, such as leaf, branch, flower, and fruit, were dried and weighed to estimated total litterfall biomass. After one year, the sampling and the analysis for trees and understory vegetation were repeated, and the increases in biomass were taken as forest NPP.

2.2.2. Measurements of Soil N

Soil sampling depths varied from a few centimeters to several meters in different studies. Considering data availability and distribution of plant roots in the soil, in our study, only the data from the depths of 0–60.0 cm in soil were used. Generally, the soil was sampled from the soil pits selected randomly in each forest ecosystem, and soil bulk density was determined by collecting samples in volumetric rings. In preparation for analysis, soil samples were air-dried and then passed through the mesh sieve for measurement of N concentration. N$_{soil}$ was calculated by multiplying the mean concentration of N in each layer with the corresponding mean soil bulk density Equation (1).

Annual litter N (N_{re}) was calculated as litter biomass multiplied by the respective N concentration, and finally, the contents of the individual fractions were summed in Equation (2).

$$N_{soil} = \sum(\text{mean concentration of N in each layer} \times \text{corresponding mean soil bulk density}) \quad (1)$$

$$N_{re} = \sum(\text{annual litter biomass of each component} \times \text{litter N concentration in each component}) \quad (2)$$

where, N_{soil} is soil N content (kg ha^{-1}), N_{re} is annual litter N (kg N ha^{-1} a^{-1}).

2.2.3. Climate Data

When investigating the relationship between climate and forest productivity, it would be better to employ climate data for particular sampling years. Hence, synchronous climate data for particular sampling years were collected. However, most sampling sites were far away from weather stations, and thus, climate data were missing. Therefore, the meteorological data from the Meteorological Database of the Chinese Ecological Research Network (CERN) Synthesis Research Center were extracted. Climate data in a grid cell where the sampling site was located was extracted as the climate data of the sampling site. MAP and MAT were the average values of 1980–2000 with a 10-day-0.1° spatial–temporal resolution [28].

2.3. Statistical Analysis

One-way analysis of variance (ANOVA) was adopted to test the differences in forest NPP among different climate zones, and was followed by Fisher's least significant difference (LSD) comparisons when the differences were significant. Linear regression analysis or dynamic curve fit was used to analyze the relationships between forest NPP and the driving forces. Considering the differences in sample size, we compared R^2 and root mean squared error (RMSE), and selected the better-fit functions that had a higher R^2 and lower RMSE. The stepwise regression was used to analyze the linear regression on forest NPP with climate and soil N. In the stepwise regression, the minimum *p*-value for a variable to be recommended for adding to and removing from the model was 0.05. Considering the results of stepwise regression, only MAT, MAP, and N_{re} were analyzed in the ensuing analysis. To identify the relative effects and interactive effects of the above three factors on forest NPP, we conducted a partial general linear model (GLM) using NPP as dependent variable, and MAT, MAP, and N_{re} as predictors. The partial regression divides the variation in response variable explained by several predictor variables into independent components (representing the independent effects of an individual explanatory variable when controlling effects of the other explanatory variables) and joint components (usually representing the collinearities between explanatory variables). The variation partitioning with three explanatory matrices leads to the identification of seven fractions in this study, i.e., independent effects of MAT, MAP, and N_{re}; interactive effects of MAT and MAP, MAT and N_{re}, MAP and N_{re}, and the interactive effect of all variables. Further details about the method were given in Heikkinen et al. (2005) [29].

Figure 1 was plotted with ArcGIS 10.1 software (Esri, Realands, CA, USA), and other graphs were performed by Sigma Plot 13.0 software (Systat Software Inc., San Jose, CA, USA). The partial GLM was performed with SAS statistical software and other analyses were conducted by SPSS 16.0 statistical software (SPSS Inc., Chicago, IL, USA).

3. Results

3.1. Latitudinal Pattern and Statistics of NPP

Forest NPP across eastern China exhibited significantly obvious latitudinal patterns ($p < 0.001$), but no clear longitudinal trend ($p = 0.20$) (Figure 2). Therefore, we did not elaborate on the longitudinal pattern of forest NPP. Generally, NPP decreased with increasing latitude when latitude was below 35.0° N, but increased when the latitude was higher than the threshold level (Figure 2). As a whole, mean total NPP was 9.5 ± 0.7 t ha^{-1} a^{-1} (mean $\pm$ SE), ranging from 0.8 to 29.6 t ha^{-1} a^{-1}, with the variability (coefficient of variation, CV) up to 64.0% (Table 1). In terms of different climate zones, tropical forests below 23.0° N had the highest forest productivity with an average NPP of 15.3 ± 1.3 t ha^{-1} a^{-1}, but the smallest spatial variability (CV = 31.6%). NPP of temperate forests was, on average, 5.5 ± 0.8 t ha^{-1} a^{-1}, which tended to be the lowest among the four climate zones. The other climate zones (subtropical forests and boreal forests) were not different from each other, and were intermediate to the others. Specifically, the average NPP of subtropical forests was 10.1 ± 0.9 t ha^{-1} a^{-1}, which was close to the value of boreal forests (12.1 ± 4.7 t ha^{-1} a^{-1}), but the latter forests exhibited a higher spatial variability in NPP from available data (CV = 78.4%) (Table 1).

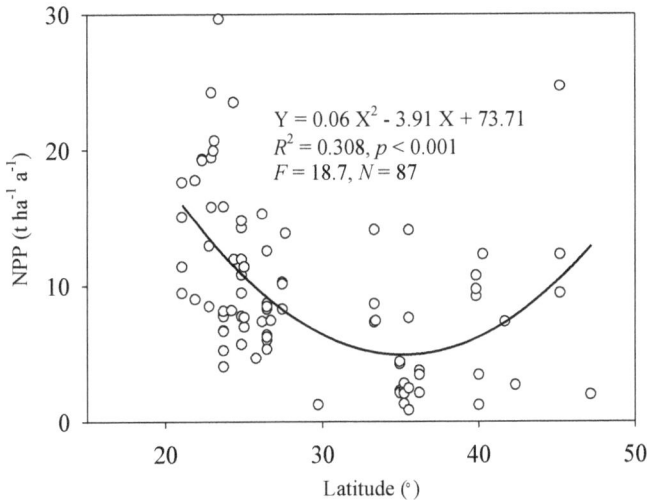

Figure 2. Trends of net primary productivity (NPP) along latitude for forest ecosystems in eastern China. N, number of observations; *F*, F Values.

Table 1. Comparisons of net primary productivity (NPP) of forest ecosystems in different climate zones. Tropical zone: <23.0° N; Subtropical zone: 23.0–33.0° N; Temperate zone: 33.0–45.0° N; and Boreal zone: >45.0° N. Number of observations (N), mean value (Mean), maximum value (Max), minimum value (Min), standard error (SE), and coefficient of variation (CV) were reported. Differences among climate zones were tested using one-way ANOVA with Fisher's LSD comparisons; differences at $p < 0.05$ were indicated with different letters.

Climate Zone	N	NPP (t ha^{-1} a^{-1})				
		Mean	Max	Min	SE	CV (%)
Boreal	4	12.1 [ab]	24.6	1.9	4.7	78.4
Temperate	28	5.5 [c]	14.1	0.8	0.8	73.2
Subtropical	42	10.1 [b]	29.6	1.2	0.9	54.7
Tropical	13	15.3 [a]	24.2	8.6	1.3	31.6
Overall	87	9.5	29.6	0.8	0.7	64.0

3.2. The Impact of Climatic Factors on the Spatial Pattern of NPP

Temperature is an important driving factor for the ecosystem carbon budget, and water is the basic material for maintaining ecosystem structure and functions. As was clearly shown in Figure 3, the plot based forest NPP grew linearly with increasing MAT, MAP, and the combinations, although variation occurred within the climate band and few sites fell out of 95.0% prediction band. On average, MAT and

MAP contributed 16.6% and 21.8% of the spatial variation of NPP, respectively (Figure 3, Table 2). We then analyzed the combined contribution of MAT and MAP to the spatial variation of NPP, and found that MAT and MAP jointly explained 24.3% of the spatial variation of NPP, which was only just 2.5% higher in the prediction of NPP than single climatic factor (MAP) alone (Figure 3, Table 2). As for different climate zones, climatic factors were also closely related to the occurrence of NPP. Generally, there was a consistent and significant shift from species with high NPP in warm climate with high MAT and MAP toward species with low NPP in low MAT and MAP conditions, while the trend in boreal forest ecosystems differed (Figure S1). Specifically, for boreal forest, higher NPP occurred with lower MAT and MAP. More speculatively, soil nutrients or other environmental factors might contribute to the variation of forest NPP in this region, and this should be taken into consideration in the ensuing analyses. It is worth noting that the number of boreal forests was quite low, hence, the reliability of the results should be considered carefully.

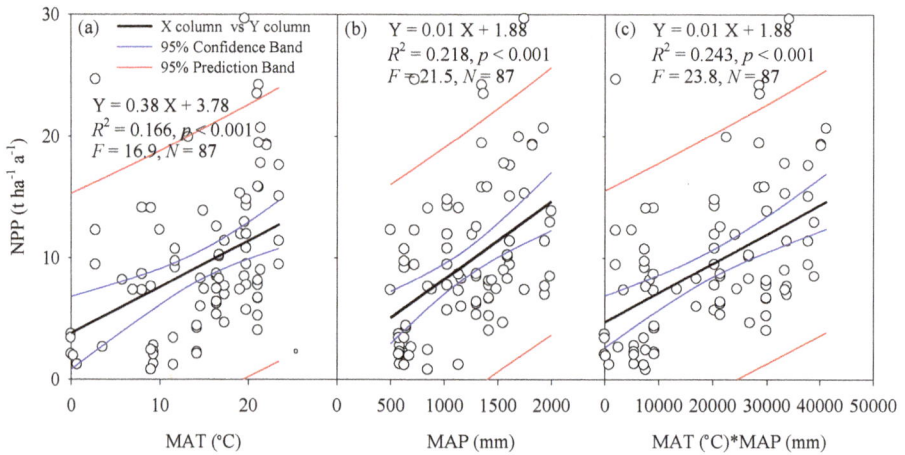

Figure 3. Relationships between net primary productivity (NPP) and mean annual temperature (MAT), mean annual precipitation (MAP), and combinations of MAT and MAP. N, number of observations; *F*, *F* Values.

Table 2. Bivariate and multivariate regression models of net primary productivity (NPP) on climatic factors and soil N. MAP, mean annual precipitation; MAT, mean annual temperature; N_{soil}, soil N content; N_{re}, annual litter N; N, number of observations; *F*, *F* Values; RMSE, root mean square error.

Variables	Model Type	N	R^2	*F*	RMSE	*p*
MAT	linear	87	0.166	16.9	5.6	<0.001
	exponential	87	0.202	22.0	5.6	<0.001
MAP	linear	87	0.218	21.5	5.4	<0.001
	exponential	87	0.237	23.2	5.3	<0.001
MAT+MAP	linear	87	0.243	23.8	4.2	<0.001
N_{soil}	linear	42	0.106	4.7	7.0	<0.05
	exponential	42	0.148	4.5	6.7	<0.05
N_{re}	linear	86	0.284	30.1	3.4	<0.001
$N_{soil} + N_{re}$	linear	42	0.382	32.7	2.2	<0.001
MAT + MAP + N_{soil} + N_{re}	linear	42	0.543	10.6	1.6	<0.001

3.3. The Impact of Soil N on the Spatial Pattern of NPP

Similar to climatic factors, our data illustrated that soil N was also closely related to forest NPP (Figure 4). Generally, positive linear function fitted well the relationship between forest NPP and soil

N, and namely, N_{soil} and N_{re} accounted for 10.6% and 28.4% of variation of NPP (Figure 4, Table 2). When examining the combined effect of N_{soil} and N_{re}, we found that N_{soil} and N_{re} in the model jointly explained 38.2% of variation in NPP, a marked improvement over N_{soil} or N_{re} alone. (Figure 4, Table 2). Although forest NPP covaried spatially with soil N, it should be noted that a statistically significant relationship did not necessarily imply causality. Moreover, the occurrence of high N_{soil} and N_{re} rather than MAT and MAP favored the increase of NPP in boreal forest ecosystems (Figure S2). Additionally, across all bivariate analyses above, R^2-value and F-value of N_{re} for NPP were almost always higher than those of NPP versus other factors involved in this study, illustrating that plant litter N played a more important role in determining forest NPP than other environmental factors.

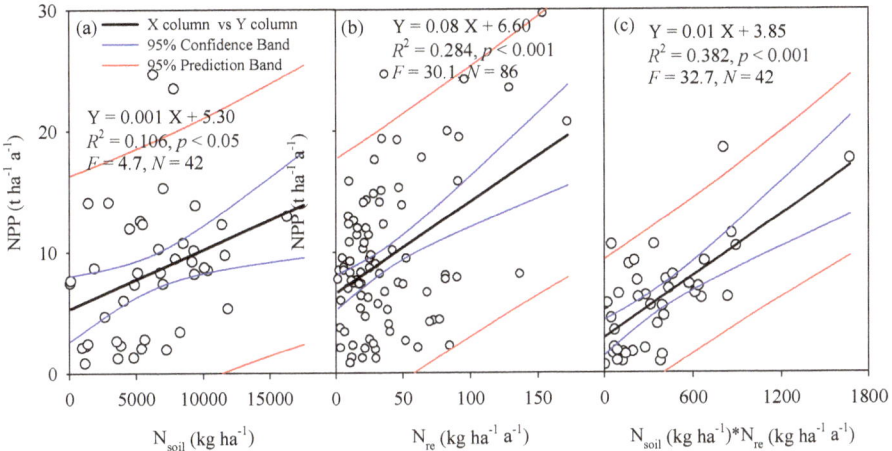

Figure 4. Relationships between net primary productivity (NPP) and soil N content (N_{soil}), annual litter N (N_{re}), and combinations of N_{soil} and N_{re}. N, number of observations; F, F Values.

3.4. The Combined Impact of Climate and Soil on the Spatial Pattern of NPP

Even though the climate-NPP and soil-NPP relationships analyzed above were statistically significant, many points were still scattered around the fitted lines, and a great deal of variability for NPP was not captured (Figures 3 and 4). Subsequently, we conducted a stepwise multiple regression to identify the effects of climatic factors and soil N on forest NPP. The results showed that N_{soil} was excluded from the linear regression model, and the explanatory power of these three factors (MAT, MAP, and N_{re}) for NPP was just 49.2%, a marginal decline over four factors together, in Equation (3).

$$NPP = -0.3387 - 0.033 \, MAT + 0.004 \, MAP + 0.073 N_{re}$$

$$R^2 = 0.492, N = 42, F = 10.2, p < 0.001 \tag{3}$$

However, there were significant collinearities between these environmental factors. Table 3 summarized correlation coefficients among these variables. Climatic factors for MAT and MAP were highly correlated, and both of them had marked correlations with N_{re}, but had non-robust correlation with N_{soil}, and mutually, N_{re} had a marginal relationship with N_{soil} (Table 3).

Table 3. Correlation matrix of independent variables. MAT, mean annual temperature; MAP, mean annual precipitation; N_{soil}, soil N content; N_{re}, annual litter N; **, $p < 0.01$.

Variables	Data Range	MAT	MAP	N_{soil}
MAT (°C)	0–23.5	1		
MAP (mm)	500.0–2000.0	0.670 **		
N_{soil} (kg ha^{-1})	115.6–163.5	0.213	0.365	
N_{re} (kg ha^{-1} a^{-1})	1.9–172.2	0.288 **	0.242 **	0.161

Considering the results of stepwise multiple regression analysis and correlation coefficients above, only MAT, MAP, and N_{re} were referred to when identifying the combined effects of the drivers on NPP in ensuing analysis. Given the significant collinearities among the three factors, their true roles for NPP could be obscured. Therefore, we used partial GLM to examine their relative causality in the control of spatial pattern of NPP. General linear model involving MAT, MAP and N_{re} could account for 84.8% of the variation in NPP, and MAT, MAP, and N_{re} explained 9.8%, 12.7%, and 35.3% of the variation in NPP, respectively; the interactive effects of MAT and MAP (ab), MAT and N_{re} (ac), MAP and N_{re} (bc), and MAT, MAP, and N_{re} (abc) represented 10.9%, 4.6%, 4.0%, and 7.5%, respectively (Figure 5). The results indicated that N_{re} was much more important in shaping forest NPP than the climatic factors.

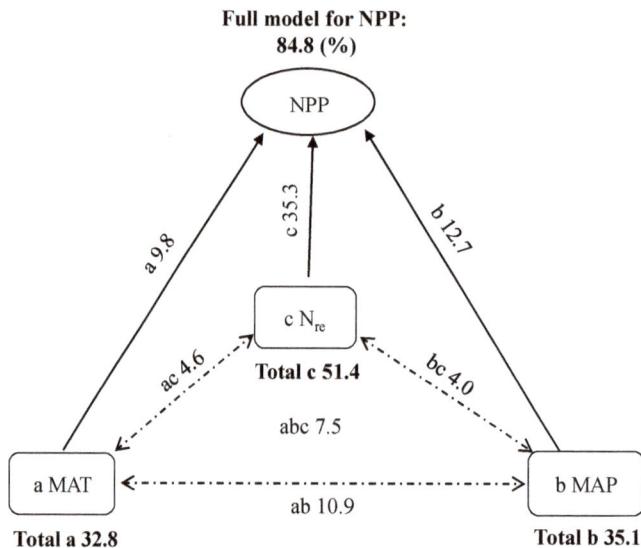

Figure 5. Summary of the partial general linear model (partial GLM) for the effects of mean annual temperature (MAT), mean annual precipitation (MAP), and annual litter N (N_{re}) on net primary productivity (NPP). In the partial GLM, a, b, and c denoted the independent effects of MAT, MAP and N_{re}, respectively; ab, ac, and bc indicated respectively the interactive effects between MAT and MAP, MAT, and N_{re}, MAP and N_{re}; and abc represented the interactive effects of the three different factors.

4. Discussion

4.1. Carbon Budget

NPP in forests in eastern China in our study were, on average, 4.3 t C ha^{-1} a^{-1}, which was similar to that in boreal forest ecosystems (4.2 t C ha^{-1} a^{-1}) [30], but lower than that in European Forest Ecosystems (6.5 t ha^{-1} a^{-1}) [31]. Actually, considering the number and spatial representation of observation sites in different regions, our results need to be further validated with more data. Based on the 7th national forest inventory data and biomass–volume relationship, we estimated that forest biomass carbon in China was approximately 840.3 Tg C a^{-1} (forest area was 195.0×10^6 ha), assuming that forests represented potential vegetation in forest regions, and the impact of human disturbance was negligible. Our findings were similar to the result reported by Ni et al. (2003) [20] (738.9 Tg C a^{-1}), but was about two times that estimated by Fang et al. (2001) [19] (461.0 Tg C a^{-1}), and the results might be ascribed to the differences in NPP estimate methods and forest area calculations. As a large country in the world, China contributes much to regional and global carbon budget. Therefore, determining distribution pattern of carbon budget of the forest ecosystems in China will be helpful for exploring the carbon cycle of the terrestrial ecosystem and addressing global warming [32].

4.2. Explanations for the Distribution Pattern of NPP

4.2.1. Climate Control

Temperature is an important factor for regulating potential photosynthetic activity of vegetation, as well as the length of the growing season [33,34], both of which jointly determine ecosystem productivity [35]. In addition to temperature, precipitation is another determinant of the spatial variation of NPP [36,37]. Numerous studies have indicated that NPP increased with both increasing temperature and precipitation, but the response rates and patterns varied in different regions [9,10,36]. Our study found that NPP in the forest ecosystems of China increased linearly at a rate of 0.38 t ha^{-1} a^{-1} for a 1.0 °C increase in MAT, and 1 t ha^{-1} a^{-1} for a 1.0 mm increase in MAP (Figure 3). Our results were similar to that of Ni et al. (2001) [38], who noted that NPP in the forest ecosystems of China increased by 0.48 kg ha^{-1} a^{-1} for 1.0 °C increase in MAT and 0.1 kg ha^{-1} a^{-1} for 1.0 mm increase in MAP. Luo et al. (2004) [10] found that NPP increased exponentially, rather than linearly, with an increase in MAT, whereas NPP increased with increasing MAP when MAP was lower than the threshold level of 1490.0 mm, then decreased when MAP was higher than the threshold level. Luyssaert et al. (2007) [36] illustrated that in a global forest study, NPP increased with increasing MAT from 5.0 to 10.0 °C, but appeared to be saturated beyond 10.0 °C. Similarly, NPP increased with increasing MAP until leveling off at 1500.0 mm. Although the responses of NPP to climatic factors were different in a certain extent, the decisive effects of temperature and precipitation on spatial variation of NPP were established in most studies.

4.2.2. N Control

Soil N is a primary factor that limits plant growth owing to the large discrepancy between demand and supply, and its vital role in plant carbon assimilation [39,40]. Significant advances have been made in the past decades toward understanding the relationship between soil N and NPP in terrestrial ecosystems, whereas substantial uncertainties persist, and discrepancies between studies remain unresolved [40–42]. Yuan et al. (2006) [43] found a significant positive linear correlation between NPP and soil inorganic N for grassland ecosystems in Mongolia, whereas Luo et al. (2004) [10] found a curved relationship between NPP and total soil N. Our field data indicated that N_{soil} alone explained nearly 10.6% of the NPP variation within the transect (Figure 4, Table 2). Hobbie (2015) [23] focused on feedback to NPP operating through litter decomposition, and reported that positive or negative effects of litter N on the later stages of litter decomposition could strengthen or weaken the positive loop of NPP. Therefore, beside N_{soil}, we introduced N_{re} to estimate the effect of soil N on forest NPP, and found that N_{re} alone accounted for 28.4% of the variation, which had higher explanation for NPP than N_{soil}, and the positive effects of N_{re} on forest NPP were confirmed (Figure 4, Table 2). The role of soil N in shaping forest NPP, in our study, was imperative for models based on mechanism and ecological process, such as CEVSA model, and BIOME-BGC model for NPP estimation. Generally, the plant growth module in CEVSA model described photosynthesis, carbon allocation, leaf area index, and litter production; the biogeochemical module simulated the transformation and decomposition of organic materials, and nitrogen inputs and outputs in soil. BIOME-BGC model was described as the function of temperature, leaf area, water, and soil nitrogen.

Soil N and litter N were involved in the key process of these models. Thus, future efforts should be focused on identifying the role of soil N, litter N, litter nutrient limiting decomposition, and litter nutrient release versus NPP.

4.2.3. Joint Control

The present study identified one geographical trend in forest productivity, and found that MAT and MAP, alone or in combination, explained 16.6–24.3% of the NPP variation within the transect, while N_{soil} and N_{re} explained 10.6–38.2% of the variation (Figures 3 and 4, Table 2). Although forest NPP covaried spatially with climatic factors and soil N, it should be noted that a statistically significant relationship did not necessarily imply causality. Importantly, we found that the trends of the relationships of NPP to soil N were more similar than that of NPP versus climatic factors, especially for boreal forests, indicating that rich soil N, rather than warm climate, stimulated the increase in productivity in this region (Figures 3 and 4). To confirm the effects of climate and soil on NPP, partial GLM was conducted, and the results showed that the overall model including MAT, MAP, and N_{re}, could account for 84.8% of the spatial variation in NPP (Figure 5). The findings suggest that the combination of temperature and precipitation-related physiology and soil substrate-related N is responsible for the observed pattern of forest productivity, and also reflects pervasive geographic pattern in the structure and function of forest ecosystems [42]. It was worth noting that in the overall model, N_{re} explained independently more than 35.0% of the total variation for NPP, while MAT and MAP explained independently less than 15.0% (Figure 5). The results above suggest that the geography of NPP was largely controlled by N_{re} rather than climate, highlighting the role of litter N in shaping the spatial pattern of forest NPP. The finding is valuable for planners and decision-makers in their attempts to evaluate the effects of nutrient status on forest ecosystems and to develop suitable strategies for forest management.

4.3. Uncertainty Analysis

In this study, we synthesized the data in the primary literature on NPP in the forest ecosystems in eastern China to produce a consistent dataset on NPP. With a wide coverage and a reasonable range, the results of this study provide a robust estimate of NPP in this region. However, there was still a great deal of variability in NPP, which might be ascribed to limitations of data and methodology. Firstly, new methods of data collection are still urgently required. Data collection should come directly from the ecologists who measure forest NPP, rather than indirectly from publications. Standard methods in measuring biomass, standard measure time, and suitable methods in estimating NPP are strongly encouraged. Secondly, uneven site distribution could be an important factor contributing to the uncertainty. In this study, the sites tended to be more concentrated in the temperate and subtropical than boreal zone; therefore, limitations in the results of the analysis were introduced by this uneven site distribution. Additionally, when analyzing the spatial variation of NPP at an annual scale, errors were likely to be produced due to different time spans of data collection. Moreover, available field measurements of NPP depend largely on the sum of the positive increments of biomass, since root production was rarely estimated. As a result, existing field-based estimates of forest NPP in our study were likely to be significant underestimates. Furthermore, it is worth noting that although MAT, MAP, and N_{re} were important factors in the spatial variation of NPP in the forests of eastern China, the total variability that was not captured was 15.2% for NPP (Figure 5). We inferred that NPP showed high spatial variability due to impacts from various environmental factors, and biological factors, such as stand age, vegetation distribution, nitrogen deposition, and disturbance (i.e., thinning harvesting irrigation, and drainage) [44–46], could weaken the climate and soil regulation on forest productivity to a certain extent. An intensive study on this aspect is expected to be carried out in the future, which is critical for expanding current analysis and including other carbon cycles related to more environmental factors.

5. Conclusions

Based on published data from Chinese literature and reports, we concluded that NPP increased from north to south within the NSTEC, and high spatial variation of NPP was found among climate zones. Spatially, climate and soil were both significantly linearly correlated to NPP, and the combined effect of MAT, MAP, and N_{re} accounted for 84.8% of the spatial variation of NPP. Considering the true roles of the controlling factors in NPP, N_{re} was the most important, followed by MAP and MAT in succession. The findings demonstrate that forest productivity was determined by climate, mainly via the status of soil N, highlighting the role of litter N in shaping the spatial pattern of forest NPP. The results are helpful for elucidating spatial variation of NPP and evaluating potential responses of forest ecosystems to global climatic change. However, further studies need to be carried out to fully understand and verify the causes of variability of NPP in forest ecosystems across wider geographical sites.

Supplementary Materials: The following materials are available online at http://www.mdpi.com/1999-4907/9/6/322/s1. Table S1: Location, mean annual temperature (MAT), mean annual precipitation (MAP), soil N content (N_{soil}), annual litter N (N_{re}) and net primary productivity (NPP) in this study. NOTES: Table S1 is a long table, and we put it in a separate file; Table S2: Summary of analysis of variance (ANOVA) of mean annual temperature (MAT), mean annual precipitation (MAP), soil N content (N_{soil}), annual litter N (N_{re}) and net primary productivity (NPP) among three classification of publication year (1980s (1983–1989), 1990s (1991–1999) and 2000s (2001–2010)). Figure S1: Comparisons of mean annual temperature (MAT), mean annual precipitation (MAP), and net primary productivity (NPP) of forest ecosystems in different climate zones. Different letters above bars indicated significant differences among climate zones, which were determined by Fisher's least significant difference (LSD) comparisons ($p < 0.05$). Figure S2: Comparisons of soil N content (N_{soil}), annual litter N (N_{re}), and net primary productivity (NPP) of forest ecosystems in different climate zones. Different letters above bars indicated significant differences among climate zones, which were determined by Fisher's least significant difference (LSD) comparisons ($p < 0.05$).

Author Contributions: Z.X. wrote the paper; G.M. collected the data; Z.T. designed the scheme and revised the paper.

Funding: This work was financially supported by National Key R & D Program of China (2017YFA0604803), 13th Five Year Informatization Plan of Chinese Academy of Sciences (XXH13505-07), and National Natural Science Foundation of China (41503078, 51509238).

Acknowledgments: The authors would like to thank all related staffs for their contributions to data collection and data analysis. Deep appreciation goes to anonymous reviewers for presenting valuable suggestions to improve this paper.

Conflicts of Interest: No potential conflict of interest was reported by the authors.

References

1. Dixon, R.K.; Brown, S.; Houghton, R.A.; Solomon, A.M.; Trexler, M.C.; Wisniewski, J. Carbon pools and flux of global forest ecosystmes. *Science* **1994**, *263*, 185–190. [CrossRef] [PubMed]

2. Niemeijer, D. Developing indicators for environmental policy: Data-driven and theory-driven approaches examined by example. *Environ. Sci. Policy* **2002**, *5*, 91–103. [CrossRef]

3. Kitayama, K.; Aiba, S.I. Ecosystem structure and productivity of tropical rain forests along altitudinal gradients with contrasting soil phosphorus pools on Mount Kinabalu, Borneo. *J. Ecol.* **2002**, *90*, 37–51. [CrossRef]

4. Raich, J.W.; Russell, A.E.; Kitayama, K.; Parton, W.J.; Vitousek, P.M. Temperature influences carbon accumulation in moist tropical forests. *Ecology* **2006**, *87*, 76–87. [CrossRef] [PubMed]

5. Keeling, H.C.; Phillips, O.L. The global relationship between forest productivity and biomass. *Glob. Ecol. Biogeogr.* **2007**, *16*, 618–631. [CrossRef]

6. Matsushita, B.; Xu, M.; Chen, J.; Kameyama, S.; Tamura, M. Estimation of regional net primary productivity (NPP) using a process-based ecosystem model: How important is the accuracy of climate data? *Ecol. Model.* **2004**, *178*, 371–388. [CrossRef]

7.   Chapin, F.S.; McFarland, J.; McGuire, A.D.; Euskirchen, E.S.; Ruess, R.W.; Kielland, K. The changing global carbon cycle: Linking plant-soil carbon dynamics to global consequences. *J. Ecol.* **2009**, *97*, 840–850.

8.   Lieth, H. Primary productivity of the biosphere. In *Modeling the Primary Productivity of the World*; Lieth, H., Whittaker, R.H., Eds.; Springer: New York, NY, USA, 1975; pp. 237–264.

9.   Knapp, A.K.; Smith, M.D. Variation among biomes in temporal dynamics of aboveground primary production. *Science* **2001**, *291*, 481–484. [CrossRef] [PubMed]

10. Luo, T.X.; Pan, Y.D.; Ouyang, H.; Shi, P.L.; Luo, J.; Yu, Z.L.; Lu, Q. Leaf area index and net primary productivity along subtropical to alpine gradients in the Tibetan Plateau. *Glob. Ecol. Biogeogr.* **2004**, *13*, 345–358. [CrossRef]

11. Del Grosso, S.; Parton, W.; Stohlgren, T.; Zheng, D.L.; Bachelet, D.; Prince, S.; Hibbard, K.; Richard Olson, A. Global potential net primary production predicted from vegetation class, precipitation, and temperature. *Ecology* **2008**, *89*, 2117–2126. [CrossRef] [PubMed]

12. Sala, O.E.; Parton, W.J.; Joyce, L.A.; Lauenroth, W.K. Primary production of the central grassland region of the United States. *Ecology* **1988**, *69*, 40–45. [CrossRef]

13. Clark, D.A.; Brown, S.; Kicklighter, D.W.; Chambers, J.Q.; Thomlinson, J.R.; Ni, J.; Holland, E.A. Net primary production in tropical forests: An evaluation and synthesis of existing field data. *Ecol. Appl.* **2001**, *11*, 371–384. [CrossRef]

14. Raich, J.W.; Russell, A.E.; Vitousek, P.M. Primary productivity and ecosystem development along an elevational gradient on Mauna Loa, Hawaii. *Ecology* **1997**, *78*, 707–721.

15. Schuur, E.A.G.; Matson, P.A. Net primary productivity and nutrient cycling across a mesic to wet precipitation gradient in Hawaiian montane forest. *Oecologia* **2001**, *128*, 431–442. [CrossRef] [PubMed]

16. Smith, M.D.; La Pierre, K.J.; Collins, S.L.; Knapp, A.K.; Gross, K.L.; Barrett, J.E.; Frey, S.D.; Gough, L.; Miller, R.J.; Morris, J.T.; et al. Global environmental change and the nature of aboveground net primary productivity responses: Insights from long-term experiments. *Oecologia* **2015**, *177*, 935–947. [CrossRef] [PubMed]

17. Scurlock, J.M.O.; Johnson, K.; Olson, R.J. Estimating net primary productivity from grassland biomass dynamics measurements. *Glob. Chang. Biol.* **2002**, *8*, 736–753. [CrossRef]

18. Field, C.B.; Behrenfeld, M.J.; Randerson, J.T.; Falkowski, P. Primary production of the biosphere: Integrating terrestrial and oceanic components. *Science* **1998**, *281*, 237–240. [CrossRef] [PubMed]

19. Fang, J.Y.; Chen, A.P.; Peng, C.H.; Zhao, S.Q.; Ci, L. Changes in forest biomass carbon storage in China between 1949 and 1998. *Science* **2001**, *292*, 2320–2322. [CrossRef] [PubMed]

20. Ni, J. Net primary productivity in forests of China: Scaling-up of national inventory data and comparison with model predictions. *For. Ecol. Manag.* **2003**, *176*, 485–495. [CrossRef]

21. Fang, J.Y.; Guo, Z.D.; Piao, S.L.; Chen, A.P. Terrestrial vegetation carbon sinks in China, 1981–2000. *Sci. China Ser. D* **2007**, *50*, 1341–1350. [CrossRef]

22. Peng, S.L.; Zhao, P.; Ren, H.; Zheng, F.Y. The possible heat-driven pattern variation of zonal vegetation and agricultural ecosystem, along the north-south of China under the global change. *Earth. Sci. Front.* **2002**, *9*, 218–226.

23. Hobbie, S.E. Plant species effects on nutrient cycling: Revisiting litter feed backs. *Trends Ecol. Evol.* **2015**, *30*, 357–363. [CrossRef] [PubMed]

24. Woodward, F.I.; Lomas, M.R.; Kelly, C.K. Global climate and the distribution of plant biomes. *Philos. Trans. Soc. B* **2004**, *359*, 1465–1476. [CrossRef] [PubMed]

25. Luo, T.X. *Patterns of Biological Production and Its Mathematical Models for Main Forest Types of China*; Committee of Synthesis Investigation of Natural Resources, Chinese Academy of Sciences: Beijing, China, 1996.

26. Pan, Y.D.; Luo, T.X.; Birdsey, R.; Hom, J.; Melillo, J. New estimates of carbon storage and sequestration in China's forests: Effects of age-class and method on inventory-based carbon estimation. *Clim. Chang.* **2004**, *67*, 211–236. [CrossRef]

27. Keith, H.; Mackey, B.G.; Lindenmayer, D.B. Re-evaluation of forest biomass carbon stocks and lessons from the world's most carbon-dense forests. *Proc. Natl. Acad. Sci. USA* **2009**, *106*, 11635–11640. [CrossRef] [PubMed]

28. Tao, B.; Cao, M.K.; Li, K.R.; Gu, F.X.; Ji, J.J.; Huang, M.; Zhang, L.M. Spatial patterns of terrestrial net ecosystem productivity in China during 1981–2000. *Sci. China Ser. D* **2007**, *50*, 745–753. [CrossRef]

29. Heikkinen, R.K.; Luoto, M.; Kuussaari, M.; Poyry, J. New insights into butterfly-environment relationships using partitioning methods. *Proc. Biol. Sci.* **2005**, *272*, 2203–2210. [CrossRef] [PubMed]

30. Gower, S.T.; Krankina, O.; Olson, R.J.; Apps, M.; Linder, S.; Wang, C. Net primary production and carbon allocation patterns of boreal forest ecosystems. *Ecol. Appl.* **2001**, *11*, 1395–1411. [CrossRef]

31. Scarascia Mugnozza, G.A.B.; Persoon, H.G.; Matteucci, A. Carbon and Nitrogen Cycling in European Forest Ecosystems. In *Tree Biomass, Growth and Nutrient Pools*; Schulze, E.D., Ed.; Springer: Berlin/Heidelberg, Germany, 2000; pp. 49–62.

32. Raupach, M.R. Carbon cycle pinning down the land carbon sink. *Nat. Clim. Chang.* **2011**, *1*, 148–149. [CrossRef]

33. Hirata, R.; Saigusa, N.; Yamamoto, S.; Ohtani, Y.; Ide, R.; Asanuma, J.; Gamo, M.; Hirano, T.; Kondo, H.; Kosugi, Y.; et al. Spatial distribution of carbon balance in forest ecosystems across East Asia. *Agric. For. Meteorol.* **2008**, *148*, 761–775. [CrossRef]

34. Saigusa, N.; Yamamoto, S.; Hirata, R.; Ohtani, Y.; Ide, R.; Asanuma, J.; Gamo, M.; Hirano, T.; Kondo, H.; Kosugi, Y.; et al. Temporal and spatial variations in the seasonal patterns of CO_2 flux in boreal, temperate, and tropical forests in East Asia. *Agric. For. Meteorol.* **2008**, *148*, 700–713. [CrossRef]

35. Chapin, F.S., III; Matson, P.A.; Mooney, H.A. *Principles of Terrestrial Ecosystem Ecology*; Springer: New York, NY, USA, 2002.

36. Luyssaert, S.; Inglima, I.; Jung, M.; Richardson, A.D.; Reichstein, M.; Papale, D.; Piao, S.L.; Schulze, E.D.; Wingate, L.; Matteucci, G.; et al. CO_2 balance of boreal, temperate, and tropical forests derived from a global database. *Glob. Chang. Biol.* **2007**, *13*, 2509–2537. [CrossRef]

37. Kato, T.; Tang, Y.H. Spatial variability and major controlling factors of CO_2 sink strength in Asian terrestrial ecosystems: Evidence from eddy covariance data. *Glob. Chang. Biol.* **2008**, *14*, 2333–2348. [CrossRef]

38. Ni, J.; Zhang, X.S.; Scurlock, J.M.O. Synthesis and analysis of biomass and net primary productivity in Chinese forests. *Ann. For. Sci.* **2001**, *58*, 351–384. [CrossRef]

39. Matson, P.; Lohse, K.A.; Hall, S.J. The globalization of nitrogen deposition: Consequences for terrestrial ecosystems. *Ambio* **2002**, *31*, 113–119. [CrossRef] [PubMed]

40. LeBauer, D.S.; Treseder, K.K. Nitrogen limitation of net primary productivity in terrestrial ecosystems is globally distributed. *Ecology* **2008**, *89*, 371–379. [CrossRef] [PubMed]

41. Finzi, A.C.; Norby, R.J.; Calfapietra, C.; Gallet-Budynek, A.; Gielen, B.; Holmes, W.E.; Hoosbeek, M.R.; Iversen, C.M.; Jackson, R.B.; Kubiske, M.E.; et al. Increases in nitrogen uptake rather than nitrogen-use efficiency support higher rates of temperate forest productivity under elevated CO_2. *Proc. Natl. Acad. Sci. USA* **2007**, *104*, 14014–14019. [CrossRef] [PubMed]

42. Cleveland, C.C.; Townsend, A.R.; Taylor, P.; Alvarez-Clare, S.; Bustamante, M.M.C.; Chuyong, G.; Dobrowski, S.Z.; Grierson, P.; Harms, K.E.; Houlton, B.Z.; et al. Relationships among net primary productivity, nutrients and climate in tropical rain forest: A pan-tropical analysis. *Ecol. Lett.* **2011**, *14*, 939–947. [CrossRef] [PubMed]

43. Yuan, Z.Y.; Li, L.H.; Han, X.G.; Chen, S.P.; Wang, Z.W.; Chen, Q.S.; Wang, Z.W.; Chen, Q.S.; Bai, Y.F. Nitrogen response efficiency increased monotonically with decreasing soil resource availability: A case study from a semiarid grassland in northern China. *Oecologia* **2006**, *148*, 564–572. [CrossRef] [PubMed]

44. Litton, C.M.; Ryan, M.G.; Knight, D.H. Effects of tree density and stand age on carbon allocation patterns in postfire lodgepole pine. *Ecol. Appl.* **2004**, *14*, 460–475. [CrossRef]

45. Wang, S.Q.; Zhou, L.; Chen, J.M.; Ju, W.M.; Feng, X.F.; Wu, W.X. Relationships between net primary productivity and stand age for several forest types and their influence on China's carbon balance. *J. Environ. Manag.* **2011**, *92*, 1651–1662. [CrossRef] [PubMed]

46. Stevens, C.J.; Lind, E.M.; Hautier, Y.; Harpole, W.S.; Borer, E.T.; Hobbie, S.E.; Seabloom, E.W.; Ladwig, L.M.; Bakker, J.D.; Chu, C.; et al. Anthropogenic nitrogen deposition predicts local grassland primary production worldwide. *Ecology* **2015**, *96*, 1459–1465. [CrossRef]

forests

MDPI

Article

Charcoal Increases Microbial Activity in Eastern Sierra Nevada Forest Soils

Zachary W. Carter [1], Benjamin W. Sullivan [1], Robert G. Qualls [1], Robert R. Blank [2], Casey A. Schmidt [3] and Paul S.J. Verburg [1,*]

[1] Department of Natural Resources and Environmental Science, University of Nevada, Reno, NV 89557, USA; zcarter@nevada.unr.edu (Z.W.C.); bsullivan@cabnr.unr.edu (B.W.S.); qualls@unr.edu (R.G.Q.)
[2] Agricultural Research Service, United States Department of Agriculture, Reno, NV 89512, USA; bob.blank@ars.usda.gov
[3] Division of Hydrologic Sciences, Desert Research Institute, Reno, NV 89512, USA; casey.schmidt@dri.edu
* Correspondence: pverburg@cabnr.unr.edu; Tel.: +1-775-784-4019

Received: 10 January 2018; Accepted: 14 February 2018; Published: 16 February 2018

Abstract: Fire is an important component of forests in the western United States. Not only are forests subjected to wildfires, but fire is also an important management tool to reduce fuels loads. Charcoal, a product of fire, can have major impacts on carbon (C) and nitrogen (N) cycling in forest soils, but it is unclear how these effects vary by dominant vegetation. In this study, soils collected from Jeffrey pine (JP) or lodgepole pine (LP) dominated areas and amended with charcoal derived from JP or LP were incubated to assess the importance of charcoal on microbial respiration and potential nitrification. In addition, polyphenol sorption was measured in unamended and charcoal-amended soils. In general, microbial respiration was highest at the 1% and 2.5% charcoal additions, but charcoal amendment had limited effects on potential nitrification rates throughout the incubation. Microbial respiration rates decreased but potential nitrification rates increased over time across most treatments. Increased microbial respiration may have been caused by priming of native organic matter rather than the decomposition of charcoal itself. Charcoal had a larger stimulatory effect on microbial respiration in LP soils than JP soils. Charcoal type had little effect on microbial processes, but polyphenol sorption was higher on LP-derived than JP-derived charcoal at higher amendment levels despite surface area being similar for both charcoal types. The results from our study suggest that the presence of charcoal can increase microbial activity in soils, but the exact mechanisms are still unclear.

Keywords: charcoal; forest soil; carbon mineralization; microbial activity; nitrification; polyphenols

1. Introduction

Wildfire is an important component of semi-arid forests of the western United States. Fire suppression and prevention during the last century have, however, contributed to increases in stand density, the number of small trees, the amount of understory shrub vegetation, and accumulation of forest floor, thereby increasing the risk of severe, stand-replacing wildfires [1]. Fire is also employed to reduce the risk of wildfire through broadcast burning or disposal of fuels following mechanical thinning through pile burning [2,3]. Fire typically reduces soil organic carbon (C) and nitrogen (N) pools due to volatilization of organic matter [4], thereby decreasing available substrate for primary decomposers. Bioavailable forms of soil N, including ammonium (NH_4^+) and nitrate (NO_3^-), can, however, increase due to chemical mineralization of organic material, deposition of ash, and lysis of soil microorganisms [5]. Elevated NH_4^+ levels can be sustained through subsequent mineralization of N. Following increases in NH_4^+, nitrification can be stimulated, resulting in an increase in soil

NO_3^- [6]. Microbial biomass typically decreases after a fire because the transfer of heat into soils is large enough to kill most organisms in the top 5 to 10 cm [3,7]. The direct effects of fire on soil chemical and microbial processes depend on fire severity, with impacts being large in high-severity wildfires or pile burns and much less pronounced in low-severity broadcast burns [8,9]. Indirect effects of fire on microbial communities can be mediated through vegetation recovery following fire, with plant colonization being important in shaping post-fire microbial community structure in high-severity fires compared to low-severity fires [10].

In addition to vegetation affecting post-fire microbial processes, the presence of charcoal formed due to the incomplete combustion of woody materials may affect microbial processes following a fire [11–14]. Charcoal is highly resistant to decay and remains chemically active for up to a century [14,15]. Charcoal may affect microbial processes through (1) an increase in soil pH affecting microbial community structure; (2) adsorption of C compounds such as polyphenols that can inhibit mineralization and nitrification; (3) adsorption of labile C compounds that can act as substrate for microbial communities; and (4) an increase in water and nutrient availability [12,13,16–18]. In addition, microorganisms may enter pore spaces smaller than 20 μm in charcoal, thus avoiding predation by larger micro-arthropods [11,13,14,17].

The biochemical composition of the vegetation from which charcoal is derived constrains its final physical and chemical properties and thus the ability of charcoal to affect microbial processes [19,20]. Previous studies used a range of pyrogenic materials including field-collected charcoal [21], activated carbon [22], grasses [12], or biochar feedstock [11]. We are aware of only one study that compared charcoal derived from different tree species. This study found that charcoal derived from ponderosa pine (*Pinus ponderosa* Lawson & C. Lawson) had a higher sorption capacity for allelopathic substances than charcoal derived from Douglas-fir (*Pseudotsuga menziesii* (Mirb.) Franco) due to a smaller tracheid diameter of the wood in ponderosa pine [23]. Despite these differences in chemistry, net nitrification rates were similar between charcoal types [23]. Other studies have shown, however, that chemical composition of charcoal can affect enzyme activity [15], thereby potentially affecting C and N fluxes in soils.

The objective of our study was to assess the response of soil microbes to charcoal that formed during wood combustion in semi-arid forests. We hypothesized that (1) microbial respiration and nitrification increases with increasing amounts of charcoal due to the increased sorption of labile C and nitrification-inhibiting compounds including polyphenols to the surface of charcoal particles, and (2) the effects of charcoal amendment on microbial activity depend on the source of charcoal, with microbial activity being greater in soils amended with charcoal that has greater surface area, porosity, and adsorption capacity.

To test our hypotheses, charcoal was derived from Jeffrey pine (*Pinus jeffreyi* Balf.) and lodgepole pine (*Pinus contorta* Douglas ex Loudon var. *murrayana* (Balf.) Engelm.), two tree species commonly found in semi-arid western forests. Charcoal was added to unburned soils collected from a site dominated by these two tree species and we measured physical and chemical properties of soils and charcoal, adsorption of polyphenolic substances, and microbial activity in amended and unamended soils. While our approach ignores the direct effects of fire on soils, this approach allowed us to separate effects of charcoal from other effects of fire on soil. In addition, our incubation conditions may represent low severity burns where direct impacts of fire on soils are small.

2. Materials and Methods

2.1. Field Site

Unburned soil and wood samples were collected from Little Valley, Nevada, approximately 30 km south of Reno, Nevada (39°14′23.76″ N, 119°52′54.62″ W). Little Valley is located in the Carson Range and the elevation ranges from 1900 to 2500 m. The overstory vegetation is dominated by Jeffrey pine and lodgepole pine. Two sites differing in overstory vegetation were selected adjacent to each other. Soils with a Jeffrey pine overstory were located on areas with a 10–15% slope and that had a sparse understory consisting of bitterbrush (*Purshia tridentata* (Pursh) DC.) and manzanita (*Arctostaphylos patula* Greene). Soils with a lodgepole pine overstory were located in an adjacent meadow on slopes of less than 5% and with an understory of mixed herbaceous species. Soils at both sites belonged to the Marla series (sandy, mixed, frigid Aquic Dystroxerepts; [24]). The climate at both sites is characterized by dry, warm summers and cold, wet winters. The mean annual air temperature is 5 °C and the mean annual precipitation is 87.5 cm with approximately half of the precipitation occurring as snow [25].

2.2. Soil Collection and Charcoal Production

We collected soil from the 0–10 cm mineral soil layer in ten locations at each site dominated by either Jeffrey pine (referred to as 'JP soil') or lodgepole pine (referred to as 'LP soil'). Each sample contained approximately 1 kg of soil. All samples were composited by vegetation type. Prior to sampling, forest floor material was removed. After sampling, we air-dried, sieved (2 mm mesh size), and homogenized the samples. Root materials present in the mineral soil were discarded.

Branch material for charcoal production was collected from multiple slash piles present at the JP and LP sites. The branches were up to 2.5 cm in diameter. We produced charcoal following the procedure of Zackrisson et al. [14]. Branch material from JP and LP was allowed to oxidize at 450 °C for 15 min in a muffle furnace. Following this treatment, all material was charred. The charcoal was then gently crushed with a mortar and pestle and sieved using a 2 mm mesh sieve. Charcoal was mixed in with the soil.

We amended unburned JP and LP soils with charcoal using four amendment levels (0.5%, 1%, 2.5%, and 5% charcoal by weight), and two charcoal sources (JP and LP). In addition, we included unamended control soils resulting in a total of 20 soil/amendment combinations. The amendment levels were based on charcoal quantification in burn scars following pile burns. These analyses revealed that mineral soils in burn scars contained approximately 2.5% charcoal (Z. Carter, unpublished data). Unburned JP soils contained 0.03% charcoal while unburned LP soils contained 0.01% charcoal (Z. Carter, unpublished data).

2.3. Soil and Charcoal Characterization

Total %C and %N of soils and charcoal were quantified in triplicate using a Leco TruSpec CN analyzer after soil and charcoal types were ground for 48 h using a Mavco vial rotator. Soil pH of unamended and amended soils was determined in triplicate using a 1:2 soil to water (g g^{-1}) ratio. Cation exchange capacity (CEC) was measured using the ammonium saturation method [26] at the Waters Agricultural Laboratories Inc., Camilla, GA, USA using a single CEC measurement for each soil and charcoal type. Surface area of charcoal was quantified using N_2 adsorption isotherms at −196 °C at the United States Department of Agriculture-Agricultural Research Service Southern Regional Research Center using a Nova 2000 Surface Area Analyzer (Quantachrome Corp., Boynton Beach, FL, USA). Specific surface area, which included micropores >2 nm, was calculated from the adsorption isotherms using the Brunner-Emmett-Teller (BET) equation. Surface area was measured on crushed samples similar to what was added to the soil. Surface area measurements were performed in duplicate.

2.4. Polyphenol Adsorption

To determine the effects of charcoal on polyphenol adsorption, we extracted unamended JP and LP soils, JP soils amended with JP-derived charcoal, and LP soils amended with LP-derived charcoal with deionized water. We used the same treatment combinations used for the potential nitrification measurements (see below). Three replicates were used for each treatment ($n = 3$), resulting in a total of 30 samples. Ten grams of soil and 30 mL of deionized water were added to 50 mL centrifuge tubes, shaken for 24 h using an overhead shaker, centrifuged at 2500 rpm, decanted, and filtered using a 0.45 μm nylon fiber filter. Polyphenol concentration in the aqueous extract was measured using the Folin-Ciocalteu reagent [27]. Absorbance was measured using a Shimadzu spectrophotometer (Shimadzu, Columbia, MD, USA) at 760 nm wavelength in quartz cuvettes and compared to tannic acid standards.

2.5. Microbial Respiration

To quantify microbial respiration, 250 mL Pyrex media storage bottles were filled with 25 g of unamended or charcoal-amended soils and deionized water was added until the soil moisture level was at 60% of field capacity by weight similar to Kolb et al. [11] and DeLuca et al. [21]. For the incubation, JP and LP soils were amended with both JP and LP charcoal, resulting in a total of 20 treatment combinations. Soil moisture at field capacity was determined by saturating the unamended and amended soils and letting water drain for 24–48 h. Moist samples were weighed prior to and after oven-drying at 105 °C until a constant mass was reached. At the start of the incubation, we weighed and capped the incubation bottles, and bottles were stored in a dark incubator at 25 °C throughout the incubation. Samples were pre-incubated for one week before any measurements were taken to minimize artifacts in response to rewetting the soil. Microbial CO_2 production was measured in all samples once a week throughout the 8-week incubation. Before each measurement, the incubation bottles were briefly opened to allow the headspace to equilibrate with atmospheric conditions. Subsequently, the bottles were closed and one headspace sample was taken at time 0 (baseline) and again after 24 h with a 250 µL syringe through a septum in the lid. The CO_2 concentration was measured using a LI-COR LI-8100A infrared gas analyzer (LI-COR Biosciences, Lincoln, NE, USA) configured for bench top measurements. Microbial respiration rates were calculated as the increase in CO_2 concentration during the 24 h measurement period. Following each respiration measurement, we weighed the bottles and re-established the target moisture content if needed. Except during the 24 h CO_2 measurement period, lids were put on the jars loosely to allow for CO_2/O_2 exchange, thus preventing the headspace and soils from becoming anoxic while minimizing water loss. Three replicate samples were used for each soil-charcoal mixture ($n = 3$), resulting in a total of 60 samples.

2.6. Potential Nitrification Rate

Potential nitrification rates were determined for unamended JP and LP soils, JP soils amended with JP-derived charcoal, and LP soils amended with LP-derived charcoal. Nitrification potential was measured using a shaken soil slurry method [28] as modified by Hart et al. [29]. We mixed 15 g of soil in a 250 mL Erlenmeyer flask with 100 mL of deionized water, 1.5 mM NH_4^+, and 1 mM PO_4^{3-} solution to ensure that nitrifiers were not limited by NH_4^+ and P availability. The slurries were shaken in an orbital shaker at 180 rpm to maintain aerobic conditions for a 24 h period at room temperature (21 °C). Subsamples (10 mL) were taken at 2, 4, 22, and 24 h from each slurry, which included both soil and solution in order to keep the ratio of soil to solution constant. The subsamples were centrifuged at 2500 rpm, decanted, filtered using a 1.5 µm glass fiber filter, and analyzed for NO_3^- concentration using a Lachat Autoanalyzer (Hach Company, Loveland, CO, USA). The nitrification rate was determined by obtaining the slope of the regression line between the NO_3^- concentration and time. The slope was subsequently corrected for temperature, water content, and soil mass in the sample. The 24-h potential nitrification was measured at the beginning of the incubation and after 4 and 8 weeks on different sets of samples. In between samplings, samples were incubated at 25 °C and maintained for water content as described for the microbial respiration measurements above. Three replicates were used for each treatment ($n = 3$), resulting in a total of 30 samples per time period. We did not include all possible treatment combinations because we were not able to extract all samples in the 24 h period without compromising the data quality. Because of the relatively slow growth of the nitrifying microbial community, and because the nitrifying community is dependent on substrate supply for maintenance and growth, the potential nitrification assay can be used to indicate relative differences in the size of the nitrifying community that are integrated over the relevant periods of the incubation (initial, 4 weeks, 8 weeks; [28–30]).

2.7. Data Analysis

Soil pH data were analyzed using Analysis of Variance (ANOVA) with soil type (JP and LP), charcoal type (JP and LP), and charcoal amendment level (0%, 0.5%, 1%, 2.5%, and 5%) as discrete independent variables. Polyphenol data were analyzed using ANOVA with soil type and charcoal amendment level as discrete independent variables. Microbial respiration was expressed as CO_2 production per gram of soil C or total C (soil C plus charcoal C). Respiration data was analyzed using repeated-measures ANOVA with soil type, charcoal type, and amendment level as discrete independent variables. Potential nitrification rate data expressed as mg N kg^{-1} soil day^{-1} were analyzed separately within each time period using ANOVA with soil type and charcoal amendment level as discrete independent variables. We tested for differences in potential nitrification between time periods using student's *t*-tests. Potential nitrification data were log-transformed for the ANOVA analyses to ensure normal distribution of the data, but untransformed data are presented for ease of interpretation. Following all ANOVA analyses, we conducted post-hoc tests to determine significant differences between treatments using a Tukey's approach. For the respiration and potential nitrification measurements, post-hoc tests were performed for each measurement time separately. All statistical analyses were conducted with JMP® 13 (Version 13, SAS Institute Inc., Cary, NC, USA). Probability levels ≤ 0.05 were considered to be significant.

3. Results

3.1. Soil and Charcoal Characteristics

Total C, N, and pH were significantly higher in LP than in JP soils (Tables 1 and 2; Figure 1). The addition of charcoal significantly increased the pH of both soils (Table 2). In LP soils, pH increased at all amendment levels, but in JP soils, pH was similar at the 0.5% and 1% amendment level. In addition, the effect of amendment level was strongest in soils amended with JP charcoal. Overall, pH was higher in soils amended with JP charcoal than in soils amended with LP charcoal. However, in JP soils, charcoal type did not affect pH, but in LP soils, pH was higher when amended with JP compared to LP charcoal. It should be noted though that differences in pH between soil and charcoal types were typically small (<0.1 unit; Figure 1). Surface area and CEC were similar in JP and LP-derived charcoal (Table 1).

Table 1. Physical and chemical properties of Jeffrey pine (JP) and lodgepole pine (LP) soils and charcoal used in this study. The pH, total C, and total N measurements were done in triplicate, surface area was measured in duplicate, and no replicates were used for cation exchange capacity (CEC) measurements.

Site	pH	Total C	Total N	CEC	Surface Area
		(%)	(%)	meq 100 g^{-1}	$m^2\,g^{-1}$
JP Soil	6.02 ± 0.02	1.59 ± 0.02	0.06 ± 0.002	9.8	-
LP Soil	6.13 ± 0.01	2.24 ± 0.21	0.08 ± 0.006	11.2	-
JP charcoal	-	93.0 ± 0.2	-	26.5	468.5 ± 14.5
LP charcoal	-	92.9 ± 5.7	-	27.7	449.1 ± 7.0

Table 2. Results from the ANOVA (degrees of freedom, df; F-values, F; and probabilities, *p*) for pH, polyphenol sorption, and respiration expressed per g soil C or total C (soil C plus charcoal C). Significant F and *p* values are highlighted in bold.

	pH			Polyphenol			Respiration Per g Soil C			Respiration Per g Total C		
	df	F	p	df	F	p	df	F	p	df	F	p
Amendment Level (AL)	4	**574.997**	**<0.001**	4	**21.845**	**<0.001**	4	**6.997**	**<0.001**	4	**24.021**	**<0.001**
Charcoal Type (CT)	1	**37.358**	**<0.001**	1	-	-	1	<0.001	0.980	1	0.030	0.863
Soil Type (ST)	1	**69.938**	**<0.001**	1	**32.040**	**<0.001**	1	**4.772**	**0.035**	1	**6.630**	**0.014**
Time (T)	-	-	-	-	-	-	8	**66.556**	**<0.001**	8	**79.612**	**<0.001**
AL × CT	4	**6.923**	**<0.001**	4	-	-	4	0.264	0.990	4	0.191	0.942
AL × ST	4	**6.333**	**<0.001**	4	**4.107**	**0.016**	4	**3.618**	**0.013**	4	**3.035**	**0.028**
AL × T	-	-	-	-	-	-	32	**2.109**	**0.002**	32	**4.215**	**<0.001**
CT × ST	1	**11.25**	**0.002**	-	-	-	1	0.021	0.886	1	0.080	0.780
CT × T	-	-	-	-	-	-	8	1.461	0.209	8	1.313	0.271
T × ST	-	-	-	-	-	-	8	**5.880**	**<0.001**	8	**5.702**	**<0.001**
AL × CT × ST	4	2.095	0.100	-	-	-	32	0.925	0.459	32	0.341	0.848
AL × CT × T	-	-	-	-	-	-	32	1.344	0.128	32	0.974	0.516
AL × ST × T	-	-	-	-	-	-	32	**1.656**	**0.027**	32	**2.160**	**<0.001**
CT × ST × T	-	-	-	-	-	-	8	**3.292**	**<0.001**	8	1.611	0.160
CT × T × ST × AL	-	-	-	-	-	-	32	**1.735**	**0.017**	32	1.097	0.349

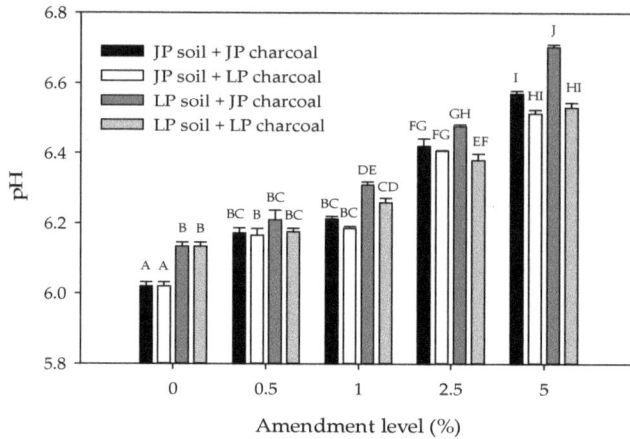

Figure 1. Soil pH as a function of charcoal type and amendment level in Jeffrey pine (JP) and lodgepole pine (LP) soils. Error bars represent standard error of the mean pH ($n = 3$). Different letters indicate significant differences ($p < 0.05$) across soil types, charcoal types, and amendment levels.

3.2. Polyphenol Adsorption

Average polyphenol concentrations in the soil extracts decreased with increasing charcoal amendment level in both JP and LP soils (Figure 2; Table 2). The post-hoc tests showed that polyphenol concentrations were similar at the 0%, 0.5%, and 1% amendment levels but consistently higher than at the 2.5% and 5% levels. The effect of amendment level varied by soil type; in LP soils, polyphenol concentrations were similar at the 0% and 0.5% amendment levels and significantly lower at the 1%, 2.5%, and 5% levels. In JP soils, polyphenol concentrations were similar at the 0%, 0.5%, and 1% levels and significantly lower at the 2.5% and 5% levels. The polyphenol concentrations were significantly higher in unamended and amended LP than in JP soils (Figure 2).

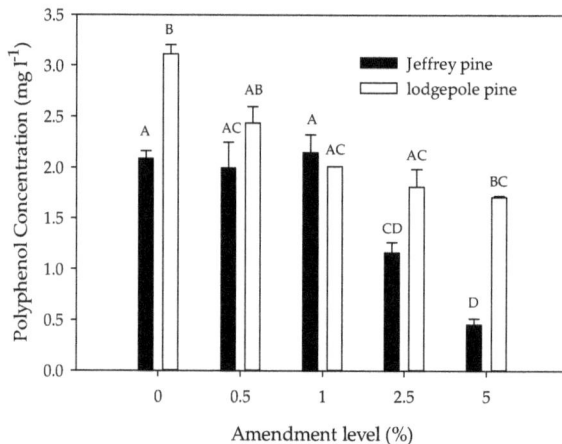

Figure 2. Polyphenol concentration in soil extracts as a function of charcoal amendment level (% by mass) in Jeffrey pine and lodgepole pine soils. Error bars represent standard error of the mean polyphenol concentration ($n = 3$). Different letters indicate significant differences ($p < 0.05$) across soil types and amendment levels.

3.3. Microbial Respiration

Overall, respiration expressed per gram of soil C was significantly higher in amended than in unamended soils (Figures 3 and 4; Tables 2 and 3), but respiration rates in the amended soils were consistently and significantly higher than in the unamended soils only at the 1% and 2.5% amendment levels (Figure 3; Tables 2 and 3). The effect of charcoal amendment level on respiration depended on the soil type. In JP soils, respiration was stimulated most at the 2.5% amendment level, whereas in LP soils, respiration was highest at the 1% level. The effect of amendment on respiration varied with time. During the first three weeks, respiration was generally similar between unamended and amended soils (Figure 3; Table 3). After three weeks, respiration was generally higher in amended than in unamended soils but was similar among amendment levels. At the 5% amendment level, respiration was only significantly higher after 7 weeks. The temporal effects of amendment also varied with soil type. When all amendment levels were combined, respiration was significantly higher in amended than in unamended soils after day 14 in LP soils (Figure 4). In JP soils, amendment only had a significant effect on respiration on the last measurement day. Respiration rates in all treatments decreased during the incubation, but temporal patterns in respiration depended on soil type, with respiration declining more rapidly with time in LP than in JP soils. Overall, respiration rates were higher in LP soils than in JP soils.

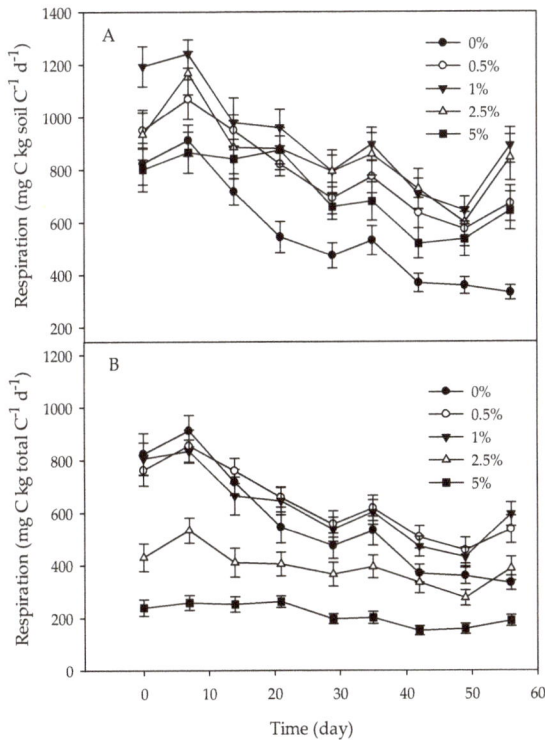

Figure 3. Microbial respiration expressed per g soil C (**A**) or total C (soil C plus charcoal C); (**B**) during a two-month long incubation. Data are pooled by soil type and charcoal type to emphasize the effect of amendment level. Day 0 represents the first measurement day following a one-week pre-incubation period. Error bars represent standard errors of the mean ($n = 12$). Results of post-hoc tests assessing differences between amendment levels for each measurement time are shown in Table 3.

Table 3. Results from post-hoc tests assessing differences in microbial respiration between amendment levels at each measurement time. Day 0 represents the first measurement day following a one-week pre-incubation period. Different letters indicate significant ($p < 0.05$) differences between amendment levels for each measurement day.

	Amendment	Day								
		0	7	14	21	29	35	42	49	56
	0%	A	A	A	A	A	A	A	A	A
	0.5%	AB	AB	AB	B	AB	AB	B	B	B
Per g soil C	1%	B	B	B	B	B	B	B	B	B
	2.5%	AB	AB	AB	B	B	B	B	B	B
	5%	A	A	AB	B	AB	AB	AB	B	B
	0%	A	A	A	AB	AB	AB	AB	AB	CD
	0.5%	A	A	A	A	A	A	A	A	AB
Per g total C	1%	A	A	A	A	A	A	AB	A	A
	2.5%	B	B	B	BC	B	BC	B	BC	BC
	5%	B	C	B	C	C	C	C	C	D

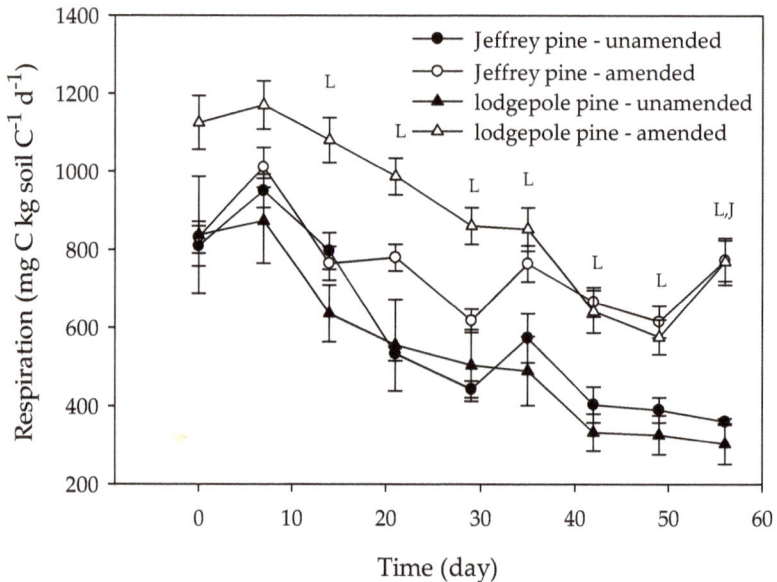

Figure 4. Microbial respiration expressed per gram of soil C in unamended and amended Jeffrey and lodgepole pine soils. Values for amended soils were pooled by amendment level and charcoal type to emphasize how soil type affects the response to charcoal amendment. Day 0 represents the first measurement day following a one-week pre-incubation period. Error bars represent standard errors of the mean ($n = 6$ for unamended soils, $n = 24$ for amended soils). The letters at different measurement times indicate significant ($p < 0.05$) differences between unamended and amended Jeffrey pine soils (J) or unamended and amended lodgepole pine soils (L) for that particular measurement time.

When respiration rates were expressed per gram of total C, respiration was similar between the 0%, 0.5%, and 1% charcoal addition levels, but respiration rates declined with increasing charcoal amendment level (Figure 3; Tables 2 and 3). Similar as when expressed per gram of soil C, the effect of amendment level on respiration varied by soil type. In JP soils, respiration was lowest at the 2.5% and

5% amendment levels. In LP soils, respiration was lowest at the 5% level, while respiration rates were similar at all other amendment levels.

Regardless of how respiration was expressed, respiration rates were not significantly affected by charcoal type. The ANOVA, however, showed significant 3- and 4-way interactions that included charcoal type when respiration was expressed per gram of soil C indicating that charcoal type affected respiration depending on measurement time and soil type (Table 2). In JP soils, charcoal type had no effect on respiration. In LP soils, average respiration was higher in soils amended with LP-derived charcoal during the first two weeks, but in the following period, respiration was higher in soils amended with JP-derived charcoal. It should be noted though that post-hoc tests conducted for each individual time period did not indicate significant differences in respiration between LP soils amended with LP- or JP-derived charcoal. None of these interactions were present when respiration was expressed per gram of total C.

3.4. Potential Nitrification

The potential nitrification rate was similar for day 0 and day 30 ($p = 0.84$) but significantly higher at day 64 compared to day 0 ($p < 0.001$) and day 30 ($p < 0.001$). Effects of amendment level and soil type depended on the measurement period. At day 0, potential nitrification was similar in the unamended soil and soils amended with 0.5%, 1%, and 2.5% charcoal, but it was significantly lower at the 5% amendment level (Figure 5; Table 4). The effects of amendment varied by soil type; in JP soils, nitrification was similar regardless of amendment level, but in LP soils, nitrification was significantly lower at the 2.5% and 5% amendment levels compared to the unamended soils (Figure 5; Table 4). Overall, potential nitrification rates were similar for JP and LP soils.

At day 30, potential nitrification rates were significantly higher in amended than the unamended soils, but rates were similar for the 0.5%, 1%, 2.5%, and 5% amendment levels (Figure 5; Table 4). The potential nitrification rate for the unamended LP soils appeared to be anomalously low, however, given that rates for unamended soils at day 0 and day 64 were much higher than for day 30. Similar to the observed patterns at day 0, the effect of charcoal addition at day 30 varied by soil type, with potential nitrification being similar for unamended and amended JP soils, but the anomalously low value for the unamended LP soils at day 30 caused the amended LP soils to have a significantly higher nitrification rate. Overall, potential nitrification was significantly higher in JP than LP soils, but when the low value for the LP soils was removed, no significant differences between soil types were observed.

At day 64, potential nitrification rates in the 2.5% charcoal amendment treatments exceeded the unamended treatment, but this was mainly caused by patterns observed in LP soils; potential nitrification rates were similar in unamended and amended JP soils, but in LP soils rates were higher in the 2.5% amendment treatment compared to the unamended soils. At day 64, potential nitrification rates were significantly higher in LP than in JP soils.

Table 4. Results from ANOVA (degrees of freedom, df; F-values, F; and probabilities, *p*) for potential nitrification for each measurement period. Day 0 represents the first measurement day following a one-week pre-incubation period. Significant F and *p* values are given in bold.

	Day 0			Day 30			Day 60		
	df	F	*p*	df	F	*p*	df	F	*p*
Amendment Level	4	**5.941**	**0.026**	4	**7.569**	**0.001**	4	**3.444**	**0.028**
Soil Type	1	0.177	0.679	1	**8.414**	**0.010**	1	**8.478**	**0.009**
Amend. Level × Soil Type	4	**3.224**	**0.034**	4	**7.230**	**0.002**	4	0.838	0.518

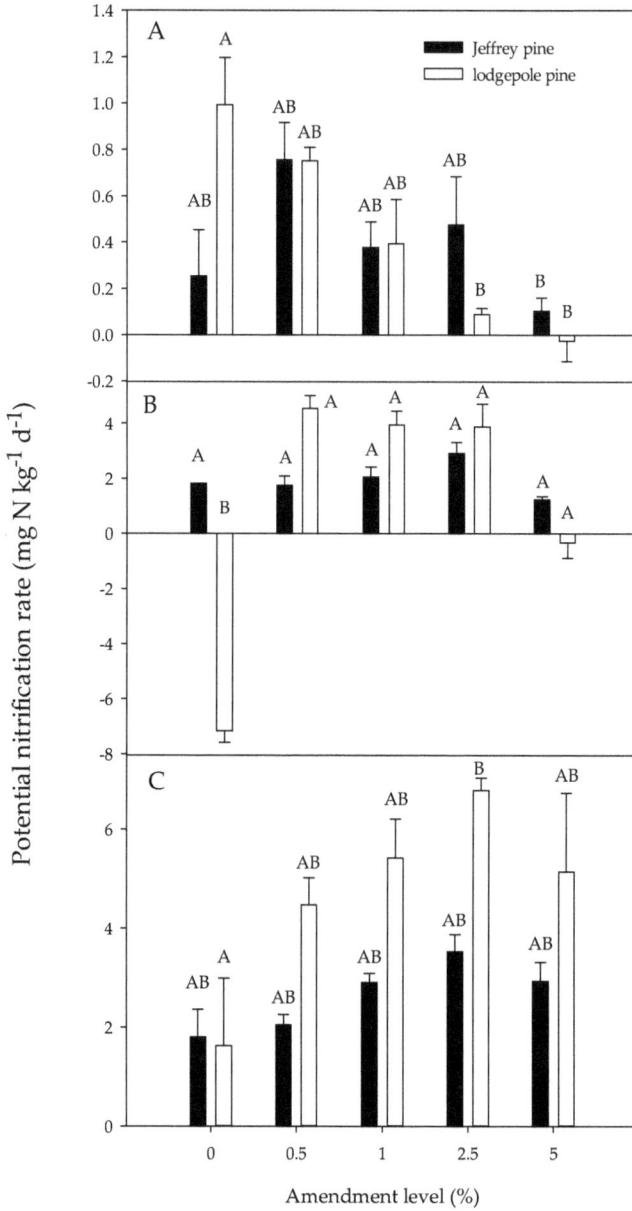

Figure 5. Potential nitrification rate as a function of the charcoal amendment level for both Jeffrey pine and lodgepole pine soil treatments at day 0 (**A**), day 30 (**B**), and day 64 (**C**). Day 0 represents the first measurement day following a one-week pre-incubation period. Error bars represent standard error of the mean (*n* = 3). Different letters indicate significant differences (*p* < 0.05) across soil types and amendment levels.

4. Discussion

Our study showed that the addition of charcoal to soils with low background levels of charcoal had mixed effects on microbial processes. Microbial respiration generally increased at intermediate amendment levels, but potential nitrification rates showed few effects of charcoal addition. Several mechanisms may explain our observed patterns. First, any increases in respiration upon addition of charcoal could have been caused by priming of native organic matter. This priming could be due to charcoal providing a refuge for microbes, thereby allowing them to avoid predation by larger soil fauna [14]. Previous studies have shown mixed effects of charcoal on priming and noted that conflicting reports regarding effects of priming may be due to soil types used, length of the incubation, and pre-incubation conditions, but no specifics were given how these factors may have affected priming [31]. Whether or not priming was important in our study could have been addressed by using isotopically labeled charcoal and measuring the isotopic signature of the respired CO_2. However, creating uniformly labeled charcoal would require a very lengthy continuous exposure to ^{13}C-enriched or depleted CO_2, which would be challenging. In addition, we could have used a soil that contained organic matter derived from C_4 vegetation to create a differential C signature between SOM and charcoal. We chose to use native soils originating from areas associated with the vegetation used to produce charcoal in order to include native microbial populations.

Second, decomposition of charcoal itself may have contributed to the increase in microbial respiration [12,32]. If this were the case, respiration expressed per gram of total C should have increased with increasing amounts of charcoal. Instead, respiration per gram of total C decreased with increasing charcoal amendment level, indicating that, even if charcoal itself decomposed, it had a lower quality of C than the native organic matter in the soil [18] and was not the main reason why respiration increased at intermediate amendment levels. Similar to priming, previous studies have shown inconsistent results with regard to the decomposability of charcoal. For instance, one study showed that charcoal derived from ponderosa pine was very resistant against decomposition by fungi [15], but a second study showed that the same charcoal type can decompose especially when charcoal had not been exposed to light or ultraviolet radiation [31]. The interaction between charcoal amendment level and soil type (Figure 3; Table 3) showed that charcoal had a larger stimulatory effect in LP than in JP soils, further indicating that other mechanisms than decomposition of charcoal itself were responsible for the observed patterns of respiration.

Third, charcoal may have adsorbed labile C compounds that were subsequently used as a substrate by microorganisms [14,17,22,33,34], thereby causing an increase in respiration. Our polyphenol data showed that polyphenol concentrations were lower in charcoal-amended than in unamended soils (Figure 2). Sorption of polyphenols was higher in JP soils at the higher amendment levels, but respiration did not increase at these higher levels, thus indicating that sorption of C compounds by itself may not explain patterns in respiration. Despite increased sorption of polyphenols, the effects of charcoal amendments on potential nitrification were small. When in solution, polyphenols can either immobilize N through complexation or cause enzymatic inhibition of nitrification, but when removed from solution, nitrification can be stimulated [17,21,35]; however, our data did not show consistent increases in nitrification potential at higher amendment levels. In addition, overall potential nitrification rates were higher in LP soils despite polyphenol sorption being lower, suggesting that the relationship between polyphenols and potential nitrification was tenuous at best.

Finally, the addition of charcoal caused an increase in soil pH (Figure 1), potentially making conditions more favorable for the microbial community and/or shifting the microbial community structure. Still, the changes in pH upon charcoal addition were relatively modest (around 0.1 unit) and unlikely to be ecologically significant. Other studies have shown, however, that the presence of charcoal can directly affect the composition of the microbial community. For instance, a recent study showed that the presence of activated charcoal increased the relative abundance of an unidentified bacterium and an Actinomycetales and decreased the relative abundance of a *Flavobacterium*, suggesting that charcoal can have an important impact on microbial community structure [36]. Still, overall, no single mechanism appeared to explain the observed patterns in respiration and potential nitrification in our study. We conducted a multiple regression analysis and the only significant correlation we observed was between polyphenol concentration and pH (r^2 = 0.52; p < 0.001). We did not find significant correlations between pH and/or polyphenol concentration and potential nitrification rates or microbial respiration.

In general, charcoal type did not differentially affect microbial respiration in the amended soils, but the ANOVA showed that for LP soils, charcoal type affected respiration depending on the time period. It appeared that LP soils amended with LP-derived charcoal initially had higher respiration rates than soils amended with JP-derived charcoal, but this pattern was reversed in the later stages of the incubation. It is not entirely clear what mechanisms were responsible for these patterns, given that chemical and physical properties of JP- and LP-derived charcoal were very similar and this effect was only present in LP soils. Still, the different charcoal types differentially affected soil pH and polyphenol sorption, suggesting that other properties besides surface area and CEC may affect charcoal behavior in soils. The interaction between charcoal type, soil type, and time was not present when respiration was expressed per gram of total (soil C plus charcoal C), potentially because differences between soil types became smaller as relative differences in total C decreased with increasing charcoal content. Our results support previous studies showing that charcoal type does not have a major impact on microbial processes. A study comparing microbial activity in soils amended with charcoal originating from *Empetrum nigrum* twigs and forest humus material and activated C mixed with pumice found that both charcoal types stimulated basal respiration similarly [17]. In addition, another study found that net nitrification rates were similar in charcoal derived from ponderosa pine bark compared to other charcoal types despite physico-chemical characteristics being different [23]. This last study suggested that the temperature at which charcoal is produced is more important in determining microbial responses than source material. The lack of response of microbial processes to differences in charcoal composition in our study and other studies is somewhat surprising given that recent studies have shown that changes in chemical composition of charcoal can affect enzyme activity [15,31]. It is important to note that we did not cross all soil and charcoal type treatments for the polyphenol and potential nitrification measurements, as soils were amended with their respective charcoal types. As a result, it is difficult to separate the effects of soil type versus charcoal type on potential nitrification and/or polyphenol sorption. Particularly the polyphenol data showed differential behavior of charcoal/soil type that can most likely be ascribed to charcoal type rather than soil type. Polyphenol sorption was much higher at the higher amendment levels in JP soils than in LP soils despite surface area being similar between charcoal types, suggesting differences in reactivity between the charcoal types.

Charcoal appeared to have a larger stimulatory effect on microbial respiration in LP than in JP soils (Figure 3), which may be related to the type of organic matter present in the soil. Both under- and overstory vegetation differed between soil types, so it is likely that, in addition to differences in %C and %N, the organic matter chemistry was different between vegetation types as well. Despite these potential differences, respiration in unamended soils expressed per gram of soil C was similar in unamended JP and LP soils (Figure 4). Potential nitrification rates at the end of the incubation were higher in LP than in JP soils, however, (Figure 5; Table 4) indicating a higher capacity of LP soils to support nitrifying microorganisms. The higher nitrification rate in LP soils was not due to lower concentrations of potentially inhibitory compounds, since polyphenol concentrations were higher in LP than in JP unamended soils (Figure 5), again suggesting that relationships between polyphenol concentrations and nitrification are weak. The higher potential nitrification rates may have been related to potentially higher rates of N mineralization and thus increased NH_4^+ availability in LP soils.

The results from our study are different from previous studies that showed that respiration, NO_3^- concentrations, and net nitrification rates increase with increasing charcoal concentration even when higher amendment levels were used [11,21,22]. In some studies, lower levels of charcoal addition were used of up to 1% [21,22], but Kolb et al. [11] used amendment levels up to 10% and observed the largest stimulation of microbial respiration at the 5% and 10% amendment levels. The reasons for the discrepancies between our study and previous studies are not entirely clear. Potentially, the type of charcoal used varied between studies. Previous studies used activated C [20] or charcoal derived from mixtures of manure and woody materials [11] although our study and other studies suggest that charcoal type may not have a large effect on microbial processes. Potentially, particle size distribution of the charcoal and associated surface area may have varied between studies. Some studies used a mixture of course (>2mm) and fine (<2 mm) charcoal [11], fine (<2 mm) charcoal only [21], or unknown particle size [22]. Finally, incubation times varied between studies ranging from 21 days [32] to about 100 days [11] and several studies, including ours, show that effects of amendments can change over time. Particularly our potential nitrification data showed temporal effects and results could have changed had we extended the incubation period.

5. Conclusions

In our study, charcoal was added to soils that had not recently been subjected to fire, similar to the approach taken in other studies [11,21,22]. By amending unburned soils with charcoal, we were able to separate the effects of charcoal from effects of fire on soils, but one potential drawback is that the results of this study may not easily apply to field conditions. However, our incubation conditions may be representative for low-severity burns where direct impacts of fire on soils are typically small [8]. Overall, the presence of charcoal appeared to stimulate microbial respiration and may benefit the recovery of microbial populations following fire events. The degree of stimulation depended on the amendment level and at high levels of charcoal (>2.5%) presence of charcoal did not affect microbial respiration. Presence of charcoal had limited effects on potential nitrification rates. Charcoal had a larger stimulatory effect on microbial respiration in LP soils than JP soils. Charcoal type had little effect on microbial processes, but polyphenol sorption was higher on LP-derived than JP-derived charcoal at higher amendment levels despite surface area being similar for both charcoal types. In addition to the direct effects of charcoal on microbial processes, charcoal may indirectly affect microbial activity by increasing the water holding capacity and nutrient availability of the soil [34,37]. These indirect effects are likely to have a positive impact on microbial processes, but these aspects were not included in our study. Despite our results showing generally positive effects of charcoal on microbial activity, the exact mechanisms responsible for these effects are still somewhat poorly understood.

Acknowledgments: Funding for this work was provided by the Department of Natural Resources and Environmental Science of the University of Nevada, Reno, and the Whittell Graduate Fellowship awarded to Zachary Carter. We thank Jonathan Birkel who assisted with the field and laboratory work. We thank two anonymous reviewers for their valuable comments that greatly helped improve the paper.

Author Contributions: Z.C. and P.V. conceived and designed the experiments; Z.C. performed the experiments; Z.C. and P.V. analyzed the data; B.S., R.Q., R.B., and C.S. contributed to the data collection and provided analytical capabilities; Z.C. and P.V. wrote the paper.

Conflicts of Interest: The authors declare no conflict of interest.

References

1. Stephens, S.L.; Collins, B.M. Fire regimes of mixed-conifer forests in the north-central Sierra Nevada at multiple spatial scales. *Northwest Sci.* **2004**, *78*, 12–23.
2. Moghaddas, E.E.Y.; Stephens, S.L. Thinning, burning, and thin-burn fuel treatment effects on soil properties in a Sierra Nevada mixed-conifer forest. *For. Ecol. Manag.* **2007**, *250*, 156–166. [CrossRef]
3. Jiménez Esquilín, A.E.; Stromberger, M.E.; Massman, W.J.; Frank, J.M.; Shepperd, W.D. Microbial community structure and activity in a Colorado Rocky Mountain forest soil scarred by slash pile burning. *Soil Biol. Biochem.* **2007**, *39*, 1111–1120. [CrossRef]
4. Johnson, D.W.; Miller, W.W.; Susfalk, R.B.; Murphy, J.D.; Dahlgren, R.A.; Glass, D.W. Biogeochemical cycling in forest soils of the eastern Sierra Nevada Mountains, USA. *For. Ecol. Manag.* **2009**, *258*, 2249–2260. [CrossRef]

5. Prieto-Fernández, Á.; Carballas, M.; Carballas, T. Inorganic and organic N pools in soils burned or heated: Immediate alterations and evolution after forest wildfires. *Geoderma* **2004**, *121*, 291–306. [CrossRef]

6. Canfield, D.E.; Glazer, A.N.; Falkowski, P.G. The evolution and future of Earth's nitrogen cycle. *Science* **2010**, *330*, 192–196. [CrossRef] [PubMed]

7. Choromanska, U.; DeLuca, T.H. Prescribed fire alters the impact of wildfire on soil biochemical properties in a Ponderosa pine forest. *Soil Sci. Soc. Am. J.* **2001**, *65*, 232–238. [CrossRef]

8. Certini, G. Effects of fire on properties of forest soils: A review. *Oecologia* **2005**, *143*, 1–10. [CrossRef] [PubMed]

9. Holden, S.R.; Rogers, B.M.; Treseder, K.R.; Randerson, J.T. Fire severity influences the response of soil microbes to a boreal forest fire. *Environ. Res. Lett.* **2016**, *11*, 035004. [CrossRef]

10. Knelman, J.E.; Graham, E.B.; Trahan, N.A.; Schmidt, S.K.; Nemergut, D.R. Fire severity shapes plant colonization effects on bacterial community structure, microbial biomass, and soil enzyme activity in secondary succession of a burned forest. *Soil Biol. Biochem.* **2015**, *90*, 161–168. [CrossRef]

11. Kolb, S.E.; Fermanich, K.J.; Dornbush, M.E. Effect of charcoal quantity on microbial biomass and activity in temperate soils. *Soil Sci. Soc. Am. J.* **2009**, *73*, 1173–1181. [CrossRef]

12. Smith, J.L.; Collins, H.P.; Bailey, V.L. The effect of young biochar on soil respiration. *Soil Biol. Biochem.* **2010**, *42*, 2345–2347. [CrossRef]

13. Wardle, D.A.; Nilsson, M.C.; Zackrisson, O. Fire-derived charcoal causes loss of forest humus. *Science* **2008**, *320*, 629. [CrossRef] [PubMed]

14. Zackrisson, O.; Nilsson, M.C.; Wardle, D.A. Key ecological function of charcoal from wildfire in the Boreal forest. *Oikos* **1996**, *77*, 10–19. [CrossRef]

15. Gibson, C.; Berry, T.D.; Wang, R.; Spencer, J.A.; Johnston, C.T.; Jiang, Y.; Bird, J.A.; Filley, T.R. Weathering of pyrogenic organic matter induces fungal oxidative enzyme response in single culture inoculation experiments. *Org. Geochem.* **2016**, *92*, 32–41. [CrossRef]

16. Alexander, M. *Introduction to Soil Microbiology*, 2nd ed.; Wiley: New York, NY, USA, 1977; 467p.

17. Pietikäinen, J.; Kiikkilä, O.; Fritze, H. Charcoal as a habitat for microbes and its effect on the microbial community of the underlying humus. *Oikos* **2000**, *89*, 231–242. [CrossRef]

18. DeLuca, T.H.; Aplet, G.H. Charcoal and carbon storage in forest soils of the Rocky Mountain West. *Front. Ecol. Environ.* **2008**, *6*, 18–24. [CrossRef]

19. Keech, O.; Carcaillet, C.; Nilsson, M.C. Adsorption of allelopathic compounds by wood-derived charcoal: The role of wood porosity. *Plant Soil.* **2005**, *272*, 291–300. [CrossRef]

20. Hatton, P.J.; Chatterjee, S.; Filley, T.; Dastmalchi, K.; Plante, A.F.; Abiven, S.; Ga, X.; Masiello, C.A.; Leavitt, S.W.; Nadelhoffer, K.J.; et al. Tree taxa and pyrolysis temperature interact to control the efficacy of pyrogenic organic matter formation. *Biogeochemistry* **2016**, *130*, 103–116. [CrossRef]

21. DeLuca, T.H.; MacKenzie, M.D.; Gundale, M.J.; Holben, W.E. Wildfire-produced charcoal directly influences nitrogen cycling in ponderosa pine forests. *Soil Sci. Soc. Am. J.* **2006**, *70*, 448–453. [CrossRef]

22. Berglund, L.M.; DeLuca, T.H.; Zackrisson, O. Activated carbon amendments to soil alters nitrification rates in Scots pine forests. *Soil Biol. Biochem.* **2004**, *36*, 2067–2073. [CrossRef]

23. Gundale, M.J.; DeLuca, T.H. Temperature and source material influence ecological attributes of ponderosa pine and Douglas-fir charcoal. *For. Ecol. Manag.* **2006**, *231*, 86–93. [CrossRef]

24. Soil Survey Staff. *Keys to Soil Taxonomy*, 12th ed.; United States Department of Agriculture-National Resources Conservation Service: Washington, DC, USA, 2014; pp. 1–360.

25. Johnson, B.G.; Johnson, D.W.; Miller, W.W.; Carroll-Moore, E.M.; Board, D.I. The effects of slash pile burning on soil and water macronutrients. *Soil Sci.* **2011**, *176*, 413–425. [CrossRef]

26. Sumner, M.E.; Miller, W.P. Cation exchange capacity and exchange coefficients. In *Methods of Soil Analysis. Part 3. Chemical Methods*; Sparks, D.L., Ed.; Soil Science Society of America: Madison, WI, USA, 1996; pp. 1201–1230.

27. Rice, E.W.; Baird, R.B.; Eaton, A.D.; Clesceri, L.S. *Standard Methods for Examination of Water and Wastewater*, 22nd ed.; American Water Works Association: Washington, DC, USA, 2012.

28. Belser, L.W.; Mays, E.L. Specific inhibition of nitrite oxidation by chlorate and its use in assessing nitrification in soils and sediments. *Appl. Environ. Microb.* **1980**, *39*, 505–510.

29. Hart, S.C.; Stark, J.M.; Davidson, E.A.; Firestone, M.K. Nitrogen mineralization, immobilization, and nitrification. In *Methods of Soil Analysis: II. Microbiological and Biochemical Properties*; Weaver, R.W., Ed.; Soil Science Society of America: Madison, WI, USA, 1994; pp. 985–1018.

30. Sullivan, B.W.; Selmants, P.C.; Hart, S.C. New evidence that high potential nitrification rates occur in soils during dry sesasons: Are microbial communities metabolically active during dry seasons? *Soil Biol. Biochem.* **2012**, *53*, 28–31. [CrossRef]

31. Wang, R.; Gibson, C.D.; Berry, T.D.; Jiang, Y.; Bird, J.A.; Filley, T.R. Photooxidation of pyrogenic organic matter (PyOM) reduces the reactive, labile C pool and the apparent soil oxidative microbial enzyme response. *Geoderma* **2017**, *293*, 10–18. [CrossRef]

32. Gundale, M.J.; DeLuca, T.H. Charcoal effects on soil solution chemistry and growth of Koeleria macrantha in the ponderosa pine/Douglas-fir ecosystem. *Biol. Fertil. Soils* **2007**, *43*, 303–311. [CrossRef]

33. DeLuca, T.H.; Nilsson, M.C.; Zackrisson, O. Nitrogen mineralization and phenol accumulation along a fire chronosequence in northern Sweden. *Oecologia* **2002**, *133*, 206–214. [CrossRef] [PubMed]

34. Wardle, D.A.; Zackrisson, O.; Nilsson, M.C. The charcoal effect in Boreal forests: Mechanisms and ecological consequences. *Oecologia* **1998**, *115*, 419–426. [CrossRef] [PubMed]

35. Ball, P.N.; MacKenzie, M.D.; DeLuca, T.H.; Holben, W.E. Wildfire and charcoal enhance nitrification and ammonium-oxidizing bacterial abundance in dry montane forest soils. *J. Environ. Qual.* **2010**, *39*, 1243–1253. [CrossRef] [PubMed]

36. Nolan, N.E.; Kulmatiski, A.; Beard, K.H.; Norton, J.M. Activated carbon decreases invasive plant growth by mediating plant–microbe interactions. *AoB Plants* **2015**, *7*, 1–11. [CrossRef] [PubMed]

37. Glaser, B.; Lehmann, J.; Zech, W. Ameliorating physical and chemical properties of highly weathered soils in the tropics with charcoal—A review. *Biol. Fertil. Soils* **2002**, *35*, 219–230. [CrossRef]

forests

MDPI

Article

Soil Organic Matter Accumulation and Carbon Fractions along a Moisture Gradient of Forest Soils

Ewa Błońska * and Jarosław Lasota

Department of Forest Soil Science, Faculty of Forestry, University of Agriculture, Al. 29 Listopada 46, 31-425 Krakow, Poland; rllasota@cyf-kr.edu.pl
* Correspondence: eblonska@ar.krakow.pl

Received: 23 October 2017; Accepted: 14 November 2017; Published: 17 November 2017

Abstract: The aim of the study was to present effects of soil properties, especially moisture, on the quantity and quality of soil organic matter. The investigation was performed in the Czarna Rózga Reserve in Central Poland. Forty circular test areas were located in a regular grid of points (100 × 300 m). Each plot was represented by one soil profile located at the plot's center. Sample plots were located in the area with Gleysols, Cambisols and Podzols with the water table from 0 to 100 cm. In each soil sample, particle size, total carbon and nitrogen content, acidity, base cations content and fractions of soil organic matter were determined. The organic carbon stock (SOCs) was calculated based on its total content at particular genetic soil horizons. A Carbon Distribution Index (CDI) was calculated from the ratio of the carbon accumulation in organic horizons and the amount of organic carbon accumulation in the mineral horizons, up to 60 cm. In the soils under study, in the temperate zone, moisture is an important factor in the accumulation of organic carbon in the soil. The highest accumulation of carbon was observed in soils of swampy variant, while the lowest was in the soils of moist variant. Large accumulation of C in the soils with water table 80–100 cm results from the thick organic horizons that are characterized by lower organic matter decomposition and higher acidity. The proportion of carbon accumulation in the organic horizons to the total accumulation in the mineral horizons expresses the distribution of carbon accumulated in the soil profile, and is a measure of quality of the organic matter accumulated. Studies have confirmed the importance of moisture content in the formation of the fractional organic matter. With greater soil moisture, the ratio of humic to fulvic acids (HA/FA) decreases, which may suggest an increase in carbon mobility in soils.

Keywords: carbon distribution index; moisture gradient; soil organic matter fraction

1. Introduction

Accumulation of organic carbon in soils is currently discussed for many reasons. With the increased pool of carbon stored in terrestrial ecosystems, a reduction of carbon dioxide emissions into the atmosphere and prevention from climate changes are observed [1,2]. In the carbon cycle on Earth, soil is its most important reservoir. It is estimated that carbon contained in soil constitutes 75% of the total pool of organic carbon, exceeding twice the resources of carbon in the atmosphere [3]. The accumulation of organic carbon in forest soils depends on many factors, but predominantly on climatic conditions [4], soil properties [5], soil moisture [6,7], the type of plant cover and type of forest management [8–11]. The site conditions and plant coverage are strongly related to each other. Here, the site conditions play the main role via driving formation of certain species composition and the structure of plant cover, thus determining the type and direction of soil formation processes [12]. Among the physical properties of the soil, clay fraction content acts in the formation of soil organic matter (SOM) accumulation. Many authors have studied the influence of clay and silt content on the protection of SOM, and have found a relationship between the content of SOM and soil texture, mainly with the content of fine fractions (silt and clay) [13,14]. According to Wosten et al. [15], the accumulation and stabilization of soil carbon are associated with the content of clay; however, it may also depend on the soil moisture. Bauer et al. [16] and Meersmaus et al. [17] reported a strong influence of soil moisture on the decomposition of SOM, and consequently, on the carbon accumulation in the soil. SOM decomposition dynamics are influenced by pH making it an important variable, as an increase in SOM content is connected with changes of other soil parameters, namely pH [18].

By specifying the total content of soil carbon (C) [19] or by estimating the available fractions of SOM [20], the quantity and quality of SOM can be determined. To evaluate the changes in soil C dynamics due to agricultural use, the use of SOM chemical fractions was more effective than the determination of total SOM in Guimarães et al.'s [21] study. The SOM fraction can be used as an indicator of changes in the soil. The amount of SOM fractions reflects the potential mobility of C in the soil and the rate of SOM decomposition [22].

Until now, a limited number of studies reporting the relationship between the forest site condition, the accumulation of SOM, and fractions of humic substances has been published. There is a lack of studies on the effect of different moisture levels, soil properties, and natural forest plant communities on the quantity and quality of SOM in temperate climates. We hypothesize that: (1) soil moisture is an important factor in the accumulation of organic carbon in the temperate zone, and the highest accumulation of carbon was connected with a swampy variant of soils; (2) moisture has an effect on the proportion of carbon accumulation in the organic horizon to the accumulation in the mineral horizons; and (3) fractional composition of the organic matter reflects the rate of SOM decomposition in forest soils and depends on soil moisture.

2. Materials and Methods

2.1. Study Area

The study was performed in the Czarna Rózga Reserve in Central Poland (Figure 1). The reserve area is 185.6 ha, and within the limits of the reserve stands is the dominant species common alder (*Alnus glutinosa*), which are of natural origin. The silver fir (*Abies alba*), pedunculate oak (*Quercus robur*), common hornbeam (*Carpinus betulus*) and common ash (*Fraxinus excelsior*) play the role of admixtures. The study area is characterized by the following climatic conditions: the average annual rainfall is 649 mm, the average daily temperature of the warmest month (July) is 17.8 °C and the coldest month (January) is −3.4 °C, and the length of the vegetation season lasts 200–210 days. Sample plots were located in the area with a predominance of fluvioglacial sand and loam with Gleysols, Cambisols and Podzols [23]. The groundwater level is the basic reason for differentiation of soil subtype and vegetations communities.

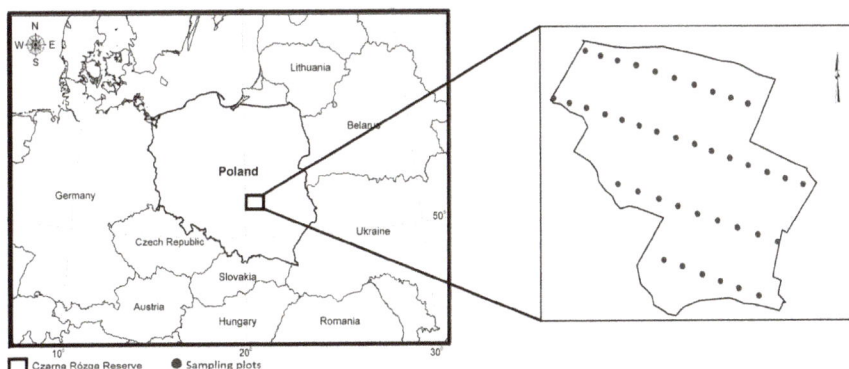

Figure 1. Study sites location (Czarna Rózga Reserve in Central Poland).

2.2. Sampling Scheme

Forty circular test areas with a surface of 0.1 hectare were established within the area of the reserve. Test surfaces were located in a regular grid of points (100 × 300 m) (Figure 1). Each plot was represented by one soil profile located at the plot's center. The surface horizons (organic—O, mineral—A, or organic/mineral—AM) and the mineral topsoil (AE or AG) horizon were sampled. The subsoil gleyic horizon (G) was sampled from the bottom of the topsoil horizon to 60 cm deep. The surface horizons' samples constituted a cumulative sample from five subsamples around the soil profile.

On each test surface (Table 1), site conditions including vegetation types, soil types, and moisture levels (based on the groundwater depth and using soil moisture monitoring sensors in the daily interval in May and June) were recorded. The obtained results revealed three types of vegetation communities (*Abietetum albae*, *Tilio-Carpinetum ficarietosum* and *Fraxino-Alnetum*) within the reserve and water table from 0 to 100 cm in soil profiles. The three variants of soil moisture content (first variant with water table 80–100 cm (WT_{80-100}), second variant with water table 40–50 cm—moist, and last variant with water table 0–30 cm—swampy) were distinguished. In further analysis, the above sequence of soil moisture content (WT_{80-100}, moist and swampy) was considered. Ten areas that represent the *Abietetum albae* from Stagnosols and Podzols have been identified for the WT_{80-100} variant. The moist variant represents the *Tilio-Carpinetum ficarietosum* (19 areas). Gleysols dominated this variant. The third swampy variant (11 areas) is represented by the *Fraxino-Alnetum* dominated by Mollic Gleysols.

Table 1. Characteristic of vegetation and soils in groups of studied areas.

Moisture Gradient	Plant Communities	Dominant Stand Species	Dominant Ground Cover Species	Type of Soil	Depth of Groundwater Level (in the Spring) (cm)	Humus Type
WT_{80-100}	*Abietetum albae*	Silver fir	*Maianthemum bifolium* *Luzula pilosa* *Thiudium tamarescinum* *Polytrichum attenuatum*	Podzols Stagnosols	80–100	mor
Moist	*Tilio-Carpinetum ficarietosum*	Common hornbeam, Common alder	*Aegopodium podgraria* *Galeobdolon luteum* *Hepatica nobilis* *Impatiens noli-tangere*	Gleysols	40–50	mull
Swampy	*Fraxino-Alnetum*	Common alder	*Ficaria verna* *Valeriana simplicifolia* *Caltha palustris* *Carex remota*	Mollic Gleysols	0–30	mull sticky

WT_{80-100}—variant with water table 80–100 cm.

2.3. Laboratory Analysis of Soil

Soil samples obtained in the field were dried and sieved through 2.0 mm mesh. Using the potentiometric method, the pH of the samples was analyzed in H_2O and KCl. By using a laser diffraction technique (Analysette 22, Fritsch, Idar-Oberstein, Germany), the soil texture was determined. Carbon (C_t) and nitrogen (N) contents were measured with an elemental analyzer (LECO CNS TrueMac Analyzer) (Leco, St. Joseph, MI, USA). Using the method of Kononowa and Bielczikowa, fractional composition of humus was determined in which extraction is performed in a mixture of 0.1 M NaOH and 0.1 M $Na_4P_2O_7$ [24]. In order to obtain a humin (Hm), humic acid (HA), and fulvic acid (FA) fraction, the chemical fraction was conducted. The relationships HA/FA and (HA+FA)/Hm were also calculated. Exchangeable Ca, Mg, K and Na were determined by inductively coupled plasma optical emission spectrometry (ICP-OES) (iCAP 6500 DUO, Thermo Fisher Scientific, Cambridge, UK) in 1 M ammonium acetate at pH 7.0 extracts. The sum of base cations (BC) and effective base saturation (BS) were calculated. Available phosphorus was determined by Bray–Kurtz's method. The hydrolytic acidity was determined in the samples after extracting 5 g of soil with 30 mL 1 mol·L^{-1} (CH$_3$COO)$_2$Ca (shaking time 1 h), followed by filtration. After extracting 5 g of soil with 30 mL KCl, exchangeable acidity was determined. Soil on the filter was washed several times by extractant solution, up to a volume of 200 mL. About 25 mL of the obtained solution was titrated by potentiometric titration (automatic titrator, Mettler Toledo Inc., Columbus, OH, USA) to a pH of 8.2. For determining the exchangeable aluminum and hydrogen, NaF was added to the KCl extract and titrated with NaOH. Total cation exchange capacity (CECt) was calculated as a sum of base cations (BC) and hydrolytic acidity (Ah); and effective cation exchange capacity (CECe) was calculated as a sum of base cations (BC) and exchangeable acidity (Aex). Using samples with intact structure collected to metal cylinders, bulk density was determined via the drying-weighing method [25].

The soil organic carbon stock (SOC$_S$) was calculated as a sum of its total content at particular soil horizons. In each of the analyzed soil samples, the SOC$_S$ was calculated in the soil block area of 1 m^2 and a depth from the surface to 60 cm. The calculation of SOC$_S$ for the particular depths was made by summing the SOC$_S$ at subsequent genetic soil horizons according to the formula:

$$SOC_S = C \times D \times m/10 \ (Mg \cdot ha^{-1}) \tag{1}$$

where C is the organic carbon content at the subsequent genetic horizon (%), D is the bulk density (g·cm^{-3}), and m is the thickness of the horizons (cm).

Moreover, a Carbon Distribution Index (*CDI*) was provided and calculated from the ratio of the carbon accumulated in organic horizons and the amount of organic carbon accumulated in the mineral horizons up to 60 cm according to the formula:

$$CDI = \frac{C \ accumulated \ in \ organic \ horizons \ \left[Mg \cdot ha^{-1}\right]}{C \ accumulated \ in \ mineral \ horizons \ to \ 60 \ cm \ depth \ \left[Mg \cdot ha^{-1}\right]} \tag{2}$$

2.4. Statistical Analysis

Basic descriptive statistics, arithmetic mean and standard deviation were used to estimate the soil properties' variability. Tukey's honest significant difference (HSD) test was used to assess differences between means for carbon accumulation in soils. To evaluate the relationships between stock of soil carbon, other soil properties, forest species, and moisture, the Principal Components Analysis (PCA) method was used. Multiple regression models were developed to describe the relationship between the estimated values of carbon content and other soil properties. The statistical significance of the results was verified at the significance level of alpha = 0.05. All statistical analyzes were performed with Statistica 12 software (2012).

3. Results

The investigated soils are characterized by diversification of properties (Table 2). The values of pH H$_2$O and pH KCl were in the range from 3.85 to 7.30 and from 3.04 to 5.78, respectively. A significantly greater pH was noted in soils of moist and swampy variants (Table 2). For cation content, a similar dependence was recorded (Table 2). The greater Ah and Aex were observed in soils with water table 80–100 cm (Table 3). The texture of the investigated soils was dominated by sand (on average 50–72% in the deeper horizons) and silt (on average 23–42% in deeper horizons) with addition of clay (on average 5–8% in the deeper horizon) (Table 3). The sand content (only in the deepest horizons) was significantly greater in WT$_{80-100}$ variant (Table 3). The silt and clay content (only in the deepest horizons) was higher in soils of moist and swampy variants (Table 3). The analyzed soils clearly differ in their moisture content (Table 4). Statistically significant higher soil moisture content was recorded in the soils of the swampy variant (mean soil moisture of the surface humus horizon was 46.85%). The lowest soil moisture content was recorded in the soils of the WT$_{80-100}$ variant (Table 4).

Table 2. Chemical properties of study soils.

Moisture Gradient	Horizon	pH H$_2$O	pH KCl	Ah	Aex	Al	P	Ca	K	Mg	Na	BC	BS	CECt	CECe
WT$_{80-100}$	O	3.85 [b] ± 0.37	3.04 [b] ± 0.35	49.68 [a] ± 18.74	10.14 [a] ± 3.43	306.98 ± 243.27	58.75 [a] ± 32.41	6.88 [b] ± 3.30	0.69 [a] ± 0.24	1.21 [a] ± 0.58	0.09 [b] ± 0.08	8.89 [b] ± 3.84	16.49 [b] ± 8.82	58.58 [a] ± 20.86	19.03 [a] ± 6.07
	AE	3.87 [c] ± 0.16	3.07 [b] ± 0.18	12.14 [a] ± 4.91	4.19 [a] ± 0.84	193.96 ± 79.65	8.67 [a] ± 4.90	0.94 [b] ± 0.32	0.05 [a] ± 0.03	0.13 [b] ± 0.09	0.02 [b] ± 0.01	1.13 [b] ± 0.41	7.15 [b] ± 4.57	13.26 [a] ± 6.07	5.32 [b] ± 1.82
	G	4.51 [b] ± 0.41	3.81 [b] ± 0.33	3.49 [a] ± 1.00	2.21 [a] ± 0.63	95.36 ± 41.21	61.46 [a] ± 46.18	0.36 [b] ± 0.19	0.01 [b] ± 0.01	0.03 [b] ± 0.02	0.01 [b] ± 0.01	0.41 [b] ± 0.37	10.03 [b] ± 6.19	3.91 [b] ± 0.87	2.62 [b] ± 0.54
Moist	A	4.94 [a] ± 0.73	3.98 [a] ± 0.69	27.57 [a,b] ± 2.10	3.04 [a] ± 1.80	69.90 ± 9.37	19.18 [b] ± 12.18	19.95 [a] ± 12.86	0.36 [a] ± 0.29	1.6 [a] ± 1.15	0.09 [a] ± 0.06	22.04 [a,b] ± 14.22	45.65 [a] ± 20.06	49.61 [a] ± 31.46	25.08 [a] ± 14.23
	AG	5.64 [b] ± 0.84	4.69 [b] ± 0.79	7.97 [a,b] ± 6.17	1.27 [b] ± 0.99	35.45 ± 68.05	3.85 [b] ± 3.69	16.07 [a] ± 14.54	0.06 [a] ± 0.04	0.93 [a] ± 0.87	0.06 [a] ± 0.04	17.12 [a] ± 12.15	60.75 [a] ± 23.38	25.10 [a] ± 19.31	18.39 [a] ± 15.02
	G	6.85 [a] ± 0.73	5.26 [a] ± 0.58	1.87 [b] ± 0.65	0.66 [b] ± 0.17	4.14 ± 5.25	1.27 [b] ± 0.72	8.30 [a] ± 5.07	0.07 [a] ± 0.05	0.60 [a] ± 0.40	0.04 [a] ± 0.02	9.01 [a] ± 5.52	76.92 [a] ± 15.04	10.88 [a] ± 5.61	9.67 [a] ± 5.44
Swampy	AM	5.71 [a] ± 0.78	4.40 [a] ± 1.50	13.01 [b] ± 11.01	1.66 [b] ± 1.01	14.99 ± 5.84	30.70 [a,b] ± 28.43	28.73 [a] ± 15.79	0.34 [b] ± 0.28	1.74 [a] ± 1.12	0.10 [a] ± 0.07	30.90 [a] ± 16.96	68.52 [a] ± 18.79	43.92 [a] ± 25.71	32.57 [a] ± 17.45
	AG	6.72 [a] ± 0.61	5.67 [a] ± 0.58	3.92 [b] ± 1.75	0.62 [b] ± 0.22	3.76 ± 6.59	8.86 [a] ± 5.88	21.73 [a] ± 12.85	0.10 [a] ± 0.08	1.09 [a] ± 0.72	0.07 [a] ± 0.05	22.99 [a] ± 13.05	82.32 [a] ± 12.04	26.92 [a] ± 14.75	23.62 [a] ± 13.45
	G	7.30 [a] ± 0.69	5.78 [a] ± 0.67	1.75 [b] ± 0.91	0.71 [b] ± 0.29	4.11 ± 1.01	5.09 [a] ± 4.47	10.59 [a] ± 7.06	0.12 [a] ± 0.07	0.70 [a] ± 0.53	0.04 [a] ± 0.02	11.40 [a] ± 7.59	83.53 [a] ± 9.73	13.16 [a] ± 8.22	12.11 [a] ± 7.47

mean ± standard deviation, small letters (a, b, c) in the upper index of the mean values mean significant differences in soil horizons between the moisture variant; WT$_{80-100}$, moist and swampy—soil moisture gradient; hydrolytic acidity (Ah) (cmol(+)·kg⁻¹); Ca, K, Mg and Na (cmol(+)·kg⁻¹); Al, P (mg·kg⁻¹); total cation exchange capacity (CECt) (cmol(+)·kg⁻¹); effective cation exchange capacity (CECe) (cmol(+)·kg⁻¹); base cations (BC) (cmol(+)·kg⁻¹); base saturation (BS) (%).

Table 3. The particle size [%] and bulk density [g·cm⁻³] of study soils.

Moisture Gradient	Horizon	Sand Fraction					Silt Fraction		Clay	Sum of Fraction		BD
		VCS	CS	MS	FS	VFS	Csi	Fsi		Sand	Silt	
WT$_{80-100}$	O	n.d.	n.d.	n.d.	n.d.	n.d.	n.d.	n.d.	n.d.	n.d.	n.d.	0.41 [b] ± 0.21
	AE	0.4 [a] ± 0.3	15.5 [a] ± 8.8	30.5 [a] ± 2.6	15.0 [a] ± 4.0	9.2 [a] ± 2.4	10.9 [a] ± 3.7	14.6 [a] ± 4.5	3.8 [a] ± 1.3	70.7 [a] ± 8.4	25.5 [a] ± 7.3	1.14 [a] ± 0.11
	G	1.3 [a] ± 0.7	24.8 [b] ± 10.5	31.5 [a] ± 3.2	10.6 [a] ± 3.8	3.5 [a] ± 1.3	7.4 [b] ± 3.0	15.9 [b] ± 6.2	4.9 [b] ± 1.8	71.6 [a] ± 9.1	23.4 [b] ± 7.6	1.34 [a] ± 0.01
Moist	A	1.1 [a] ± 1.0	10.9 [a] ± 8.4	19.7 [a] ± 11.4	18.9 [a] ± 8.1	11.26 [a] ± 5.3	11.6 [a] ± 7.2	11.6 [a] ± 9.1	2.7 [a] ± 2.0	62.0 [a] ± 30.1	23.3 [a] ± 15.7	0.83 [a] ± 0.25
	AG	0.5 [a] ± 0.2	8.9 [a] ± 6.1	20.1 [b] ± 10.6	17.5 [a] ± 3.5	10.8 [a] ± 3.7	16.8 [a] ± 6.8	21.2 [a] ± 9.5	4.3 [a] ± 2.1	57.0 [a] ± 17.5	37.9 [a] ± 15.9	1.21 [a] ± 0.08
	G	0.3 [a] ± 0.1	7.8 [a] ± 5.7	23.7 [a] ± 11.5	14.4 [a] ± 3.2	4.9 [a] ± 2.0	14.0 [a] ± 6.5	27.1 [a] ± 10.9	7.7 [a] ± 2.9	51.1 [b] ± 18.8	41.1 [a] ± 16.3	1.35 [a] ± 0.02
Swampy	AM	1.3 [a] ± 0.8	13.0 [a] ± 3.0	19.3 [a] ± 12.9	16.3 [a] ± 10.8	7.1 [a] ± 6.0	6.9 [a] ± 5.1	6.9 [a] ± 6.3	1.9 [a] ± 1.0	64.7 [a] ± 33.3	15.1 [a] ± 11.0	0.69 [a,b] ± 0.31
	AG	n.d.	9.1 [a] ± 7.1	20.3 [a,b] ± 12.4	18.2 [a] ± 5.5	9.5 [a] ± 3.9	14.8 [a] ± 7.3	21.9 [a] ± 12.9	5.3 [a] ± 2.9	57.8 [a] ± 22.2	36.8 [a] ± 19.4	1.07 [a] ± 0.19
	G	n.d.	5.7 [a] ± 2.6	22.7 [a] ± 12.6	15.4 [a] ± 6.8	5.6 [a] ± 3.4	12.9 [a] ± 8.6	28.7 [a] ± 11.1	7.8 [a] ± 3.5	49.5 [b] ± 19.8	41.6 [a] ± 18.7	1.31 [a] ± 0.08

n.d. no determine, mean ± standard deviation, small letters (a, b, c) in the upper index of the mean values mean significant differences in soil horizons between the moisture variant, WT$_{80-100}$, moist and swampy—soil moisture gradient; VCS—very coarse sand, CS—coarse sand, MS—medium sand, FS—fine sand, VFS—very fine sand, Csi—coarse silt, Fsi—fine silt; BD—bulk density.

Table 4. Humus substances, organic carbon and nitrogen content, moisture of study soils.

Moisture Gradient	Horizon	C	N	C/N	HA	FA	Hm	HA/HF	HA+HF/Hm	SOCs	CDI	Moisture
WT$_{80-100}$	O	241.87 a ± 74.21	12.45 a ± 5.28	19.6 b ± 2.8	11.9 a ± 6.78	4.43 a ± 2.28	n.d.	2.74 a ± 1.12	n.d.			16.43 c ± 3.42
	AE	35.07 a ± 19.41	1.75 a ± 1.15	21.1 b ± 3.1	3.08 a ± 1.44	0.63 a ± 0.61	0.47 a ± 1.03	4.70 a ± 1.38	8.00 a ± 5.03	14.09 a,b ± 2.98	0.837	
	G	4.87 a,b ± 2.01	0.24 a ± 0.07	20.3 b ± 2.5	0.81 a ± 0.44	0.82 a ± 0.45	0.3 a ± 0.03	1.05 a ± 1.10	54.30 a ± 22.90			
Moist	A	101.10 b ± 49.29	6.80 a ± 3.79	14.2 a ± 1.7	3.15 b ± 3.04	3.41 a ± 3.55	9.79 a ± 10.41	1.80 a ± 2.31	0.68 b ± 0.84			30.08 b ± 4.86
	AG	24.29 a ± 13.80	1.84 a ± 1.04	13.2 a ± 1.2	1.37 b ± 0.73	1.69 a ± 2.10	1.89 a ± 2.53	2.25 b ± 1.96	1.63 b ± 1.25	12.61 b ± 3.71	0.299	
	G	3.95 b ± 2.30	0.33 a ± 0.18	11.8 a ± 1.1	0.99 a ± 1.12	0.72 a ± 0.94	0.25 a ± 0.26	3.31 a ± 1.18	11.47 a ± 8.98			
Swampy	AM	135.96 a,b ± 107.11	10.09 a ± 7.02	12.8 a ± 1.6	4.67 a,b ± 3.55	2.91 a ± 3.72	5.51 a ± 7.63	3.84 a ± 2.41	1.40 a,b ± 0.68			46.85 a ± 3.28
	AG	44.13 a ± 34.29	3.31 a ± 2.39	13.1 a ± 1.2	2.26 a,b ± 1.56	1.34 a ± 1.65	1.08 a ± 1.47	2.51 b ± 1.06	3.40 b ± 2.20	19.02 a ± 1.82	0.329	
	G	8.51 a ± 4.31	0.64 a ± 0.51	12.8 a ± 1.6	0.79 a ± 0.72	0.99 a ± 0.90	0.31 a ± 0.28	2.98 a ± 1.47	16.18 a ± 12.45			

mean ± standard deviation, small letters (a, b, c) in the upper index of the mean values mean significant differences in soil horizons between the moisture variant; WT$_{80-100}$, moist and swampy—soil moisture gradient; C and N (g·kg^{-1}); HA carbon of humic acids (% C), FA carbon of fulvic acids (% C), Hm carbon of humin (% C); SOCs soil organic carbon stock (Mg·ha^{-1}); CDI carbon distribution index; moisture (%); n.d.—no determine.

The differences in the carbon content of the examined soils were clearly noted. A statistically higher carbon content in the surface horizons was noted in the soils of the WT_{80-100} variant. The average content of carbon in the surface horizon of this variant was 241.87 g·kg^{-1}, and was 101.10 g·kg^{-1} and 135.96 g·kg^{-1} in the soils of the moist and swampy variants, respectively (Table 4). The highest carbon content in the lower lying horizons was recorded for soils of the swampy variant. The mean carbon content at these horizons was 44.13 g·kg^{-1} and 8.51 g·kg^{-1}, respectively. In the case of WT_{80-100} and moist variant of soils in the deeper horizons, the carbon content was 35.07–4.87 g·kg^{-1} and 24.29–3.95 g·kg^{-1}, respectively. Similar dependencies concerned nitrogen content (Table 4). The lowest C/N, which expresses the degree of organic matter distribution, was recorded in the soils of the swampy and moist variants (the C/N ratio ranged from 11.8 to 14.2). The highest C/N ratio, from 19.6 to 20.3 depending on the horizon, was recorded in the WT_{80-100} variant of soils (Table 4). Soil organic matter (SOCs) concentration in the 0–60 cm soil horizon differed significantly among moisture gradients and followed the order: swampy > WT_{80-100} > moist variants. The average SOCs up to a depth of 60 cm in the soils of moist variant is lower and is estimated at 12.61 Mg·ha^{-1}. The higher accumulation was found in the soils of WT_{80-100} variant, where the average SOCs was 14.09 Mg·ha^{-1}. A significantly greater accumulation was noted in the soils of swampy variant (19.02 Mg·ha^{-1}). The ratio between carbon accumulation in the organic horizon and carbon accumulation in the mineral horizons, up to 60 cm of soils (CDI) of estimated variants, is notably diverse (average 0.299–0.837). The lowest values of CDI are found in the soils of moist variant in which the CDI indicator is 0.299. The soils of swampy variant are characterized by higher values of the CDI index—0.329, and the CDI index of soils of WT_{80-100} variant approached 1.0 and is 0.837 (Table 4).

Concentrations of C in fulvic acid (FA), humic acid (HA) and humin fractions in the different soil horizons of different moisture variants are presented in Table 4. Mean values of HA, FA and humin varied from 11.9% to 0.79%, 4.43% to 0.63%, and 9.79% to 0.25%, respectively. In the fractions of SOM, humic acids dominate over fulvic acids and humins. Soils of the WT_{80-100} variant are characterized by the highest content of humic and fulvic acid (Table 4). Significant differences between the soils in terms of humic acid were observed in two surface horizons (swampy variant = WT_{80-100} variant > moist variant = swampy variant). No significant differences between soil variant in terms of fulvic acid and humin were noted (Table 4). All SOM fractions decreased significantly with depth for all variants of soil. The HA/FA ratio ranged from 4.70 to 1.05. The lowest HA/FA ratio regardless of depth was in the soil of WT_{80-100} variant, and the highest in the soil of moist and swampy variant. The relationship of the sum of C humic and fulvic acids to C humin (HA+FA/Hm) assumes the values of 54.3–0.68 (Table 4). In the soil of WT_{80-100} variant, the highest HA/FA ratio and HA+FA/Hm, as compared to moist and swampy variant, was recorded and was statistically significant (Table 4).

Multiple regression models explained 89% of the variance in the C content (Table 5). C content was largely dependent on hydrolytic acidity and moisture. The relationships between carbon content, hydrolytic acidity and moisture are shown in Figure 2. The increase in C accumulation in soils correlated with high moisture and high acidity.

A projection of the variables on the factor-plane clearly demonstrated a correlation between the variables and soil moisture (supplementary variables). Two main factors had a significant total impact (60.62%) on the variance of the variables in soil. Factor 1 explained 35.76% of the variance of the examined properties and Factor 2 explained 24.86% of the variance (Figure 3). The soil of moist and swampy variants had a higher pH. In contrast, the soils of WT_{80-100} variant were characterized by a high C/N ratio. The soils of WT_{80-100} variant contained more humic acids. The fulvic acids and humins were connected with nitrogen content.

Table 5. Multiple regression analysis for carbon content based on soil characteristic. R^2 describes the percentage of explained variance, β is the regression coefficient for the given equation parameter and *p* is the significance level for the equation parameter.

	R^2	Equation Parameter	β	*p*
C	89%	Ah	5.257	0.000001
		Moisture	1.897	0.000001

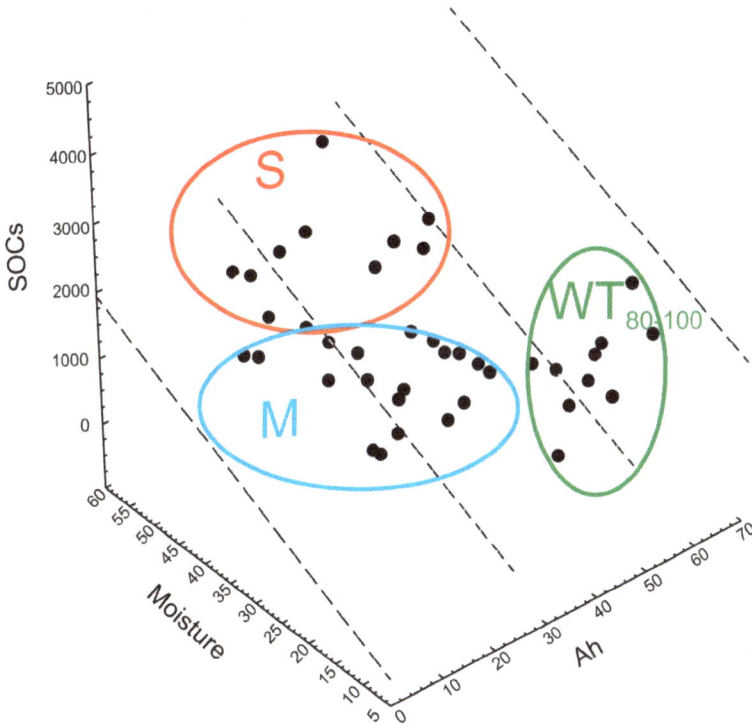

Figure 2. Three-dimensional plot of the relationship between soil organic carbon stock (SOCs), moisture and hydrolytic acidity (Ah) of soils (circles with WT80–100—group of plots with water table 80–100 cm, circles with M—group of moisture plots, circles with S—group of swampy plots); regression plane marked by black lines.

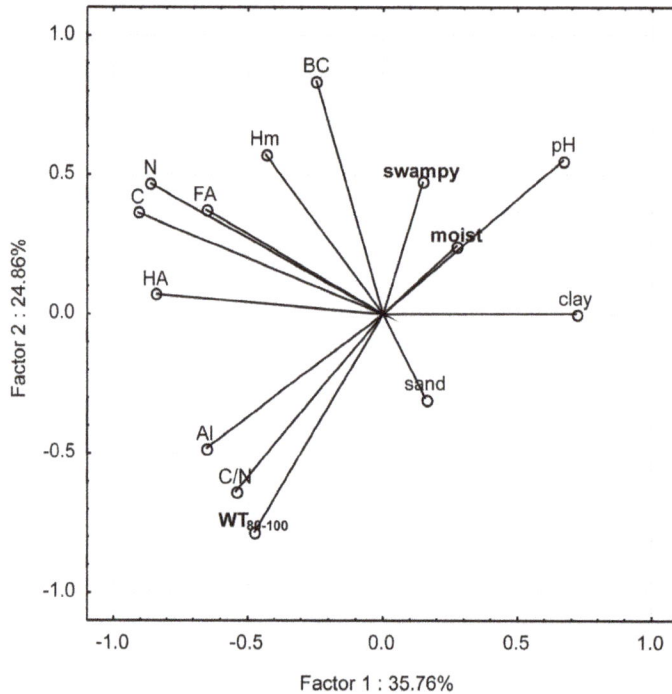

Figure 3. Diagram of Principal Components Analysis (PCA) with projection of variables on a plane of the first and second factor in the surface horizon (WT$_{80-100}$, moist and swampy—soil moisture gradient; HA—humic acids, FA—fulvic acids, Hm—humins).

4. Discussion

Results present in this study demonstrate that soil moisture is a factor that most strongly influences the carbon accumulation in forest soils. Compared to the soil with water table from 80 to 100 cm and moist soils, those with the highest moisture content (swampy variant) are characterized by the higher C accumulation. Soil organic matter (SOCs) concentration differed significantly among moisture gradient and followed the order: swampy > WT$_{80-100}$ > moist variants. The greater SOM accumulation in the soils is affected not only by the high moisture, but also periodic deficiency of water. Our results show the high SOC stock on the sites with low moisture (WT$_{80-100}$ variant) and in swampy variant, and that site condition promoted lower rate decomposition. The high SOC stock in the soils of WT$_{80-100}$ variant is affected by greater acidity (surface horizons are not enriched due to low groundwater level). In the soils of swampy variant, there is an effect of anaerobic processes: the soils better aerated in the moist variants are characterized by faster and more efficient decomposition of SOM. Jandl et al. [26] and Stendahl et al. [27] related the accumulation of SOC with site conditions. A large stock of carbon can be associated with the accumulation of organic matter in acidic environments. A similar situation was noted for the soils of the WT$_{80-100}$ variant where surface horizons were under coniferous species influence. Tree species affect pH, and the specific composition of organic matter may lead to the modification of soil acidity and nutrient pools [28–30]. pH is an important variable that influences SOM decomposition dynamics [31]. Our results confirmed the importance of moisture and acidity of soils on soil organic matter accumulation. In our study, 89% of the variance in the C content was explained by the hydrolytic acidity and moisture of the soils. Li et al.'s [32] results indicated that moisture level had an effect on carbon accumulation. At the same time, the increase of soil's acidity results in deterioration of the conditions of decomposition of dead organic matter, slowing down their humification rate and microbial activity. On the other hand, the increase in organic matter accumulation results in an increase in sorption capacity, which, under acidic forest soil condition and lower level of groundwater, includes aluminum and hydrogen ions. The increase in soil moisture was favorable for the accumulation of water soluble carbon. In our study, soils in the swampy variant are characterized by the highest accumulation of C, which can be combined with the accumulation of organic debris in the peat-forming process. These studies confirm earlier findings, that although the peat covers only 3% of the Earth's surface, it constitutes an important reservoir of carbon [33,34]. Under anaerobic conditions, 2–6 times less CO_2 was produced than under aerobic conditions [35]. Furthermore, the presence of water in the soil profile facilitates dissolution of minerals, and during the capillary transport, also their transportation to higher horizons where they can form stable conjunctions to the humic substances. In addition, higher soil moisture results in higher biomass production of trees and plant cover of lower layers of the stand, which results in increased accumulation of organic material into the forest floor and soil.

In our study, CDI was used to evaluate the distribution of carbon accumulated in the soil profile and as a measure of quality of the accumulated organic matter. The advantage of accumulation of carbon in the mineral horizons over accumulation in the organic horizons indicates a greater efficiency of the processes of organic matter decomposition. In numerous studies [36], the greater resistance of SOM, in combination with soil minerals, is observed. In comparison to the accumulation occurring in the organic horizons, the predominance of carbon accumulation in the mineral horizons can be observed in soils of moist and swampy variants. The soils of moist variants were occupied by a very diverse species composition; the presence of admixture species—mainly hornbeam, alder and maple—provides abundant input of easily decaying organic matter to the soil. The positive impact of hornbeam on the accumulation of SOM was also described by other authors [31,37]. Stand mixing improves soil properties, especially the quality of SOM and biochemical properties [38]. The dominance of coniferous species in the forest (WT_{80-100} variant) changes the CDI index. In coniferous forests with fir, the C accumulation in the mineral horizons is reduced in relation to the total accumulation in the whole soil profile. The reason for the higher accumulation of C at the organic horizons is undoubtedly the slow decomposition of organic debris. Litter decomposition is particularly influenced by decomposer organisms [39]. The conifer litter decomposes slower than that from hardwood species [40]. The soils of coniferous forests are characterized by low pH, and the decomposition of litter is mainly conducted by fungal organisms. Fungi dominate in acidic soils, whilst bacteria in soils are characterized by a neutral or alkaline pH [41].

Forest soils in the studied moisture content gradient differ in the accumulation of carbon and the fractional composition of the soil organic matter. In the studied soils, a large proportion of humic acids were found, which prevailed in the moist variants. This is in line with Tavares and Nahas' study [42]. More humic acids than fulvic acids indicate the potentially low mobility of carbon accumulated in the surface soil horizons [21]. In our studies, the ratio of humic to fulvic acids (HA/FA) changes as moisture content of soils increases, suggesting an increase in carbon mobility in these soils. More humic acids than fulvic acids indicate high humification rates. The differences in soil moisture contents influences the amounts and chemical composition of the soil humic fraction [32]. The results implied that a high moisture level was unfavorable for the accumulation of humic fraction in the soil. In our experiment, we used the ratio between the sum of humic and fulvic acids and humin ((HA+FA)/humin) as an indicator of loss of SOM through the soil profile. The highest ratio ((HA+FA)/humin) was determined in the soils with the lowest moisture. As the soil moisture content increased, the ratio decreased. The lower ratio ((HA+FA)/humin) indicates the sufficient humin level in the soils with the highest moisture content.

5. Conclusions

In the soils under study, in the temperate zone, moisture is an important factor in the accumulation of organic carbon in the soil. The highest accumulation of carbon was observed in soils of swampy variant, while the lowest in the soils of moist variant. In the soil of swampy variant, the anaerobic conditions affect the organic matter decomposition leading to slower processes. The increased accumulation of C in soils of WT_{80-100} variant is caused by other factors. Large accumulation of C in the soils of the WT_{80-100} variant results from the thick organic horizons that are characterized by lower organic matter decomposition and higher acidity. In the WT_{80-100} variant, the higher acidity is simultaneously an effect of the stand species composition and lack of water neutralization. The proportion of carbon accumulation in the organic horizons to the total accumulation in the mineral horizons expresses the distribution of carbon accumulated in the soil profile, and is a measure of quality of the organic matter accumulated. The advantage of accumulation at the mineral horizons over accumulation at the organic horizons confirms a greater efficiency of decomposition processes of organic matter, which, thanks to more advanced transformations and connecting with mineral soil, can be considered stable. Studies have confirmed the importance of moisture content in the formation of the fractional organic matter. The increase in soil moisture content changes the ratio of humic to fulvic acids (HA/FA), which may suggest an increase in carbon mobility in soils.

Acknowledgments: The project was financed by the National Science Centre, Poland: decision no. DEC-2016/21/D/NZ9/01333.

Author Contributions: E.B. conceived and designed the experiments, performed the experiments, wrote the paper; J.L. analyzed the data.

Conflicts of Interest: The authors declare no conflict of interest.

References

1. Dorrepaal, E.; Sylvia, T.; van Logtestijn, R.S.P.; Swart, E.; Van de Weg, M.J.; Callaghan, T.V.; Aerts, R. Carbon respiration from subsurface peat accelerated by climate warming in the subarctic. *Nature* **2009**, *460*, 616–619. [CrossRef]
2. Fornara, D.A.; Steinbeiss, S.; McNamara, N.P.; Gleixner, G.; Oakley, S.; Poulton, P.R. Increases in soil organic carbon sequestration can reduce the global warming potential of long-term liming to permanent grassland. *Glob. Chang. Biol.* **2011**, *17*, 2762. [CrossRef]
3. Farquhar, G.D.; Fasham, M.J.R.; Goulden, M.L.; Heimann, M.; Jaramillo, V.J.; Kheshgi, H.S.; Le Quere, C.; Scholes, R.J.; Wallace, D.W.R. Climate change. The Scientific Basis IPCC. Chapter 3. In *The Carbon Cycle and Atmospheric Carbon Dioxide*; Cambridge University Press: Cambridge, UK, 2001; pp. 183–237.
4. Stergiadi, M.; Van der Peck, M.; de Nijs, T.C.M.; Bierkens, M.F.P. Effect of climate change and land management on soil organic carbon dynamics and carbon leaching in northwestern Europe. *Biogeosciences* **2016**, *13*, 1519–1536. [CrossRef]
5. Martin, M.P.; Wattenbach, M.; Smith, P.; Meersmans, J.; Jolivet, C.; Boulonne, L.; Arrouays, D. Spatial distribution of soil organic carbon stock in France. *Biogeosciences* **2011**, *8*, 1053–1065. [CrossRef]
6. Parajuli, P.B.; Duffy, S. Evaluation of Soil Organic Carbon and Soil Moisture Content from Agricultural Fileds in Mississipi. *Open J. Soil Sci.* **2013**, *3*, 81–90. [CrossRef]
7. Hobley, E.U.; Wilson, B. The depth distribution of organic carbon in the soils of eastern Australia. *Ecosphere* **2016**, *7*, e01214. [CrossRef]
8. Degórski, M. Influence of forest management into the carbon storage. *Monit. Environ.* **2005**, *6*, 75–83.
9. Zwydak, M.; Brożek, S.; Lasota, J.; Małek, S. Reserve of Organic Carbon in Forest Soils of Lowlands in Poland. *Pol. J. Environ. Stud.* **2008**, *17*, 632–637.
10. Ostrowska, A.; Porębska, G.; Kanafa, M. Carbon accumulation and Distribution in Profiles of Forest Soils. *Pol. J. Environ. Stud.* **2010**, *19*, 1307–1315.
11. Stavi, I. Biochar use in forestry and tree-based agro-ecosystems for increasing climate change mitigation and adaptation. *Int. J. Sustain. Dev. World Ecol.* **2013**, *20*, 166–181. [CrossRef]
12. Lasota, J.; Błońska, E. *Forest Site Science in the Polish Lowlands and Highlands*; Scientific Papers; University of Agriculture in Krakow: Kraków, Poland, 2013.
13. Hassink, J. The capacity of soils to preserze organic C and N by their association with Clay and silt particles. *Plant Soil* **1997**, *191*, 77–87. [CrossRef]

14. Six, J.; Paul, E.; Paustian, K. Stabilization Maechanisms of Soil Organic Matter: Implications for C—Saturation of Soils. *Plant Soil* **2002**, *241*, 155–176. [CrossRef]

15. Wosten, J.H.; Lilly, A.; Nemes, A.; Le Bas, C. Development and use of a database of hydraulic properties of European soils. *Geoderma* **1999**, *90*, 169–185. [CrossRef]

16. Bauer, J.; Herbst, M.; Huisman, J.A.; Weihermuller, L.; Vereecken, H. Sensitivity of simulated soil heterotrophic respiration to temperature and moisture reduction functions. *Geoderma* **2008**, *145*, 17–27. [CrossRef]

17. Meersmans, J.; De Ridder, F.; Canters, F.; De Baets, S.; Van Molle, M. A multiple regression approach to assess the spatial distribution of Soil Organic Carbon (SOC) at the regional scale (Flanders, Belgium). *Geoderma* **2008**, *143*, 1–13. [CrossRef]

18. Frouz, J.; Kalčík, J. Accumulation of soil organic carbon in relation to other soil characteristic during spontaneous succession in non-reclaimed colliery spoil heaps after brown coal mining near Sokolov (the Czech Republic). *Ekológia* **2006**, *25*, 388–397.

19. Haynes, R.J. Labile organic matter fractions as central components of the quality of agricultural soils: An overview. *Adv. Agron.* **2005**, *85*, 221–268.

20. Dębska, B.; Długosz, J.; Piotrowska-Długosz, A.; Banach-Szott, M. The impact of a bio-fertilizer on the soil organic matter status and carbon sequestration—Results from a field-scale study. *J. Soils Sediments* **2016**, *16*, 2335–2343. [CrossRef]

21. Guimarães, D.V.; Gonzaga, M.I.S.; da Silva, T.O.; da Silva, T.L.; Dias, N.D.; Matias, M.I.S. Soil organic matter pools and carbon fractions in soil under different land uses. *Soil Till. Res.* **2013**, *126*, 177–182. [CrossRef]

22. Błońska, E.; Lasota, J.; Piaszczyk, P.; Wiecheć, M.; Klamerus-Iwan, A. The effect of landslide on soil organic carbon stock and biochemical properties of soil. *J. Soil Sendiments* **2017**. [CrossRef]

23. World Reference Base (WRB). *World Reference Base for Soil Resource*; FAO: Rome, Italy, 2014.

24. Dziadowiec, H.; Gonet, S. Przewodnik metodyczny do badań materii organicznej gleb. *Prace Komisji Naukowych PTG* **1999**, *20*, 42–43.

25. Grossman, R.B.; Reinsch, T.G. Bulk density and linear extensibility. In *Methods of Soil Analysis, Part 4*; Dane, J.H., Topp, G.C., Eds.; Soil Science Society of America: Madison, WI, USA, 2002; pp. 201–225.

26. Jandl, R.; Lindner, M.; Vesterdal, L.; Bauwens, B.; Baritz, R.; Hagedorn, F.; Johnson, D.W.; Minkkinen, K.; Byrne, K.A. How strongly can forest management influence soil carbon sequestration? *Geoderma* **2007**, *137*, 253–268. [CrossRef]

27. Stendhal, J.; Johansson, M.B.; Eriksson, E.; Langvall, O. Soil organic carbon in Swedish spruce and pine forests-differences in stock levels and regional patterns. *Silva Fennica* **2010**, *33*, 5–21. [CrossRef]

28. Prescott, P.C.; Grayston, S.J. Tree species influence on microbial communities in litter and soil: Current knowledge and research needs. *For. Ecol. Manag.* **2013**, *309*, 19–27. [CrossRef]

29. Gałka, B.; Kabała, C.; Łabaz, B.; Bogacz, A. Influence of stands with diversed share of Norway spruce in species structure on soils of various forest habitats in the Stołowe Mountains. *Sylwan* **2014**, *158*, 684–694.

30. Gruba, P.; Mulder, J. Tree species affect cation exchange capacity (CEC) and cation binding properties of organic matter in acid forest soils. *Sci. Total Environ.* **2015**, *511*, 655–662. [CrossRef] [PubMed]

31. Błońska, E.; Lasota, J.; Gruba, P. Effect of temperate forest tree species on soil dehydrogenase and urease activities in relation to other properties of soil derived from loess and glaciofluvial sand. *Ecol. Res.* **2016**, *31*, 655–664. [CrossRef]

32. Li, C.; Gao, S.; Zhang, J.; Zhao, L.; Wang, L. Moisture effect on soil humus characteristics in a laboratory incuba-tion experiment. *Soil Water Res.* **2016**, *11*, 37–43. [CrossRef]

33. Turunen, J.; Tomppo, E.; Tolonen, K.; Reinikainen, A. Estimating carbon accumulation rates of undrained mires in Finland–application to boreal and subarctic regions. *Holocene* **2002**, *12*, 69–80. [CrossRef]

34. Wellock, M.L.; Reidy, B.; Laperle, C.M.; Bolger, T.; Kiely, G. Soil organic carbon stocks of afforested peatlands in Irleand. *Forestry* **2011**, *84*, 441–451. [CrossRef]

35. Walz, J.; Knoblauch, C.; Böhme, L.; Pfeiffer, E.M. Regulation of soil organic matter decomposition in permafrost-affected Siberian tundra soils—Impact of oxygen availability, freezing and thawing, temperature, and labile organic matter. *Soil Biol. Biochem.* **2017**, *110*, 34–43. [CrossRef]

36. Mikutta, R.; Kleber, M.; Torn, M.S.; Reinhold, J. Stabilization of Soil Organic Matter: Association with Minerals or Chemical Recalcitrance? *Biogeochemistry* **2006**, *77*, 25. [CrossRef]

37. Błońska, E. *Effect of Stand Species Composition on the Enzyme Activity and Organic Matter Stabilization in Forest Soil*; Scientific Papers of University of Agriculture in Krakow No. 527; University of Agriculture in Krakow: Kraków, Poland, 2015; Volume 404.

38. Błońska, E.; Lasota, J.; Zwydak, M.; Piaszczyk, W. Stand mixing effect on enzyme activity and other soil properties. *Soil Sci. Ann.* **2016**, *67*, 173–178. [CrossRef]

39. Osono, T.; Ono, Y.; Takeda, H. Fungal ingrowth on forest floor and decomposing needle litter of *Chamaecyparis obtusa* in relation to resource availability and moisture condition. *Soil Biol. Biochem.* **2003**, *35*, 1423–1431. [CrossRef]

40. Jurgensen, M.; Ree, D.; Page-Dumroese, D.; Laks, P.; Collins, A.; Mroz, G.; Degórski, M. Wood strength loss as a measure of decomposition in northern forest mineral soil. *Eur. J. Soil Biol.* **2006**, *42*, 23–31. [CrossRef]

41. Rousk, J.; Bååth, E. Growth of saprotrophic fungi and bacteria in soil. *FEMS Microbiol. Ecol.* **2011**, *78*, 17–30. [CrossRef] [PubMed]

42. Tavares, R.L.M.; Nahas, E. Humic fractions of forest, pasture and maize crop soils resulting from microbial activity. *Braz. J. Microbial.* **2014**, *45*, 963–969. [CrossRef]

![forests logo] *forests*

MDPI

Article
Interstorm Variability in the Biolability of Tree-Derived Dissolved Organic Matter (Tree-DOM) in Throughfall and Stemflow

Daniel H. Howard [1], John T. Van Stan [1,*], Ansley Whitetree [1], Lixin Zhu [2,3] and Aron Stubbins [2,4]

[1] Geology and Geography, Georgia Southern University, Statesboro, GA 30458, USA;
    dh05256@georgiasouthern.edu (D.H.H.); aw08547@georgiasouthern.edu (A.W.)
[2] Skidaway Institute of Oceanography, University of Georgia, Savannah, GA 31411, USA;
    lixinzhu0305@hotmail.com (L.Z.); a.stubbins@northeastern.edu (A.S.)
[3] State Key Laboratory of Estuarine and Coastal Research, East China Normal University,
    Shanghai 200062, China
[4] Departments of Marine and Environmental Sciences, Civil and Environmental Engineering,
    and Chemistry and Chemical Biology, Northeastern University, Boston, MA 02115, USA
*   Correspondence: jvanstan@georgiasouthern.edu; Tel.: +1-912-478-8040

Received: 29 March 2018; Accepted: 27 April 2018; Published: 1 May 2018

Abstract: Dissolved organic matter (DOM) drives carbon (C) cycling in soils. Current DOM work has paid little attention to interactions between rain and plant canopies (including their epiphytes), where rainfall is enriched with tree-derived DOM (tree-DOM) prior to reaching the soil. Tree-DOM during storms reaches soils as throughfall (drip through canopy gaps and from canopy surfaces) and stemflow (rainwater drained down the trunk). This study (1) assessed the susceptibility of tree-DOM to the consumption by microbes (biolability); (2) evaluated interstorm variability in the proportion and decay kinetics of biolabile tree-DOM (tree-BDOM), and (3) determined whether the presence of arboreal epiphytes affected tree-BDOM. Tree-BDOM from *Juniperus virginiana* L. was determined by subjecting throughfall and stemflow samples from five storms to 14-day microbial incubations. Tree-DOM was highly biolabile, decreasing in concentration by 36–73% within 1–4 days. Tree-BDOM yield was 3–63 mg-C m^{-2} mm^{-1} rainfall, which could represent 33–47% of annual net ecosystem exchange in Georgia (USA) forests. Amount and decay kinetics of tree-BDOM were not significantly different between throughfall versus stemflow, or epiphyte-covered versus bare canopy. However, epiphyte presence reduced water yields which reduced tree-BDOM yields. Interstorm proportions, rates and yields of tree-BDOM were highly variable, but throughfall and stemflow consistently contained high tree-BDOM proportions (>30%) compared to previously-published litter and soil leachate data (10–30%). The high biolability of tree-DOM indicates that tree-BDOM likely provides C subsidies to microbial communities at the forest floor, in soils and the rhizosphere.

Keywords: biolability; tree-DOM; dissolved organic matter (DOM); carbon; dissolved organic carbon (DOC); stemflow; throughfall

1. Introduction

Dissolved organic carbon (DOC) has long been known to exert significant influence over soil processes, from soil formation [1] to the preservation of soil organic matter [2] and pollutant transfer [3], as well as the cycling of other nutrients, e.g., nitrogen [4] and phosphorous [5]. Dissolved organic matter (DOM) is usually quantified as the concentration of DOC, which constitutes approximately 50% of DOM dry mass [6]. In vegetated landscapes, soils receive DOM during storms from multiple sources: rain falling through canopy gaps and dripping from canopy surfaces (throughfall), droplets that drain down the stem (stemflow) and litter leachates. The fate and transport of these DOM sources in and through soils depends, in large part, on the portion that is degradable by microbial communities [7]. For litter leachates, this "biolability" has received decades of extensive research attention [8,9] compared to tree-derived DOM in throughfall and stemflow (tree-DOM) whose biolability has rarely been assessed [7].

The paucity of tree-DOM biolability research is surprising, because it is highly concentrated (10–480 mg-C L^{-1}), yields substantial C from canopies (10–90 g-C m^{-2} year^{-1}), and its role in soils and downstream ecosystems is currently poorly understood [10]. In fact, total DOM yields (per unit canopy area), when converted to inputs per unit infiltration area, can exceed 300 g-C m^{-2} year^{-1}, for example [11], which, if processed in soils, can be approximately one-quarter of annual soil respiration, 1248 g-C m^{-2} year^{-1} [12]. Arguably, a key endeavor to deciphering the role that tree-DOM plays in shaping the form and function of receiving ecosystems is the determination of how much may be consumed by soil microbial communities. Biolabile portions of tree-DOM (tree-BDOM) represent a mechanism preventing DOM loss from the ecosystem [7]. The remaining, non-biolabile tree-DOM may then be sequestered through mineral complexation deeper in the soil profile [13] or be flushed into streams and rivers along preferential pathways in the soil [14]. Thus, the first objective of this study was to quantify the biolabile portion and decay coefficient for tree-DOM at a subtropical cedar (*Juniperus virginiana* L.) forest site in southeast Georgia (USA) and compare these values for throughfall and stemflow. Forests cover 31% of the global land surface [15] and *Juniperus* vegetation can be found throughout forest types, from arid [16] to humid conditions [17].

Tree-BDOM has been found to vary significantly across a growing season (May through August) for weekly throughfall samples [7]. However, no work known to the authors has examined the interstorm variability of tree-BDOM. This minimal examination of temporal variability in tree-BDOM constrains our understanding of the respective relevance of decomposition versus adsorption in the removal of DOM from soils. To address this knowledge gap, our second objective was to evaluate the variability of throughfall and stemflow tree-BDOM quantities and decay coefficients between individual storm events.

Forest canopies can host substantial lichen, bryophyte and/or vascular epiphyte assemblages that trap detritus and aerosols [18], but the authors are unaware of any research on epiphyte influences over tree-BDOM. Epiphytes, although particularly concentrated in the tropics and subtropics, are found in all forests and warming of the temperate regions is expected to extend the range of vascular epiphytes poleward [19]. Thus, the role of epiphyte communities in forest biogeochemistry is gaining increased attention [18]. Epiphyte cover can increase mean DOC concentrations relative to bare canopies for cedar throughfall (54 ± 38 versus 20 ± 13 mg-C L^{-1}) and stemflow (63 ± 40 versus 36 ± 29 mg-C L^{-1}) at the study site [11]. However, the molecular signatures of tree-DOM derived from epiphyte-covered and bare cedar canopies were similar—containing over two-thirds carbohydrate formulas and being rich in highly unsaturated formulas [20]. These molecular signatures indicate the potential for high biolability whether epiphytes are present or not. However, an individual highly unsaturated molecular formula may represent many different structural isomers of correspondingly diverse biogeochemical functions, including biolability [21]. As a result, our third objective was to compare biolability of throughfall and stemflow from bare and epiphyte-covered cedar canopies. As tree-DOM represents the first, and in some cases the largest, enrichment of DOC in rainwater [10], and is large from cedar trees at our site: 5–48 g-C m^{-2} year^{-1} [14], accomplishing these objectives advances our understanding of a key DOM supply to the forest floor, soil and other downstream ecosystems.

2. Materials and Methods

2.1. Study Site Description

The study was conducted in a forest on Skidaway Island, Georgia, USA, located at $31.9885°$ N, $81.0212°$ W (Figure 1a) along Georgia's coast. Climate is humid subtropical (Köppen *Cfa*). Thirty-year mean annual temperature and precipitation (exclusively rainfall) was 19.3 °C and

975 mm [22]. The forest site (Figure 1b) contained 162 stems ha^{-1} of *Juniperus virginiana* L. (eastern red cedar, hereafter "cedar"). Cedars on site host substantial epiphyte biomass, *Tillandsia usneoides* L. (Spanish moss) (Figure 1c), yet some individual trees were bare. Further site information and a complete site inventory can be found in [11].

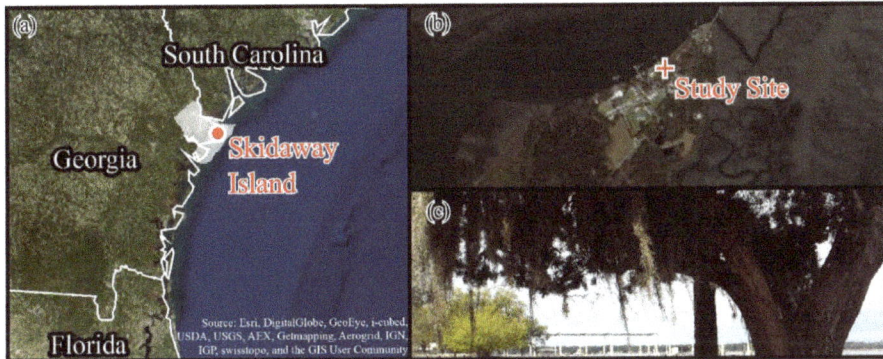

Figure 1. Location of (**a**) Skidaway Island in Chatham County (shaded area), Georgia, USA; and (**b**) the study site location on the Skidaway Institute of Oceanography campus; (**c**) Photograph provided showing an example cedar canopy hosting the epiphyte, *Tillandsia usneoides* L.

2.2. Rainfall, Throughfall and Stemflow Sampling

Sampling of rainfall, throughfall and stemflow was performed for five storms during the 2017 growing season: 14 May, 3 June, 30 July, 28 August and 23 October following published methods [11]. Three bulk rainfall and twenty throughfall samplers consisting of 0.18 m^2 high density polyethylene (HDPE) bins were deployed immediately before each storm. Rainfall was collected in an open area immediately beside the forest site and throughfall samplers were evenly split between bare (*n* = 10) and epiphyte-covered (*n* = 10) cedar canopies. Ten individual cedar trees were selected for stemflow sampling, five each with bare and epiphyte-covered canopies (individual tree characteristics provided in Table S1). Stemflow samplers consisted of collars made from polyethylene tubing cut longitudinally, wrapped around the stem at 1.4 m height, fixed to the stem with aluminum nails, sealed to the bark with silicone, and connected to a 120 L HDPE bin. All samplers were pre-cleaned with pH 2 (using trace clean 6 N HCl) ultrapurified water (Milli-Q), then triple-rinsed with ultrapure water, air dried, and covered until the start of a storm. Sample volumes were measured manually with graduated cylinders. Samples for DOC quantification were collected within hours after a storm, filtered to 0.2 μm through hydrophilic polypropylene syringe filters (Acrodisc) into precombusted glass vials, acidified, then stored at 4 °C in the dark until quantification of DOC concentration (hold period no longer than one month). All sampling materials were precleaned with acidified ultrapure water, triple-rinsed with ultrapure water, then triple-rinsed with sample water.

2.3. Bioincubations

Bioincubations lasting 14 days and based on a modified protocol [7,23] were performed to estimate the biolabile portion and decay coefficients for tree-DOM in throughfall and stemflow for each storm. Immediately after each storm, in addition to the sample taken for DOC analysis in Section 2.2, four 20 mL samples per stemflow sampler and from a composite sample of every two throughfall samplers were filtered to 0.2 μm and placed into precombusted glass vials. A bacterial inoculum was added (2 mL) to each vial along with 2 mL of Nitrogen-Phosphorous-Potassium (NPK) nutrient solution (10:10:10) to prevent nutrient limitation from constraining biodegradation. As throughfall and stemflow at this site contain 10^4–10^6 bacteria mL^{-1} [24], the inoculum was prepared for each storm from a volume-weighted composite of freshly collected throughfall and stemflow samples filtered through a 50 μm mesh to remove microbial grazers and coarse particulates. Caps were placed loosely on the bottles to allow air movement, then samples were incubated for 1, 2, 4, and 14 days at 25 °C in the dark on a shaker table (60 rpm). After bioincubation, each sample was filtered to 0.2 μm into a new precombusted glass vial, acidified, then stored at 4 °C in the dark until quantification of DOC concentration (hold period no longer than one month).

2.4. Dissolved Organic Carbon Concentrations

Concentrations of DOC were determined as nonpurgable organic C using a total organic carbon (TOC)-VCPH analyzer with an ASI-V autosampler (Shimadzu, Columbia, MD, USA). Calibration curves were made with potassium hydrogen phthalate stock solution. Instrument reproducibility was checked against deep seawater reference material from the Consensus Reference Material (CRM) Project. CRM analyses were <5% from reported (http://yyy.rsmas.miami. edu/groups/biogeochem/Table1.htm). This configuration has a minimum DOC detection limit of 0.034 ± 0.004 mg L^{-1} with typical standard errors for DOC concentration being $1.7 \pm 0.5\%$ [25].

2.5. Data Analysis

First-order decay curves were fit to the DOC concentrations quantified after bioincubations:

$$\%DOC\ remaining = be^{-kt} + c \qquad (1)$$

where b is the biolabile proportion, k is the decay coefficient, and c is the recalcitrant proportion of tree-DOM that would theoretically resist biodegradation indefinitely under these experimental conditions. First-order decay curves were only fit to sample data where DOC concentrations stopped decreasing between two consecutive measurements. Tree-BDOM yield (mg-C m^{-2} mm^{-1} rainfall) for each storm from each flux/cover type was computed as the product of b and total tree-DOM yield. Total tree-DOM yield was calculated as DOC concentration (mg-C L^{-1}) × water volume (L)/canopy area (m^2) and rainfall amount (mm). Two-way Analysis of Variance (ANOVA) with a Tukey's Honest Significant Differences (HSD) test was performed to compare initial DOC concentrations between fluxes with and without epiphyte cover using Statistica 13.2 (Statsoft, Tulsa, OK, USA). The threshold for significance was $p < 0.05$ unless otherwise noted and variability about the mean is expressed in the text as standard deviation.

3. Results

3.1. Hydrometeorology for Sampled Storms

Sampled storms ranged in magnitude from 8 mm to almost 50 mm, while storm duration ranged from 8.5 to 59.3 h (Table 1). Rainfall intensity for sampled storms varied from 0.7 to 1.8 mm h^{-1} (Table 1). Throughfall volumes were generally larger beneath bare compared to epiphyte-covered cedar canopies, except for the largest storm (14 May 2017; Table 1). Stemflow volumes from the sampled bare cedar trees were 2–5 times greater than observed beneath the epiphyte-covered cedars (Table 1). The sum of throughfall and stemflow exceeded total rainfall for the 30 July storm (9.2 mm net rainfall versus 8.9 mm gross rainfall), which is a common artifact observed when throughfall drip points are oversampled [26], rainfall is undersampled, or wind conditions permit greater three-dimensional rainfall capture area than represented by two-dimensional projected canopy areas [27].

Table 1. Rainfall conditions and throughfall and stemflow (mm across canopy area) for the five storms sampled for biolability testing. Throughfall and stemflow as percent rainfall provided in parentheses.

Condition	14 May 2017	3 June 2017	30 July 2017	28 August 2017	23 October 2017
Magnitude, mm	48.3	8	8.9	26.1	30.8
Duration, h	59.3	11.5	8.5	30	17.3
Intensity, mm h^{-1}	0.8	0.7	1.0	0.9	1.8
Throughfall					
Bare, mm (%)	26.4 (55%)	4.7 (58%)	7.5 (84%)	19.0 (73%)	25 (81%)
Epiphyte, mm (%)	28.0 (58%)	2.1 (26%)	2.9 (33%)	13.8 (53%)	20.0 (65%)
Stemflow					
Bare, mm (%)	8.0 (17%)	1.0 (12%)	1.7 (19%)	3.9 (15%)	4.4 (14%)
Epiphyte, mm (%)	4.0 (8%)	0.2 (3%)	0.4 (4%)	0.9 (4%)	1.6 (5%)

3.2. Initial DOM Concentrations

DOC concentrations in rainwater (<7 mg-C L^{-1}) were always significantly lower than in throughfall and stemflow (entire dataset provided in Table S2). The highest mean concentration across sample types was found in stemflow from epiphyte-covered canopies, 85 ± 38 mg-C L^{-1}, followed by epiphyte-covered throughfall, 64 ± 34 mg-C L^{-1}, then bare-canopy stemflow, 55 ± 34 mg-C L^{-1}, and bare-canopy throughfall, 35 ± 19 mg-C L^{-1} (Table S2). The maximum initial tree-DOM concentration observed was 143 mg-C L^{-1} from epiphyte-covered stemflow (Table S2). Significant differences between sample types were found for bare-canopy throughfall versus epiphyte-covered throughfall, bare-canopy stemflow versus epiphyte-covered stemflow, and bare-canopy throughfall versus epiphyte-covered stemflow ($p < 0.001$).

3.3. Interstorm Tree-DOM Biolability

Tree-DOM concentrations declined over the bioincubation experiments, conforming to first-order decay models (Figure S1 and Table S3). Greater tree-BDOM proportions generally related to larger decay coefficients (Figure 2). Mean tree-BDOM proportion was greatest during the smallest magnitude storm (3 June, 8.0 mm) for all sample types except bare-canopy throughfall, $70 \pm 20\%$ (Figure 2). Eighty-eight percent of tree-DOM, on average, in stemflow from bare and epiphyte-covered canopies and epiphyte-covered throughfall was biolabile for the 3 June storm (Figure 2). The 3 June storm also had the largest range in mean decay coefficients across sample types, 2.4–6.7 day^{-1} for bare-canopy versus epiphyte-covered throughfall, respectively (Figure 2). The proceeding storm, 30 July, had the largest range in mean tree-BDOM across sample types, $34 \pm 12\%$ for epiphyte-covered stemflow versus $73 \pm 9\%$ for bare-canopy throughfall, but the mean decay coefficient was generally similar (Figure 2). For the 28 August storm, all sample types had similar mean tree-BDOM proportions, 49–55%, but decay coefficients were as low as 1.4 day^{-1} for epiphyte-covered stemflow and over double this value, 3.7 day^{-1} for bare-canopy throughfall (Figure 2). The 23 October and 14 May storms were both large magnitude storms (30.8 and 48.3 mm, respectively), and tree-BDOM proportions were generally the lowest on average, barring epiphyte-covered throughfall on 23 October (Figure 2). Tree-BDOM regardless of epiphyte cover or flux type produced the lowest decay coefficients, 0.2–0.8 day^{-1}, for the largest, 14 May, storm (Figure 2).

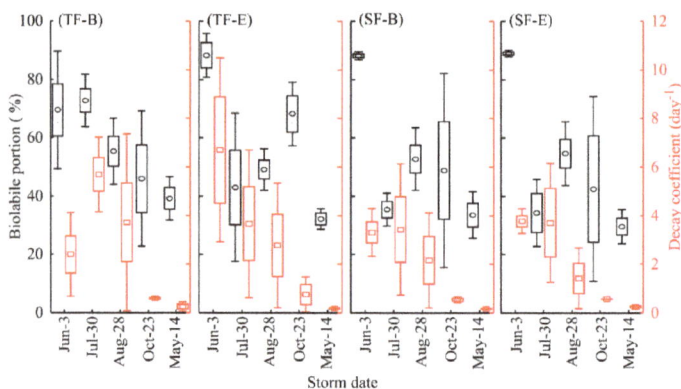

Figure 2. Mean, standard error, and standard deviation for biolabile proportion (black) and the decay coefficient (red) across the five sampled storms during 2017 arranged by increasing storm magnitude for bare-canopy throughfall (TF-B), epiphyte-covered throughfall (TF-E), bare-canopy stemflow (SF-B), and epiphyte-covered stemflow (SF-E).

3.4. Tree-BDOM Yield

For individual storms, mean tree-BDOM yield from any cover ranged over an order of magnitude, from 3.1 to 62.6 mg-C m^{-2} mm^{-1} of rainfall (Table 2). Greater water yields of throughfall from bare canopies (Table S2) compared to all other fluxes resulted in generally larger tree-BDOM yields for smaller storms (3 June, 30 July: Table 1) compared to throughfall from epiphyte-covered canopies (Table 2). Under rain events large enough to saturate the epiphyte-covered canopy areas (14 May, 28 August, 23 October), throughfall yielded greater tree-BDOM beneath epiphyte-covered canopy than bare canopy (Table 2). For stemflow, tree-BDOM yields were consistently greater from bare compared to epiphyte-covered canopies for all storms (Table 2) regardless of storm conditions (Table 1) and despite higher initial DOM concentrations from epiphyte-covered canopies (Table S2).

Table 2. Mean and standard deviation of biolabile tree-DOM (tree-BDOM) yields for throughfall (TF) and stemflow (SF) samples in each storm.

	Biolabile Yield (mg-C m^{-2} mm^{-1} Rainfall)				
	14 May 2017	3 June 2017	30 July 2017	28 August 2017	23 October 2017
TF, bare	10.3 ± 2.0	33.3 ± 15.2	62.6 ± 32.5	18.9 ± 6.1	16.3 ± 3.5
TF, epiphyte	12.2 ± 1.5	32.8 ± 18.8	25.5 ± 10.1	45.0 ± 17.1	28.6 ± 6.6
SF, bare	5.6 ± 1.8	10.6 ± 5.5	16.1 ± 8.7	7.4 ± 2.5	4.9 ± 1.1
SF, epiphyte	3.9 ± 2.2	3.7 ± 1.9	5.4 ± 3.0	3.3 ± 1.7	3.1 ± 0.6

4. Discussion

4.1. Hydrometeorology

A previous, more comprehensive sampling of storms at the study site, but for fewer cedar trees indicates that the sampled storms (Table 1) nearly represent the full range of typical rain event magnitudes and durations (7–74 mm [11]). However, rain event intensities for sampled storms (0.7–1.8 mm h^{-1}) were low compared to previous studies at the site (1–31 mm h^{-1} [11]). Throughfall water yields for these storms were comparable to previous work on the same cedar species [11,28,29]. Stemflow water yields from epiphyte-covered cedar trees were similar to past studies; however, bare stemflow volumes were higher, >10% for all storms (Table 1), than reported elsewhere,

5–7% of rainfall [28,29], and for two previously-monitored cedar trees at this site, 4.2% of rainfall [11]. Epiphytes intercepting rainwater and disrupting stemflow pathways on the epiphyte-covered trees likely diminished stemflow yield compared to the bare canopies. Elevated stemflow yield from bare cedar trees at the study site compared to other studies on the same species, however, may be a function of meteorological variables not measured (i.e., wind conditions [27]) or neighborhood conditions along the forest edge that improved individual trees' ability to entrain rainfall as stemflow [30].

4.2. Tree-DOM Concentration and Biolability

Concentrations of tree-DOM measured immediately after storm sampling were within the range reported by past throughfall and stemflow research, 10–480 mg-C L^{-1} [10]. Epiphyte presence enriched both stemflow and throughfall with tree-DOM compared to samples collected from bare canopies. The tangled, pendulous morphology of *T. usneoides* may account for increased tree-DOM concentrations, because it (1) can capture greater aerosols [31] and (2) trap and decompose materials (litter, insect frass, etc.) [32]. For stemflow and throughfall, regardless of the presence of epiphytes, tree-DOM concentrations varied between storms where increasing rainfall amount appears to dilute the limited store of tree-DOM along flowpaths through the canopy (Table S2), as reported previously at this site [11].

Tree-DOM in throughfall and stemflow was highly and rapidly biolabile, with an interquartile range of 36–73% biodegradation over 1–4 days (Figure S1), compared to other fluxes along the rainfall-to-runoff pathway: 14–33% in litter leachates, soil solution and stream water over several days [7]. Throughfall samples from Qualls and Haines [7] decayed less rapidly than observed in this study, exhausting the BDOM proportion after 1–2 weeks. The range of tree-BDOM percentages in cedar throughfall and stemflow agrees with values observed in broadleaved throughfall, ~60% [7]. Results agree with the optical character and molecular signatures of cedar tree-DOM from previous work at the site [20]. Thus, tree-DOM delivered to soils via throughfall or stemflow may be susceptible to significant biological alterations. When water yields are prodigious and highly localized (i.e., stemflow yields over 10% of rainfall for bare cedars), tree-DOM may rapidly infiltrate along root channels [33,34] to bypass biodegradation processes in the immediate soil. Throughfall, which is generally an area-diffuse flux, likely supplies a subsidy of tree-BDOM to the soil, up to 0.5 g-C m^{-2} in a single storm (i.e., on 14 May), that prevents DOM loss from the ecosystems and contributes to respiration when consumed.

There was a substantial range in tree-BDOM percentage, 2–96%, rate, 0.1–11.0 days^{-1}, and yield, 3–63 mg-C m^{-2} mm^{-1} of rainfall between the sampled storms (Figure 2; Table 2), that exceeded previously-reported ranges for throughfall tree-BDOM, 30–60% [7]. As indicated by the similar molecular formulas of epiphyte versus bare canopy tree-DOM [20], the presence or absence of epiphyte cover did not appear to influence the tree-BDOM percentage or rate (Figure 2). There was, however, a marked effect on tree-BDOM storm yields due to the epiphytes' reduction of water yields (Table 2). Interstorm variability in tree-BDOM characteristics was high and not significantly correlated to any storm data available in this study (rainfall amount, intensity, duration) (Table S2). This prevents the use of discrete storm data to estimate annual yields of tree-BDOM from throughfall and stemflow. Therefore, it is recommended that future work examine what controls may drive the high interstorm variability in tree-BDOM observed in this study. Canopy structural variability across phenological events was not measured during this study, thus we suggest future work assess the connection between canopy structure and tree-BDOM. Further information is also needed to scale spatially, including data on the interstorm variability in tree-BDOM in other forest types and whether the temporal drivers vary geographically. Perhaps optical metrics, many of which are now able to be automatically collected in the field [35], can progress large-scale data collection efforts on tree-DOM and its biolability.

5. Conclusions

Little information has been reported on the proportion, rate and yield of tree-BDOM in throughfall and stemflow. In this study, bioincubation experiments on tree-DOM from epiphyte-covered and bare *Juniperus virginiana* (cedar) at Skidaway Island (Georgia, USA) found throughfall and stemflow to be largely biolabile, 36–73% (interquartile range). Cedar tree-BDOM was consumed rapidly (within 1–4 days) and yielded 3–63 mg-C m^{-2} mm^{-1} of rainfall. A simple estimate of tree-BDOM annual yield using the mean biolability proportion for each hydrologic flux and total tree-DOM annual yields previously computed at this site [14] equates to 21.3–30.2 g-C m^{-2} year^{-1} of tree-BDOM, which is 33–47% of the 65 g-C m^{-2} year^{-1} of net ecosystem exchange estimated for Georgia (USA) forests [31]. Tree-BDOM proportions and biodegradation rates were not significantly different between throughfall and stemflow, or under epiphyte versus bare canopy conditions. However, differences in water yield between fluxes, and due to epiphyte presence, influenced tree-BDOM yields per storm. The proportion, rate and yield of tree-BDOM varied markedly between storms and was not explained by storm amount, duration, or intensity. It is recommended that future work seek to characterize drivers of interstorm variability in tree-BDOM across forest types. This will enhance understanding of a key DOM supply to soil ecosystems by enabling temporal and spatial scaling of the first, and potentially most biolabile, enrichment of rainwater with DOM along the rainfall-to-runoff pathway.

Supplementary Materials: The following are available online at http://www.mdpi.com/2079-4991/9/5/ 236/s1, Figure S1: Example tree-DOM decomposition curves, Table S1: Structural characteristics of stemflow trees, Table S2: Raw dataset, Table S3: First order decay kinetics.

Author Contributions: J.T.V.S. and A.S. conceived and designed the experiments; D.H.H. and A.W. performed the experiments; D.H.H. and L.Z. analyzed the data; A.S. contributed reagents/materials/analysis tools; D.H.H. wrote the paper with input from all other authors.

Acknowledgments: D.H.H., A.W. and J.T.V.S. acknowledges support from NSF-1518726 and the Research Scholar Award from Georgia Southern University's College Office of Undergraduate Research.

Conflicts of Interest: The authors declare no conflict of interest.

References

1. Dawson, H.; Ugolini, F.; Hrutfiord, B.; Zachara, J. Role of soluble organics in the soil processes of a podzol, central cascades, washington. *Soil Sci.* **1978**, *126*, 290–296. [CrossRef]
2. Guggenberger, G.; Kaiser, K. Dissolved organic matter in soil: Challenging the paradigm of sorptive preservation. *Geoderma* **2003**, *113*, 293–310. [CrossRef]
3. Temminghoff, E.J.; Van der Zee, S.E.; de Haan, F.A. Copper mobility in a copper-contaminated sandy soil as affected by ph and solid and dissolved organic matter. *Environ. Sci. Technol.* **1997**, *31*, 1109–1115. [CrossRef]
4. Michalzik, B.; Kalbitz, K.; Park, J.-H.; Solinger, S.; Matzner, E. Fluxes and concentrations of dissolved organic carbon and nitrogen–a synthesis for temperate forests. *Biogeochemistry* **2001**, *52*, 173–205. [CrossRef]
5. Donald, R.G.; Anderson, D.W.; Stewart, J.W. Potential role of dissolved organic carbon in phosphorus transport in forested soils. *Soil Sci. Soc. Am. J.* **1993**, *57*, 1611–1618. [CrossRef]
6. Dittmar, T.; Stubbins, A. *12.6—Dissolved organic matter in aquatic systems. Treatise on Geochemistry*, 2nd ed.; Elsevier: Oxford, UK, 2014; pp. 125–156.
7. Qualls, R.G.; Haines, B.L. Biodegradability of dissolved organic matter in forest throughfall, soil solution, and stream water. *Soil Sci. Soc. Am. J.* **1992**, *56*, 578–586. [CrossRef]
8. Cummins, K.; Klug, J.; Wetzel, R.; Petersen, R.; Suberkropp, K.; Manny, B.; Wuycheck, J.; Howard, F. Organic enrichment with leaf leachate in experimental lotic ecosystems. *BioScience* **1972**, *22*, 719–722. [CrossRef]
9. Hagedorn, F.; Bruderhofer, N.; Ferrari, A.; Niklaus, P.A. Tracking litter-derived dissolved organic matter along a soil chronosequence using 14c imaging: Biodegradation, physico-chemical retention or preferential flow? *Soil Biol. Biochem.* **2015**, *88*, 333–343. [CrossRef]
10. Van Stan, J.T.; Stubbins, A. Tree-dom: Dissolved organic matter in throughfall and stemflow. *Limnol. Oceanogr. Lett.* **2018**. [CrossRef]

11. Van Stan, J.T.; Wagner, S.; Guillemette, F.; Whitetree, A.; Lewis, J.; Silva, L.; Stubbins, A. Temporal dynamics in the concentration, flux, and optical properties of tree-derived dissolved organic matter in an epiphyte-laden oak-cedar forest. *J. Geophys. Res. Biogeosci.* **2017**, *122*, 2982–2997. [CrossRef]

12. Tan, Z.H.; Zhang, Y.P.; Liang, N.; Song, Q.H.; Liu, Y.H.; You, G.Y.; Li, L.H.; Yu, L.; Wu, C.S.; Lu, Z.Y. Soil respiration in an old-growth subtropical forest: Patterns, components, and controls. *J. Geophys. Res. Atmos.* **2013**, *118*, 2981–2990. [CrossRef]

13. Aitkenhead-Peterson, J.A.; McDowell, W.H.; Neff, J.C. Sources, production, and regulation of allochthonous dissolved organic matter inputs to surface waters. In *Aquatic Ecosystems*; Elsevier: Cambridge, MA, USA, 2003; pp. 25–70.

14. Johnson, M.S.; Lehmann, J. Double-funneling of trees: Stemflow and root-induced preferential flow. *Ecoscience* **2006**, *13*, 324–333. [CrossRef]

15. Keenan, R.J.; Reams, G.A.; Achard, F.; de Freitas, J.V.; Grainger, A.; Lindquist, E. Dynamics of global forest area: Results from the fao global forest resources assessment 2015. *For. Ecol. Manag.* **2015**, *352*, 9–20. [CrossRef]

16. Briggs, J.M.; Hoch, G.A.; Johnson, L.C. Assessing the rate, mechanisms, and consequences of the conversion of tallgrass prairie to juniperus virginiana forest. *Ecosystems* **2002**, *5*, 578–586. [CrossRef]

17. Elias, R.B.; Dias, E. Gap dynamics and regeneration strategies in juniperus-laurus forests of the azores islands. *Plant Ecol.* **2009**, *200*, 179–189. [CrossRef]

18. Van Stan, J.T., II; Pypker, T.G. A review and evaluation of forest canopy epiphyte roles in the partitioning and chemical alteration of precipitation. *Sci. Total Environ.* **2015**, *536*, 813–824. [CrossRef] [PubMed]

19. Zotz, G.; Bader, M. Epiphytic plants in a changing world-global: Change effects on vascular and non-vascular epiphytes. In *Progress in Botany*; Springer: Berlin/Heidelberg, Germany, 2009; pp. 147–170.

20. Stubbins, A.; Silva, L.M.; Dittmar, T.; Van Stan, J.T. Molecular and optical properties of tree-derived dissolved organic matter in throughfall and stemflow from live oaks and eastern red cedar. *Front. Earth Sci.* **2017**, *5*, 22. [CrossRef]

21. Stubbins, A.; Spencer, R.G.; Chen, H.; Hatcher, P.G.; Mopper, K.; Hernes, P.J.; Mwamba, V.L.; Mangangu, A.M.; Wabakanghanzi, J.N.; Six, J. Illuminated darkness: Molecular signatures of congo river dissolved organic matter and its photochemical alteration as revealed by ultrahigh precision mass spectrometry. *Limnol. Oceanogr.* **2010**, *55*, 1467–1477. [CrossRef]

22. *Climate Average from 1948 to 2016 Climate Name GA54.CLI*; University of Georgia: Athens, GA, USA, 2012.

23. Fellman, J.B.; D'Amore, D.V.; Hood, E.; Boone, R.D. Fluorescence characteristics and biodegradability of dissolved organic matter in forest and wetland soils from coastal temperate watersheds in southeast Alaska. *Biogeochemistry* **2008**, *88*, 169–184. [CrossRef]

24. Bittar, T.B.; Pound, P.; Whitetree, A.; Dean Moore, L.; Van Stan, J.T. Estimation of throughfall and stemflow bacterial flux in a subtropical oak-cedar forest. *Geophys. Res. Lett.* **2018**, *45*, 1410–1418. [CrossRef]

25. Stubbins, A.; Dittmar, T. Low volume quantification of dissolved organic carbon and dissolved nitrogen. *Limnol. Oceanogr. Methods* **2012**, *10*, 347–352. [CrossRef]

26. Crockford, R.; Richardson, D. Partitioning of rainfall into throughfall, stemflow and interception: Effect of forest type, ground cover and climate. *Hydrol. Process.* **2000**, *14*, 2903–2920. [CrossRef]

27. Herwitz, S.R.; Slye, R.E. Three-dimensional modeling of canopy tree interception of wind-driven rainfall. *J. Hydrol.* **1995**, *168*, 205–226. [CrossRef]

28. Owens, M.K.; Lyons, R.K.; Alejandro, C.L. Rainfall partitioning within semiarid juniper communities: Effects of event size and canopy cover. *Hydrol. Process.* **2006**, *20*, 3179–3189. [CrossRef]

29. Zou, C.B.; Caterina, G.L.; Will, R.E.; Stebler, E.; Turton, D. Canopy interception for a tallgrass prairie under juniper encroachment. *PLoS ONE* **2015**, *10*, e0141422. [CrossRef] [PubMed]

30. Metzger, J.C.; Germer, S.; Hildebrandt, A. Factors impacting stemflow generation in a european beech forest: Individual tree versus neighborhood properties. In Proceedings of the EGU General Assembly Conference Abstracts, Vienna, Austria, 23–28 April 2017; p. 8410.

31. Gay, T.E.; Van Stan, J.T.; Moore, L.D.; Lewis, E.S.; Reichard, J.S. Throughfall alterations by degree of tillandsia usneoides cover in a southeastern us quercus virginiana forest. *Can. J. For. Res.* **2015**, *45*, 1688–1698. [CrossRef]

32. Turner, E.C.; Snaddon, J.L.; Johnson, H.R.; Foster, W.A. The impact of bird's nest ferns on stemflow nutrient concentration in a primary rain forest, Sabah, Malaysia. *J. Trop. Ecol.* **2007**, *23*, 721–724. [CrossRef]

33. Backnäs, S.; Laine-Kaulio, H.; Kløve, B. Phosphorus forms and related soil chemistry in preferential flowpaths and the soil matrix of a forested podzolic till soil profile. *Geoderma* **2012**, *189*, 50–64. [CrossRef]
34. Spencer, S.; Meerveld, H. Double funnelling in a mature coastal british columbia forest: Spatial patterns of stemflow after infiltration. *Hydrol. Process.* **2016**, *30*, 4185–4201. [CrossRef]
35. Etheridge, J.R.; Birgand, F.; Osborne, J.A.; Osburn, C.L.; Burchell, M.R.; Irving, J. Using in situ ultraviolet-visual spectroscopy to measure nitrogen, carbon, phosphorus, and suspended solids concentrations at a high frequency in a brackish tidal marsh. *Limnol. Oceanogr. Methods* **2014**, *12*, 10–22. [CrossRef]

forests

MDPI

Review

Enrichment Planting and Soil Amendments Enhance Carbon Sequestration and Reduce Greenhouse Gas Emissions in Agroforestry Systems: A Review

Bharat M. Shrestha [1,2], Scott X. Chang [1,*], Edward W. Bork [3] and Cameron N. Carlyle [3]

[1]   Department of Renewable Resources, University of Alberta, 442 Earth Science Building, Edmonton,
     AB T6G 2E3, Canada; bshresth@ualberta.ca
[2]   Alpine Environmental and Forestry Technical Services Inc., 103 Portrush Ave, Ottawa, ON K2J 5J2, Canada
[3]   Department of Agricultural, Food and Nutritional Science, University of Alberta, 410 Agriculture/Forestry
     Centre, Edmonton, AB T6G 2H1, Canada; ebork@ualberta.ca (E.W.B.); cameron.carlyle@ualberta.ca (C.N.C.)
*   Correspondence: sxchang@ualberta.ca; Tel.: +1-780-492-6375

Received: 1 April 2018; Accepted: 4 June 2018; Published: 20 June 2018

Abstract: Agroforestry practices that intentionally integrate trees with crops and/or livestock in an agricultural production system could enhance carbon (C) sequestration and reduce greenhouse gas (GHG) emissions from terrestrial ecosystems, thereby mitigating global climate change. Beneficial management practices such as enrichment planting and the application of soil amendments can affect C sequestration and GHG emissions in agroforestry systems; however, such effects are not well understood. A literature review was conducted to synthesize information on the prospects for enhancing C sequestration and reducing GHG emissions through enrichment (i.e., in-fill) tree planting, a common practice in improving stand density within existing forests, and the application of organic amendments to soils. Our review indicates that in agroforests only a few studies have examined the effect of enrichment planting, which has been reported to increase C storage in plant biomass. The effect of adding organic amendments such as biochar, compost and manure to soil on enhancing C sequestration and reducing GHG emissions is well documented, but primarily in conventional crop production systems. Within croplands, application of biochar derived from various feedstocks, has been shown to increase soil organic C content, reduce CO_2 and N_2O emissions, and increase CH_4 uptake, as compared to no application of biochar. Depending on the feedstock used to produce biochar, biochar application can reduce N_2O emission by 3% to 84% as compared to no addition of biochars. On the other hand, application of compost emits less CO_2 and N_2O as compared to the application of manure, while the application of pelleted manure leads to more N_2O emission compared to the application of raw manure. In summary, enrichment planting and application of organic soil amendments such as compost and biochar will be better options than the application of raw manure for enhancing C sequestration and reducing GHG emissions. However, there is a shortage of data to support these practices in the field, and thus further research on the effect of these two areas of management intervention on C cycling will be imperative to developing best management practices to enhance C sequestration and minimize GHG emissions from agroforestry systems.

Keywords: climate change; manuring; manure pelleting; northern temperate; pyrolysis; information review

1. Introduction

Agriculture is the second largest emitter of greenhouse gases (GHG) after the energy sector, and is responsible for about 30% of global GHG emissions [1]. Agroforestry, the intentional integration of trees and/or shrubs with herbaceous crops and/or livestock in a production system, is a popular beneficial management practice (BMP) that can mitigate climate change by sequestering carbon (C) and reducing greenhouse gas (GHG) emissions [2–8]. The Intergovernmental Panel on Climate Change (IPCC) has recognized both afforestation and reforestation as important activities supporting C sequestration [9]. Agroforestry systems include many different permutations such as alley cropping, silvopasture, riparian buffers, savanna, forest farming, home-gardens, and woodlots, as well as other similar integrated land-use systems [10]. In all cases, agroforestry systems are recognized as a land use management framework that simultaneously integrates the dual goals of ecological conservation and socio-economic development [9–12].

The environmental service of sequestering C and reducing GHG emissions provided by agroforestry systems are relatively well-documented globally at various management systems as summarized in Section 3 below. However, data quantifying the specific role of management interventions to improve such benefits are rare, with only a few studies reporting on the potential benefits of enrichment planting [13,14] and organic amendment of soils [2,15]. Enrichment planting is commonly used for increasing the density of desired tree species in degraded (secondary) forests, particularly where these forests are low in density or occupied by less-desirable (i.e., non-productive) tree species. On the other hand, the addition of organic amendments to soils, including mulch, manure, or the application of other organic by-products from a feedlot or modified organic materials (such as biochar, composts, and manure pellets), is widely practiced in sole cropping systems but rare in agroforestry systems.

Manure is a widely available by-product from livestock production systems, particularly those involving the confined feeding of animals in large-scale livestock operations (poultry, swine, beef, dairy, etc.). Due to its high nutrient and C content, manure management and its application to soil plays a critical role in GHG emissions, including CH_4 and N_2O [16]. The Food and Agriculture Organization of the United Nations (FAO) estimated the CH_4 and N_2O emissions associated with manure storage and processing contribute 4.3% and 5.2% respectively, and N_2O emissions from the field applied and deposited manure contribute 16.4% of the GHG emissions in the global livestock supply chains [17]. Composting, despite emitting GHGs during storage, and dried pelleting, are two methods of conserving nutrients in manure and facilitating their slow release into the soil [18]. In the process, these methods reduce GHG emissions compared with raw manure application if applied to the field in an appropriate time such as avoiding wet conditions of soils [19]. Biochar is pyrolysed biomass consisting of around 50% or more recalcitrant organic material [20]. It is a promising soil amendment that improves physical and chemical properties of soils [21], as well as provides better environmental services such as improved nutrient cycling, increased C sequestration and reduce GHG emissions. However, studies on the effects of biochars in agroforestry systems are limited [22].

This paper reviews the current state of knowledge regarding opportunities to enhance C sequestration and reduce GHG emissions through two potential management interventions within agroforestry systems. The first is enrichment tree planting, and the second is the use of organic soil amendments. It starts with an overview of the impact of agroforestry in C sequestration and GHG emissions as documented by previous different empirical studies, and reviews reports at different temporal and spatial scales. Subsequent sections then review the role of enrichment planting and organic soil amendments and discuss the prospect of applying these interventions to agroforestry systems in order to enhance C sequestration and reduce GHG emissions. Finally, conclusions are drawn and areas of further research needs are identified on these two practices in order to further mitigate GHG emissions and promote climate change adaptation using agroforestry systems.

2. Methods of Literature Collection

A wide range of published literature was collected through searches using Google Scholar and ISI Web of Science with a Boolean defined by logical strings containing "and/or" with keywords "agroforestry", "environmental service", "enrichment planting", "greenhouse gas emission", "carbon sequestration", "biochar", "manure", "manure pellet", "composting", and "secondary forest".

More than 200 publications, both referred and non-reviewed, were found; they were further sorted with criteria of "carbon sequestration and agroforestry" and "greenhouse gas emission and agroforestry", "enrichment planting and secondary forest or agroforestry", "composting and greenhouse gas emission", "biochar and soil C sequestration", "raw manure and composted manure" and "manure pellet and soil carbon". With these criteria, the number of publications selected was reduced to 94. Among them, five publications were on enrichment planting, six on manure pelleting, 33 on manure, 10 on compost and management, and 40 on biochar. Key results found from these studies were compared focusing on C sequestration in vegetation and soils, as well as GHG emissions. A total of 82 publications closely related to the subject matter were used in this paper.

3. Role of Agroforestry in C Sequestration and Reducing Greenhouse Gas Emissions

Several studies have documented that the C sequestration potential of agroforestry systems varies depending on environment and specific management systems (Table 1). Estimated global C sequestration potential of agroforestry ranges between 12 and 228 Mg C ha^{-1}, leading to a net C sequestration potential of 1.1 to 2.2 Pg (1 Pg = 10^{15} g) over 50 years [23]. It is also estimated that improved management alone within existing agroforestry systems could sequester an additional 0.3 Mg C ha^{-1} y^{-1}, while undertaking land use changes from conventional cropland to agroforestry (crops combined with forests) could sequester an additional 3.1 Mg C ha^{-1} y^{-1} [24]. The SOC sequestration rate varies among agroforestry systems across different regions, ranging from 0.1 to 4.2 Mg C ha^{-1} y^{-1} depending upon the age of agroforests, and soil depths considered in the estimation [25]. In the early stage of practicing agroforestry, soil C can be lost from top soils; for example, a multi-strata agroforestry system in Ghana lost 0.4 Mg C ha^{-1} y^{-1} until 15 years after establishment, after which a small amount of SOC was stored (0.06 Mg C ha^{-1} y^{-1}) through 25 years age within the 0–15 cm soil layer (cited in [25]). Furthermore, the vegetation C sequestration rate differs by forest types. As an example, within windbreak systems, broadleaved trees demonstrated an almost double C storage capacity (4.39 ± 1.74 Mg C ha^{-1} y^{-1}) than conifer trees (2.45 ± 0.42 Mg C ha^{-1} y^{-1}) in a study involving nine ecoregions across the USA [26].

Table 1. Carbon sequestration rate, carbon stocks and greenhouse gas (GHG) emissions in different types of agroforestry systems. A negative flux shows consumption of GHGs.

Agroforestry/Management Activities/Location	Carbon/GHG Data	Reference
Above- and below-ground vegetation C sequestration rate (Mg C ha^{-1} y^{-1})		
Fodder bank, West Africa (7.5 years)	0.3	
Tree-based inter-cropping, Canada (13 years)	0.8	
Agroforest, Western Oregon, USA (11 years)	1.1	
Agrisilviculture, India (5 years)	1.3	
Silvopasture, India (5 years)	6.6	
Home gardens, Togo (23 years)	4.3	[9]
Shaded coffee, Togo (13 years)	6.3	
Home gardens, Indonesia (13 years)	8.0	
Cacao agroforest, Cameroon (26 years)	5.9	
Cacao Agroforest, Costa Rica (5 years)	10.3	
Cacao Agroforest, Costa Rica (10 years)	11.1	
Woodlots, Puerto Rico (4 years)	12.0	
Median C storage in different ecoregions		
Semi-arid (5 years)	2.6	
Sub-humid (8 years)	6.1	[27]
Humid (5 years)	10.0	
Temperate (30 years)	3.9	
Windbreak in U.S. ecoregions		
Conifers	2.0–2.9	[26]
Broadleaved	2.7–6.1	

Table 1. *Cont.*

Agroforestry/Management Activities/Location	Carbon/GHG Data	Reference
Shifting cultivation * in Peruvian Amazon and Indonesia	3.5	[28]
Improved fallow **		
12-month-old fallow	5.3–13.2	
18-month-old fallow	17.4–31.9	[4]
22-month-old fallow	21.3–30.5	
Vegetation C stock (Mg C ha^{-1})		
Improved fallow in Mediterranean	70	
Potential C storage in six continents		
Africa, agrosilvicultural	29–53	
South America, agrosilvicultural	39–195	
Southeast Asia, agrosilvicultural	12–228	[28,29]
Australia, silvopastoral	28–51	
North America, silvopastoral	90–198	
Northern Asia, silvopastoral	15–18	
Different agroforestry systems in sub-Saharan Africa:		
Arid and semi-arid silvopastoral:		
pastoral/fruit	0.8–3.9	
pastoral/fuelwood	3.9–19.4	
pastoral/shelterbelt	1.7–1.8	
Humid silvopastoral:		
pastoral/fruit	2.0–8.6	
pastoral/fuelwood	5.1–24.7	[30]
pastoral/shelterbelt	2.8–6.5	
Fruit/fuelwood	4.6–23.0	
Fruit/timber	33.3–71.3	
Fruit/shelterbelt	2.4–5.4	
Fuelwood/timber	36.4–86.8	
Fuelwood/shelterbelt	5.5–20.9	
Soil C sequestration rate (Mg C ha^{-1} y^{-1})		
Alley cropping, France (equivalent mass basis)		
26–29 cm	0.25	[31]
93–98 cm	0.35	
Improved fallow in Mediterranean	1.6	[28]
Soil C stock (Mg C ha^{-1})		
Three agroforestry systems, Alberta, Canada		
0–10 cm		
Hedgerow (natural forest + crop)	77	[32]
Shelterbelt (planted forest + crop)	67	
Silvopasture (natural forest + grassland)	101	
0–30 cm		
Hedgerow (natural forest + crop)	178	
Shelterbelt (planted forest + crop)	163	
Silvopasture (natural forest + grassland)	201	
Inter-cropping in sub-tropical China, 0–80 cm		
Tree + shrub	93	
Tree + legume & cereal	79	[33]
Tree + Oilseed & legume	74	
Humid tropics, 0–20 cm	25	[28]
Different agroforestry systems in Canada		
Alley cropping, 0–40 cm (13–25 years)	71.1–125.4	
Alley cropping, 0–30 cm (8–9 years)	43.5–113.2	[5]
Shelterbelt, 0–30 cm (various ages)	15–208	

Table 1. *Cont.*

Agroforestry/Management Activities/Location	Carbon/GHG Data	Reference
GHG emission rates (kg ha^{-1} y^{-1})		
Different agroforestry systems in Peruvian Amazon and Indonesia		
Shifting cultivation		
N$_2$O-N emission		
CH$_4$-C flux	0.8	
CO$_2$-C emission	−2.0	
Multi-strata agroforestry	5.9	
N$_2$O-N emission		[28]
CH$_4$-C flux	0.5	
CO$_2$-C emission	−2.0	
Peach-palm agroforestry	5.5	
N$_2$O-N emission		
CH$_4$-C flux	0.9	
CO$_2$-C emission	−1.5	
	5.8	
CO$_2$-C emission from agroforestry systems in Alberta		
Hedgerow (natural forest + crop)	16,425	
Shelterbelt (planted forest + crop)	10,950	[34]
Silvopasture (natural forest + grassland)	13,505	
CH$_4$-C emission		
Hedgerow (natural forest + crop)	−2.2	
Shelterbelt (planted forest + crop)	−1.8	
Silvopasture (natural forest + grassland)	−2.9	
Different agroforestry systems in Canada		
Alley Cropping		
CO$_2$-C emission	4900–6240	
Shelterbelts (combined with annual crops)		[5]
CO$_2$-C emission	1900–4000	
CH$_4$-C efflux	−0.15–−0.9	
N$_2$O-N efflux	0.25–3.0	
Shade coffee agroforestry, Sumatra		
N$_2$O-N	16	[4]
CH$_4$-C	−1.0	

* Plants are cut and burned to create the farming field—also called slash-and-burn system; ** rotation between cereal crops and tree-legume fallow.

Levels of C sequestration in vegetation and soils are known to vary among ecoregions. In the humid tropics, 70 and 25 Mg C ha^{-1} can be sequestered within vegetation and the top 20 cm of soil, respectively [28]. In Mediterranean regions, total C sequestration rates in vegetation and soils of different agroforestry systems can be up to 1.3 Mg C ha^{-1} y^{-1} [35]. In temperate climates, the potential C sequestration by aboveground vegetation of agroforestry systems could be as large as 2.1 × 10^9 Mg C y^{-1}, while in tropical regions, it could be 1.9 × 10^9 Mg C y^{-1} [36]. Collectively, these examples show that C sequestration rates and resulting C stocks vary widely, reflecting marked variation in climatic conditions, soil properties, vegetation types, and the ongoing management of agroforests. However, observed variation in estimates of C might also be due to the use of different methods for estimating soil C sequestration potential under contrasting conditions, coupled with the inherently high natural variability of soil C stocks within agroforestry systems associated with divergent agro-ecological zones [37].

Trees are also known to help reduce CH$_4$ and N$_2$O emissions, particularly in relation to neighboring cropland [5]. In the sub-tropics, agroforestry systems combining trees and inter-cropped shrubs store more C in vegetation and soils compared to systems with only trees or trees grown with legume or cereals as inter-cropped systems [38,39]. Similarly, windbreak and riparian forest buffers store significant amounts of C, in addition to providing other valuable ecosystem services such as improved water quality, biodiversity, and biomass feedstock availability [40]. Sequestered C in low-till croplands with adjacent treed windbreaks was 75% greater than in low-till lands without adjacent windbreaks [40]. Compared to sole herbland pastures, the presence of trees in the former leads to greater topsoil and subsoil C content, and larger litter inputs result in higher free and occluded organic matter (OM) fractions, and ultimately higher levels of stabilized SOM fractions [41]. A study in central Alberta, Canada showed that silvopastoral systems had higher SOC and lower GHG emissions compared to agroforestry systems containing either hedgerows or shelterbelts combined with annual cropland [32,34]. Mean SOC in the bulk soil at 0–10 cm depth was 81, 48 and 63 g kg^{-1} in the silvopasture, shelterbelt and hedgerow systems, respectively. Soil C in the more stable fine fraction (<53 μm) of the soil was higher in hedgerow systems (34 g kg^{-1}) compared to both shelterbelt and silvopasture systems (29 and 29 g kg^{-1}, respectively). Within each agroforestry system, total SOC and the SOC concentration within each size fraction was consistently greater in the forested land-use compared to the adjacent agricultural herbland [32]. The SOC stock in both the 0–10 cm and 10–30 cm soil layers were greater within the forested land cover type than in the adjacent herbland [42]. In terms of GHG emissions, the silvopasture system had 15% greater CH_4 uptake and 44% lower N_2O emission compared with the shelterbelt and silvopasture systems [34].

Despite their potential to mitigate GHG emissions, agroforestry systems can be a significant sink or source of GHGs depending upon management practices. In the humid tropics, agroforestry mitigated N_2O and CO_2 emissions from soils, and increased CH_4 uptake, compared to sole cropping systems [28]. While N_2O emission in the agroforestry system was as low as 3%, CO_2 emissions were 70% of the high input cropping systems, and CH_4 uptake was almost double that of the low input cropping system [33]. In contrast, management practices that disturbed soil and vegetation, such as tillage, burning of biomass, fertilization, and manuring, lead to net emissions of GHGs from soils and vegetation to the atmosphere [43]. Among different agroforestry systems, multi-strata systems reduce N_2O emissions and CH_4 oxidation, but emit similar CO_2 compared to shifting cultivation, crop/rubber agroforestry and short fallow systems [28].

4. Management Intervention to Enhance C Sequestration and Reduce GHG Emissions

4.1. Impacts of Enrichment Planting on C Sequestration and GHG Emissions

Enrichment planting, also known as in-fill or gap-planting, is commonly practiced to increase the density of desired tree species in degraded (secondary) forests [13,44], including those found in shelterbelts or silvopastoral plantations of agroforestry systems. Enrichment planting enables newly establishing trees to utilize available resources, including light, moisture and nutrients [45]. Agricultural systems that include trees generally store more C in deeper soil layers compared to treeless systems, and higher SOC content in the former has been associated with greater species richness of trees and tree density [4,46]. Enrichment planting in old fallow fields is beneficial in sequestering C, improving over-story tree diversity, and enhancing social, cultural, and ecosystem services [13]. The improved C storage observed after enrichment planting in eastern Panama was around 113 Mg C ha^{-1}, which is comparable to that in industrial teak plantations and primary forests [13].

The choice of tree species within plantations affects C storage in phytomass, necromass, and underlying soils. For example, after 40 years of growth in plantations, conifers had higher biomass and litter C, while broad-leaved forests had considerably more soil C [47]. The greater decomposition rate of broadleaf litter contributed favorably to soil C sequestration compared to that from conifer litter, but due to relatively steady photosynthetic rates throughout the year and high drought tolerance, conifers had more stable live biomass [47]. Understory vegetation biomass was also negatively correlated to tree-biomass, with conifer stands leading to less understory C mass due to increased canopy closure and associated light limitations [47].

Natural regeneration of vegetation in abandoned pasture land is known to sequester C in a manner similar to planted vegetation, although the rate of sequestration can be slower [48]. The aboveground C accumulation rate in 12–14 year-old forests was 5.6 Mg C ha^{-1} y^{-1}, and SOC accumulation rates were 1.49 Mg ha^{-1} y^{-1} [48]. These results indicate that natural regeneration of tree species can mimic enrichment planting after pasture abandonment.

4.2. Impact of Organic Soil Amendment on C Sequestration and Greenhouse Gas Emissions

Soil amendment with organic input is a common practice in conventional agricultural practices (e.g., on annual cropland or forage land), and involves manuring, mulching, green manuring and biochar addition. Agroforestry systems can provide various types of feedstock for bulking agents such as residues from annual crops, small woody biomass from pastures, leaf litter, as well as twigs, branches and woody biomass from trees for use in composting, pelleting or biochar production. In this section, we review relevant literature and summarize potential impacts of these amendments on C sequestration and GHG emissions in agroforestry.

4.2.1. Impacts of Biochar Applications

Biochar has a slow decomposition rate and its application to soils can sequester SOC compared to non-amended soils [49]. For example, 1.4 times higher total soil C was found in hardwood biochar amended soils compared to non-amended soils [49]. Biochar has been tested in different cropping systems to assess its impact on enhancing C sequestration and reducing GHG emissions. A review of a wide range of agro-ecosystems with biochar application showed that despite using a variety of feedstocks and crops, the resultant impact of biochar application was a decrease in GHG emissions by up to 66% in CO_2 and up to 50% in N_2O emissions [2]. In addition, biochar addition led to reduced leaching of plant nutrients and contamination of downstream water sources [2]. However, some biochar amended soils increased CO_2 emissions, which was attributed to increased soil porosity, lowered bulk density and higher pH, all of which may favor microorganism activity [2]. Biochar from wood and herbaceous feedstocks performed the best in reducing emissions (ca. −60%), while manure-based biochar was less effective, the latter of which altered N_2O emissions by −46% to +39% [50]. Biochar feedstock, pyrolysis conditions, and C/N ratios were key factors influencing the emissions of N_2O [50].

A previous meta-analysis of published data obtained from laboratory and field experiments to explore the effects of biochar on N_2O and CH_4 emissions reported nearly 50% less N_2O emissions across different soil types [51]. This same study indicated a potential to increase the uptake of CH_4 due to enhanced methanothrophy following biochar addition [51]. Effects of biochar on N_2O emissions also varied among feedstocks, with both woody and crop residue biochars decreasing emissions, while biochars derived from other feedstocks (e.g., manures, bio-solids, paper mill residues) had no significant effects [51]. A laboratory incubation study conducted across ten different soils in the USA and receiving the same hardwood biochar found no significant differences in the emissions of CO_2 and CH_4. However, this same study reported a decrease in N_2O emission (up to 63% less) across all soils after biochar applications [52]. Similarly, biochar produced from pine sawdust at 500 °C with or without steam activation decreased CO_2 and N_2O emission (up to 32% in forest soils), though no differences in CH_4 uptake were detected [53]. Pine biochar reduces GHG emissions by decreasing microbial and enzyme activities [53]. Moreover, by changing the physical (gas diffusivity, aggregation, water retention), chemical (e.g., pH, redox potential, availability of organic and mineral N and dissolved organic C, organo-mineral interactions), and biological properties (e.g., microbial community structure, microbial biomass and activity, macro faunal activity, N cycling enzymes) of soils, biochar influences N mineralization-immobilization, turnover, and nitrification or denitrification processes, all of which ultimately affect N_2O emissions [51].

The impact of biochar on GHG emissions within amended soils is dependent on both biochar and soil properties [20]. Relationships between the biochar and soil N dynamics revealed that adsorption of NH_4^+ and NO_3^- in biochar during the pyrolysis process decreased N loss during composting and after manure application, thus offering a mechanism for the slow release of fertilizer in the field [54]. Higher pyrolysis temperatures during the manufacture of biochar from manure and bio-solids also result in biochars with decreased hydrolysable organic N and increased aromatic N [54]. Short-term N_2O emissions are therefore likely to decrease following biochar application, though no clear information exists on the long-term effects of this practice. In summary, biochar input to agroecosystems represents a potential mitigation strategy for environmentally detrimental N losses, specifically as N_2O [54].

Biochar can also enhance the process of composting manure and reduce GHG emissions during composting and subsequent field applications. The impact of biochar addition in conjunction with composting, including their application to soils with manure and manure pellets, on GHG emissions and C sequestration, are summarized in Table 2. Application of biochar during the composting of chicken manure increased peak CO_2 emission, while emissions of both CH_4 and N_2O decreased [55,56]. Composting of cattle manure with added biochar increased aeration, and hence the activity of methanogens, which reduced CH_4 emission [15]. Biochar reduced N_2O and CH_4 emissions during field applications due to a change in the microenvironment for the microbial population, including soil water content, and availability of oxygen, N, and C [57]. In calcareous soils, biochar application alone increased total organic C stocks by 1.4 fold, while the application of biochar mixed with manure increased C levels by 1.7 fold [49].

Impacts of biochar on GHG emissions have shown mixed results depending on soil type, feedstock type and season of the application [58,59]. Biochar addition to upland soil increased CH_4 emissions by 37% during the summer, but had no effect in winter, while decreasing N_2O emissions up to 54% and 53% during the summer and winter seasons, respectively [58]. In Chernozemic soils amended with straw, and its biochar reduced N_2O emission but there were no significant effects on CH_4 or CO_2 emission compared with the unamended soils [59]. A soil-column experiment using non-treated soils and those amended with biochar prepared by pyrolysis of pig manure and spruce sawdust at 600 °C found no differences in N_2O, CH_4 and CO_2 emissions between the two treatments until they received an application of fresh pig manure during the 10th week [60]. After 10 weeks, cumulative GHG emissions were higher from soils amended with biochars and manure for up to four weeks compared to the non-treated soil. However, this same study found NO_3^--N leaching was 51% and 43% lower in pig manure biochar amended soils and wood biochar amended soils, respectively, compared to the pig manure only-amended soils [60].

Life cycle analysis is an emerging tool to link the full C footprint of products from their origin via different stages of the product supply chain [61–63]. A life cycle analysis of biochar systems comparing biochar produced from three feedstocks, namely corn stover, yard waste (waste from industrial-scale composting) and switchgrass energy crops, found that the net energy provided by corn stover and yard waste was negative (−864 and −885 kg CO_2e per Mg dry feedstock, respectively) while switchgrass was a net emitter (+36 kg CO_2e per Mg dry feedstock) [63]. These findings indicate that careful selection of biochar feedstock is required to avoid unintended environmental consequences, such as indirect increases in GHG emissions elsewhere in the global C and N cycles.

Table 2. Review of previous studies examining the effects of biochar and/or compost manure addition on relative changes in GHG emissions. Negative value shows reduction in GHG emissions.

Location	Experiment/Analysis	Variable	CO$_2$ (%)	CH$_4$ (%)	N$_2$O (%)	Reference	Remarks
Australia	Corn cropped red ferrosol amended with poultry litter (PL), PL biochar (PLB) and urea. Measured cumulative emissions for 57 days.	PLB Raw PL Urea PLB + urea	1.3% 2.0% 1.4% 1.6%	NA *	0.04% 0.27% 0.16% 0.10%	[64]	Estimated CO$_2$-C and CH$_4$-C as percentage of TOC and N$_2$O-N as % of TN in 20 cm plough layer after biochar and manure addition.
Ireland	Pig manure (PM) added to soil and further amended with biochar from pig manure (PMB) or spruce wood (WB). Cumulative emissions evaluated over 28 days 10 weeks after PM addition.	PM + PMB PM + WB PM	2.3% 2.1% 5.5%	0.02% 0.01% 0.06%	3.8% 4.1% 2.1%	[60]	Estimated CO$_2$-C and CH$_4$-C as percentage of TOC and N$_2$O-N as % of TN in 20 cm plough layer after biochar and manure addition.
Global	Meta-analysis on role of biochar from different feedstocks in regulating N$_2$O emissions in laboratory or field conditions. Feedstocks– biowaste [1] (BW), biosolids [2] (BS), manures or manure-based materials (MM), wood (W), herbaceous (H), lignocellulosic waste (LW).	Mean BW BS MM W H LW	NA	NA	-60 to 48% -40% NS -46 to +39% -60% -60% -40%	[50]	[1] Biowaste = Municipal solid waste; [2] Biosolids = sewage sludge from water treatment plants.
Global	Meta-analysis of published data on biochar application in soils from laboratory or field experiments.	Lab results Field results Mean	NA	NA	-60% -40% -50%	[51]	
USA	Fast pyrolysed (550 °C) Oak biochar (BC) applied to temperate soils from Colorado, Iowa, Michigan, and Minnesota, and 2 years incubation study for GHG emissions. Comparisons are to control treatments (lacking BC).	1% BC 5% BC 10% BC 20% BC	8% 36% 88% 226%	NA	-53.9% -72.4% -76.3% -83.5%	[65]	
USA	Biochar produced at 550 °C from hardwood sawdust applied to soils from various locations.	Forest soils Agricultural soils	120% 75%	4.2% -0.9%	-58.2% -54.4%	[52]	% of change in GHG emissions after biochar addition compared to controlled soil with no biochar; Forest soils, N = 2; Agricultural soils, N = 8.
Canada	Biochars produced at 300 and 550 °C with and without steam activation (BC-S) applied to forest and grassland soils at 1.5% mass basis. Comparisons are to control soils.	Forest soils BC300 BC300-S BC500 BC500-S Grassland soils BC300	 -0.1% 4.2% -16.4% -5.7% -2.7%	 0.7% 12.6% 18.1% 15.1% 1.0%	 -3.0% -30.1% -27.5% -31.5% -3.3%	[53]	

Table 2. *Cont.*

Location	Experiment/Analysis	Variable	CO$_2$ (%)	CH$_4$ (%)	N$_2$O (%)	Reference	Remarks
Germany	Manure compost from organic household wastes applied to 115-year old Norway Spruce plantation in silty and sandy soils at the rate of 6.3 kg m^{-2}.	BC300-S	−2.4%	−0.4%	−7.4%	[66]	% CO$_2$ emissions in fertilized area compared to control plots. Annual CO$_2$ emissions from control plots were 5.1 and 4.2 Mg C ha^{-1} y^{-1} in silty and sandy soils, respectively, in year 1, and 5.0 and 4.0 in year 2.
		BC500	−4.3%	4.3%	−14.8%		
		BC500-S	−2.2%	3.5%	−11.7%		
		Year 1					
		Silty soils	24%				
		Sandy soils	67%				
		Year 2					
		Silty soils	20%				
		Sandy soils	45%				
Brazil	Sewage sludge compost (SSC), sewage sludge (SS), mineral fertilizer (MF), and control (Ctrl) at the rate of 20 kg of available N ha^{-1}.	SSC	90%		85%	[67]	% GHG emissions in fertilized area compared to control plots. CO$_2$ emission from control plots were 31.1 kg CO$_2$-C and 0.005 N$_2$O N (Mg ha^{-1}).
		SS	60%	NS **	37%		
		MF	13%		9%		

* NA = not available; ** NS = not significant.

Table 3. Summary of the previously documented effects of applying raw manure, composted manure, or manure pellets on subsequent C sequestration and GHG emissions in croplands and grasslands.

Location	Experiment	Treatment	CO$_2$ (kg ha^{-1})	CH$_4$ (kg ha^{-1})	N$_2$O (kg ha^{-1})	Reference	Remarks
Saskatchewan, Canada	Surface application, direct injection, and injection with soil aeration of swine effluent at 200 kg N ha^{-1} in no-till corn grain production.	Surface application	6900	1.2	7.3	[68]	Cumulative emissions for 141 days.
		Direct injection	8470	2.6	4.7		
		Combination with soil aeration	7370	2.1	6.9		
Quebec, Canada	Pig slurry applied to agricultural soils at 200 kg N ha^{-1} in spring and fall.	Fall	997	NA *	10.2	[69]	Seasonal cumulative measurement.
		Spring	1874		18.8		
Quebec, Canada	Pig slurry (PS) applied for 19-year in loamy soil at 60 (PS60) or 120 (PS120) Mg ha^{-1} y^{-1}.	PS60	2820	NA	4.9	[70,71]	12-month cumulative.
		PS120	6079		13.1		
Germany	Soil amendments (50 mg N kg^{-1}) with cow manure (CM), poultry manure (PM), sheep and wheat straw compost (SWC), bio-waste compost (BWC) or calcium ammonium nitrate (CAN) in a laboratory experiment.	CM	8118		0.4	[72]	
		PM	2706		0.4		
		SWC	1804	NA	0.1		
		BWC	6314		0.1		
		CAN	3608		0.1		
		Controlled	1624		0.0		

Table 3. *Cont.*

Location	Experiment	Treatment	CO$_2$ (kg ha^{-1})	CH$_4$ (kg ha^{-1})	N$_2$O (kg ha^{-1})	Reference	Remarks
Japan	Poultry manure (PM) and pelleted poultry (PP) manure application in Andisol at 120 kg N ha^{-1} in field and lab incubation experiment at two different water filled porosity (WFP) levels. Cumulative emission for 365 days.	Field condition		NA		[19]	Converted efflux to kg ha^{-1} at 5 cm soil depth (bulk density = 0.56 g cm^{-3})
		PM	NA		1.3		
		PP			5.0		
		Incubation—0.3 WFP					
		Intact PP	549		0.9		
		Ground PP	634		4.4		
		Incubation—0.5 WFP					
		Intact PP	882		10.1		
		Ground PP	1060		67.8		
USA	Incubation experiment with fine poultry manure (FPM) and pelleted poultry (PP) manure application in Cecil loamy sand at 55% and 90% of water filled porosity (WFP). N application rate was 307 kg N ha^{-1} equivalent.	55% WFP				[73]	Converted efflux to kg ha^{-1} at 15 cm soil depth (bulk density = 1.33 g cm^{-3}).
		FPM	6584		53.3		
		PP	6584		65.8		
		90% WFP					
		FPM	5267		1.6		
		PP	4316		15.7		
Scotland	Combination of dry pelleted and composted sewage sludge compared with liquid cattle slurry mixed with digested sewage sludge. Treatments were broadcasted sewage sludge pellet (DP): 15–17.5 t ha^{-1}, broadcasted compost sewage sludge (CP): 52–63.4 t ha^{-1}, injected digested liquid sewage sludge (LS) 60–120 t ha^{-1}, injected cattle slurry (CS) 5.9–10 t ha^{-1}.	Spring application				[18]	Manuring rate in grassland soils varied from year to year (kg N ha^{-1}); DP: 508–510, CP: 462–615, LS: 15–116, CS: 190–240.
		DP	10,633		0.8		
		CP	11,367		3.5		
		LS	11,367		2.5		
		CS	13,200		9.1		
		Summer application					
		DP	18,700		3.8		
		CP	22,000		5.0		
		LS	20,900		3.1		
		CS	22,000		9.7		

* NA = not available.

4.2.2. Impacts of Raw Manure, Composted Manure, and Manure Pellets

Swine slurry (primarily liquid), farmyard manure (primarily from large mammals), and poultry manure are common by-products of livestock production that are recycled in the field as nutrient input to agricultural plants. Inorganic N content, labile C content and the water content in manure provide essential substrates to micro-organisms that affect GHG emissions from soil. However, GHGs can be produced and emitted to the atmosphere in each step from livestock confinement, to manure storage and treatment (i.e., handling and transport), and ultimately during application to the land [74]. Composting is a well-established manure management process because it utilizes livestock manure and residual biomass of livestock feed and bedding, and produces manure that has reduced pathogens and weed seeds [75]. On the other hand, manure pelleting, a physical method of densification, increases manure bulk density, reduces storage space requirements, reduces subsequent transportation costs, and makes these materials easier to handle. Cattle manure with 50% moisture content, and processed at a temperature of about 40 °C and a pressure of 6 MPa, resulted in maximum pellet durability [76].

In the field, slurry and manure application methods and their forms affect GHG emissions [68,77]. Effects of raw farmyard manure, compost and pelleted manure application on soil C sequestration and GHG emissions are summarized in Table 3. Conventional injected pig slurry emitted greater CH_4 compared to injection with soil aeration, while manure spread on the surface emitted higher N_2O than both types of injection [68]. Slurry application season, method and rate all affected CO_2 and N_2O emissions in the field [69–71].

The inclusion of compost into soil provides better nutrient input compared to raw manure from the perspective of C sequestration [77]. For example, four years after application about 36% of applied compost remained in the soil as sequestered C, as compared to only 25% of applied raw manure [77]. Composting increases the aromatic bonds and reduces the soluble C/N ratio in manure compost [75], which leads to the slow release of nutrients. Composted manures are more effective in reducing N_2O emissions than raw manures for soil amendments [73]. In general, N_2O is produced through the denitrification process of organic fertilizers [78] while nitrification is the most important process for inorganic fertilizer. From one to five percent of total N applied from organic manure was emitted with emission rates depending largely on soil nitrate levels, dissolved organic carbon (DOC) content and aeration, as well as soil temperature, moisture and pH [79].

Effects of manure from different livestock and composts also vary in GHG emissions from soils. Raw cattle manure emitted higher amounts of CO_2 (~1 g kg^{-1} dry matter), followed by bio-waste composts, poultry manure, and sheep waste compost, within arable soils in Germany [72]. Application of cattle manure and straw mixed together as a compost enhanced C sequestration and reduced N_2O emissions; thus, composting of manure containing high lignin, such as rice-husk or wheat straw, is beneficial [72]. Such compost reduces soil pH, which slows down the nitrification process and reduces N_2O emissions [72].

Pellets derived from a mix of manure and urea enhanced nutrient use efficiency via the slow-release of nutrients and led to increased crop yields [80]. Unlike composting, pelleted manures are less effective in reducing GHG emissions than raw (i.e., untreated) manure [19]. Annual cumulative emission of N_2O from pelleted poultry manure applied in the field was almost four times higher than that from raw poultry manure. Similarly, higher CO_2 emissions were detected from soils amended with intact pelleted poultry manure compared to the application of ground pelleted poultry manure under anaerobic incubation [19]. N_2O emission was 154 mg N kg^{-1} dry soils from intact pelleted manure-amended soils, which was almost seven times higher than from ground pelleted manure-amended soil [19]. Soils emitted significantly higher N_2O when treated with pelletized poultry litter (6.8% of applied N) than for fine-particle litter (5.5%) at 55% of water field capacity (WFP). In contrast, at 90% of WFP, fine-particle litter treated soils emitted higher N_2O (3.4%) than soils receiving pelletized litter (1.5%), indicating GHG responses to pellet application depended on moisture, with pellets leading to more GHG if moisture is low [73]. Reported CO_2 emissions ranged from 29 to

43 g C kg^{-1} across moisture levels, though they were not statistically different. Results indicate that N$_2$O emissions, but not CO$_2$ emissions, from soils treated with poultry litter depend on its physical characteristics of litter and soil water regime. Diminishing rates of N$_2$O emission after the application of manure pellets to soil are attributed to the polymer chain reaction defined by the specific type of nitrite reductase encoded by the *nirS* gene, which fluctuated with time; however, the *nirK* gene remains relatively stable, making *nirS* responsible for the denitrification process of N in manure pellets [15].

In forest ecosystems, application of organic and inorganic fertilizer has shown different effects on GHG emissions depending upon geographic location. In Germany the application of composted household waste manure increased CO$_2$ emission by 24% in silty soils and by 66% in sandy soils compared to control plots [66]. On the other hand the application of organic and inorganic fertilizer to tropical forest in Brazil increased CO$_2$ emission by 90%, and 60% in composted sewage sludge, and raw sewage sludge amended sites, respectively, compared to emissions from controlled plots [67]. Surprisingly, N$_2$O emissions were 85 and 37 fold higher in the composted sewage sludge amended and raw sewage sludge amended plots compared to control plots [67].

5. Conclusions

Agroforestry has emerged as a holistic land use practice creating a win-win scenario for environment and society [81]. Combining woody vegetation with cropping and livestock production via agroforestry systems increases total production, enhances food and nutrition security and mitigates the effects of climate change [81,82]. Carbon sequestration and GHG emissions in agroforestry systems are complex and depend on various biophysical factors such as climatic conditions, soil properties, water regime, vegetation characteristics, and the site-specific management practices undertaken, including inputs. The ability of an agroforestry system to enhance C sequestration and reduce GHG emissions depends on the region-specific biophysical condition. Several estimates showed that agroforestry systems in temperate regions have higher C pools than other climatic regions [28,36]. Silvopastoral systems were found to be superior in terms of both C sequestration and reducing GHG emissions [32,34,41] compared to agroforestry systems that included annual cropland. Interventions like enrichment planting and organic amendment of soils to slow down nutrient release are also site-specific in regulating their effectiveness [13]. Our review indicates that broadleaved tree species used in enrichment planting contribute towards more soil C, while conifers sequester more in their biomass over the long-term [47]. Results of this literature review showed that the effects of enrichment planting in agroforestry are not studied widely. The paucity of literature on this topic limits the drawing of conclusions with respect to the type of enrichment planting that will be most effective in optimizing ecosystem goods and services from agroforestry.

Livestock manure applied to soils in the form of pellets, compost or biochar, can play a significant role in increasing C sequestration and reducing GHG emissions. Soil amendment with biochar increases soil porosity, and aids water and nutrient retention, thereby creating a favorable situation for nutrient uptake by plants. By enhancing biomass production, biochar can play an important role in sequestering C in vegetation and soils. Additionally, decreased emissions of N$_2$O and CO$_2$, and increased uptake of CH$_4$, have been reported in the literature after biochar application. However, many of these studies were carried out in annual cropping systems, leaving a substantial knowledge gap with respect to their effectiveness in agroforestry systems. Further studies on the specific effects of organic amendments to soils within either the treed area or adjacent cropland may provide a better idea on how agroforestry systems can be collectively managed to achieve greater C sequestration and reduce GHG emissions. In general, raw manure management and field applications of manure were found to be sources of CO$_2$, CH$_4$, and N$_2$O emissions, although the magnitude of these GHG emissions varied with application season, methods and amounts. Composting and pelleting of manures can reduce GHG emissions while making manure more convenient to store and use. A more thorough study is warranted to better understand the relationship between different types of feedstocks and their capacity to enhance C sequestration and reduce GHG emissions within agroforestry systems.

Overall, this review found that enhancement of C sequestration and reducing GHG emissions in agroforestry are possible through management interventions. Enrichment planting practiced in secondary forestry management and organic amendment of soils in conventional cropping systems should be further explored within an agroforestry management framework, as potential interventions to enhance C sequestration and reduce GHG emissions. Further studies will provide better evidence of such beneficial practices for environment, economy, and society.

Author Contributions: B.M.S. collected the literature and conducted the systematic review of available data on the subject matter. He wrote the first draft of the manuscript. S.X.C. conceived the idea and provided guidance to the senior author about data collection and conducting the review. He revised the first draft and provided further guidance. E.W.B. reviewed and revised various versions of the manuscript. C.N.C. reviewed the manuscript and provided initial feedback to improve the quality of the manuscript.

Acknowledgments: We gratefully acknowledge Agriculture and Agri-Food Canada (AAFC) for its financial support through its Agricultural Greenhouse Gas Program (AGGP) that funded the research project-BMPs to enhance carbon sequestration and reduce greenhouse gas emissions from agroforestry systems. Assistance received from Jinhyeob Kwak, Sishir Gautam and Shuva Gautam, is thankfully acknowledged. We would like to acknowledge anonymous reviewers whose comments and feedback helped a lot to improve the paper.

Conflicts of Interest: The authors declare no conflict of interest. The funder had no role in the design of the study, in the collection, analyses, or interpretation of data, in the writing of the manuscript, or in the decision to publish the results.

References

1. McIntyre, B.D.; Herren, H.R.; Wakhungu, J.; Watson, R.T. *International Assessment of Agricultural Knowledge, Science and Technology for Development (IAASTD): Global Report*; Island Press, The Center for Resource Economics: Washington, DC, USA, 2009.
2. Stavi, I.; Lal, R. Agroforestry and biochar to offset climate change: A review. *Agron. Sustain. Dev.* **2013**, *33*, 81–96. [CrossRef]
3. Schoeneberger, M.; Bentrup, G.; de Gooijer, H.; Soolanayakanahally, R.; Sauer, T.; Brandle, J.; Zhou, X.; Current, D. Branching out: Agroforestry as a climate change mitigation and adaptation tool for agriculture. *J. Soil Water Conserv.* **2012**, *67*, 128A–136A. [CrossRef]
4. Verchot, L.V.; Van Noordwijk, M.; Kandji, S.; Tomich, T.; Ong, C.; Albrecht, A.; Mackensen, J.; Bantilan, C.; Anupama, K.V.; Palm, C. Climate change: Linking adaptation and mitigation through agroforestry. *Mitig. Adapt. Strateg. Glob. Chang.* **2007**, *12*, 901–918. [CrossRef]
5. Baah-Acheamfour, M.; Chang, S.X.; Bork, E.W.; Carlyle, C.N. The potential of agroforestry to reduce atmospheric greenhouse gases in Canada: Insight from pairwise comparisons with traditional agriculture, data gaps and future research. *For. Chron.* **2017**, *93*, 180–189. [CrossRef]
6. Jose, S. Agroforestry for ecosystem services and environmental benefits: An overview. *Agrofor. Syst.* **2009**, *76*, 1–10. [CrossRef]
7. Mbow, C.; Smith, P.; Skole, D.; Duguma, L.; Bustamante, M. Achieving mitigation and adaptation to climate change through sustainable agroforestry practices in Africa. *Curr. Opin. Environ. Sustain.* **2014**, *6*, 8–14. [CrossRef]
8. Newaj, R.; Chaturvedi, O.P.; Handa, A.K. Recent development in agroforestry research and its role in climate change adaptation and mitigation. *Indian J. Agrofor.* **2016**, *18*, 1–9.
9. Nair, P.K.R.; Nair, V.D.; Kumar, B.M.; Showalter, J.M. Carbon sequestration in agroforestry systems. *Adv. Agron.* **2010**, *108*, 237–307.
10. Kumar, B.M.; Nair, P.K.R. *Carbon Sequestration Potential of Agroforestry Systems: Opportunities and Challenges*; Springer: Berlin, Germany, 2011; Volume 8.
11. Montagnini, F.; Nair, P.K.R. Carbon sequestration: An underexploited environmental benefit of agroforestry systems. *Agrofor. Syst.* **2004**, *61*, 281.
12. Nuberg, I.; Reid, R.; George, B. *Agroforestry as Integrated Natural Resource Management*; CSIRO Publishing: Collingwood, Australia, 2009.
13. Paquette, A.; Hawryshyn, J.; Senikas, A.; Potvin, C. Enrichment planting in secondary forests: A promising clean development mechanism to increase terrestrial carbon sinks. *Ecol. Soc.* **2009**, *14*, 31. [CrossRef]

14. Suryanto, P.; Putra, E.T.S. Traditional enrichment planting in agroforestry marginal land Gunung Kidul, Java, Indonesia. *J. Sustain. Dev.* **2012**, *5*, 77. [CrossRef]

15. Sonoki, T.; Furukawa, T.; Mizumoto, H.; Jindo, K.; Aoyama, M.; Sanchez-Monedero, M.A. Impacts of biochar addition on methane and carbon dioxide emissions during composting of cattle manure. In Proceedings of the Asia Pacific Biochar Conference, Kyoto, Japan, 15–18 September 2011.

16. Zhongqi, H.E.; Pagliari, P.H.; Waldrip, H.M. Applied and Environmental Chemistry of Animal Manure: A Review. *Pedosphere* **2016**, *26*, 779–816.

17. Gerber, P.J.; Steinfeld, H.; Henderson, B.; Mottet, A.; Opio, C.; Dijkman, J.; Falcucci, A.; Tempio, G. *Tackling Climate Change Through Livestock—A Global Assessment of Emissions and Mitigation Opportunities*; Food and Agriculture Organization of the United Nations (FAO): Rome, Italy, 2013.

18. Ball, B.C.; McTaggart, I.P.; Scott, A. Mitigation of greenhouse gas emissions from soil under silage production by use of organic manures or slow-release fertilizer. *Soil Use Manag.* **2004**, *20*, 287–295. [CrossRef]

19. Hayakawa, A.; Akiyama, H.; Sudo, S.; Yagi, K. N$_2$O and NO emissions from an Andisol field as influenced by pelleted poultry manure. *Soil Biol. Biochem.* **2009**, *41*, 521–529. [CrossRef]

20. Spokas, K.A.; Reicosky, D.C. Impacts of sixteen different biochars on soil greenhouse gas production. *Ann. Environ. Sci.* **2009**, *3*, 4.

21. Hansen, V.; Müller-Stöver, D.; Munkholm, L.J.; Peltre, C.; Hauggaard-Nielsen, H.; Jensen, L.S. The effect of straw and wood gasification biochar on carbon sequestration, selected soil fertility indicators and functional groups in soil: An incubation study. *Geoderma* **2016**, *269*, 99–107. [CrossRef]

22. Stavi, I. Biochar use in forestry and tree-based agro-ecosystems for increasing climate change mitigation and adaptation. *Int. J. Sustain. Dev. World Ecol.* **2013**, *20*, 166–181. [CrossRef]

23. Albrecht, A.; Kandji, S.T. Carbon sequestration in tropical agroforestry systems. *Agric. Ecosyst. Environ.* **2003**, *99*, 15–27. [CrossRef]

24. Watson, R.T.; Noble, I.R.; Bolin, B.; Ravindranath, N.H.; Verardo, D.J.; Dokken, D.J. *Land Use, Land-Use Change and Forestry. A Special Report of the Intergovernmental Panel on Climate Change (IPCC)*; Cambridge University: Cambridge, UK, 2000.

25. Lorenz, K.; Lal, R. Soil organic carbon sequestration in agroforestry systems. A review. *Agron. Sustain. Dev.* **2014**, *34*, 443–454. [CrossRef]

26. Possu, W.B.; Brandle, J.R.; Domke, G.M.; Schoeneberger, M.; Blankenship, E. Estimating carbon storage in windbreak trees on U.S. agricultural lands. *Agrofor. Syst.* **2016**, *90*, 889–904. [CrossRef]

27. Schroeder, P. Agroforestry systems: Integrated land use to store and conserve carbon. *Clim. Res.* **1993**, *3*, 53–60. [CrossRef]

28. Mutuo, P.K.; Cadisch, G.; Albrecht, A.; Palm, C.A.; Verchot, L. Potential of agroforestry for carbon sequestration and mitigation of greenhouse gas emissions from soils in the tropics. *Nutr. Cycl. Agroecosyst.* **2005**, *71*, 43–54. [CrossRef]

29. Kandji, S.T.; Verchot, L.V.; Mackensen, J.; Boye, A.; Van Noordwijk, M.; Tomich, T.P.; Ong, C.K.; Albrecht, A.; Palm, C.A.; Garrity, D.P. Opportunities for linking climate change adaptation and mitigation through agroforestry systems. In *World Agroforestry into the Future*; World Agroforestry Centre, ICRAF: Nairobi, Kenya, 2006; pp. 113–121.

30. Unruh, J.; Houghton, R.; Lefebvre, P. Carbon storage in agroforestry: An estimate for sub-Saharan Africa. *Clim. Res.* **1993**, *3*, 39–52. [CrossRef]

31. Cardinael, R.; Chevallier, T.; Barthès, B.G.; Saby, N.P.A.; Parent, T.; Dupraz, C.; Bernoux, M.; Chenu, C. Impact of alley cropping agroforestry on stocks, forms and spatial distribution of soil organic carbon—A case study in a Mediterranean context. *Geoderma* **2015**, *259*, 288–299. [CrossRef]

32. Baah-Acheamfour, M.; Carlyle, C.N.; Bork, E.W.; Chang, S.X. Trees increase soil carbon and its stability in three agroforestry systems in central Alberta, Canada. *For. Ecol. Manag.* **2014**, *328*, 131–139. [CrossRef]

33. Wang, G.; Welham, C.; Feng, C.; Chen, L.; Cao, F. Enhanced Soil Carbon Storage under Agroforestry and Afforestation in Subtropical China. *Forests* **2015**, *6*, 2307–2323. [CrossRef]

34. Baah-Acheamfour, M.; Carlyle, C.N.; Lim, S.-S.; Bork, E.W.; Chang, S.X. Forest and grassland cover types reduce net greenhouse gas emissions from agricultural soils. *Sci. Total Environ.* **2016**, *571*, 1115–1127. [CrossRef] [PubMed]

35. Chenu, C.; Cardinael, R.; Chevallier, T.; Germon, A.; Jourdan, C.; Dupraz, C.; Barthès, B.; Bernoux, M. The contribution of agroforestry systems to climate change mitigation—Assessment of C storage in soils in a Mediterranean context. In *Our Common Future under Climate Change*; CFCC: Paris, France, 2015.

36. Oelbermann, M.; Voroney, R.P.; Gordon, A.M. Carbon sequestration in tropical and temperate agroforestry systems: A review with examples from Costa Rica and southern Canada. *Agric. Ecosyst. Environ.* **2004**, *104*, 359–377. [CrossRef]

37. Atangana, A.; Khasa, D.; Chang, S.; Degrande, A. Carbon Sequestration in Agroforestry Systems. In *Tropical Agroforestry*; Springer: Dordrecht, The Netherlands, 2014; pp. 217–225.

38. He, Y.; Qin, L.; Li, Z.; Liang, X.; Shao, M.; Tan, L. Carbon storage capacity of monoculture and mixed-species plantations in subtropical China. *For. Ecol. Manag.* **2013**, *295*, 193–198. [CrossRef]

39. Paquette, A.; Bouchard, A.; Cogliastro, A. Survival and growth of under-planted trees: A meta-analysis across four biomes. *Ecol. Appl.* **2006**, *16*, 1575–1589. [CrossRef]

40. Schoeneberger, M.M. Agroforestry: Working trees for sequestering carbon on agricultural lands. *Agrofor. Syst.* **2009**, *75*, 27–37. [CrossRef]

41. Hoosbeek, M.R.; Remme, R.P.; Rusch, G.M. Trees enhance soil carbon sequestration and nutrient cycling in a silvopastoral system in south-western Nicaragua. *Agrofor. Syst.* **2016**, *92*, 263–273. [CrossRef]

42. Baah-Acheamfour, M.; Chang, S.X.; Carlyle, C.N.; Bork, E.W. Carbon pool size and stability are affected by trees and grassland cover types within agroforestry systems of western Canada. *Agric. Ecosyst. Environ.* **2015**, *213*, 105–113. [CrossRef]

43. Dixon, R.K. Agroforestry systems: Sources of sinks of greenhouse gases? *Agrofor. Syst.* **1995**, *31*, 99–116. [CrossRef]

44. Montagnini, F.; Eibl, B.; Grance, L.; Maiocco, D.; Nozzi, D. Enrichment planting in overexploited subtropical forests of the Paranaense region of Misiones, Argentina. *For. Ecol. Manag.* **1997**, *99*, 237–246. [CrossRef]

45. McGuire, J.P.; Mitchell, R.J.; Moser, E.B.; Pecot, S.D.; Gjerstad, D.H.; Hedman, C.W. Gaps in a gappy forest: Plant resources, longleaf pine regeneration, and understory response to tree removal in longleaf pine savannas. *Can. J. For. Res.* **2001**, *31*, 765–778. [CrossRef]

46. Nair, P.K.R.; Nair, V.D.; Kumar, B.M.; Haile, S.G. Soil carbon sequestration in tropical agroforestry systems: A feasibility appraisal. *Environ. Sci. Policy* **2009**, *12*, 1099–1111. [CrossRef]

47. Gao, Y.; Cheng, J.; Ma, Z.; Zhao, Y.; Su, J. Carbon storage in biomass, litter, and soil of different plantations in a semiarid temperate region of northwest China. *Ann. For. Sci.* **2014**, *71*, 427–435. [CrossRef]

48. Feldpausch, T.R.; Rondon, M.A.; Fernandes, E.; Riha, S.J.; Wandelli, E. Carbon and nutrient accumulation in secondary forests regenerating on pastures in central Amazonia. *Ecol. Appl.* **2004**, *14*, 164–176. [CrossRef]

49. Lentz, R.D.; Ippolito, J.A. Biochar and manure affect calcareous soil and corn silage nutrient concentrations and uptake. *J. Environ. Qual.* **2012**, *41*, 1033–1043. [CrossRef] [PubMed]

50. Cayuela, M.L.; Van Zwieten, L.; Singh, B.P.; Jeffery, S.; Roig, A.; Sánchez-Monedero, M.A. Biochar's role in mitigating soil nitrous oxide emissions: A review and meta-analysis. *Agric. Ecosyst. Environ.* **2014**, *191*, 5–16. [CrossRef]

51. Van Zwieten, L.; Kammann, C.; Cayuela, M.; Singh, B.P.; Joseph, S.; Kimber, S.; Donne, S.; Clough, T.; Spokas, K.A. Biochar effects on nitrous oxide and methane emissions from soil. In *Biochar for Environmental Management: Science, Technology and Implementation*; Routledge: New York, NY, USA, 2015.

52. Thomazini, A.; Spokas, K.; Hall, K.; Ippolito, J.; Lentz, R.; Novak, J. GHG impacts of biochar: Predictability for the same biochar. *Agric. Ecosyst. Environ.* **2015**, *207*, 183–191. [CrossRef]

53. Pokharel, P.; Kwak, J.-H.; Ok, Y.S.; Chang, S.X. Pine sawdust biochar reduces GHG emission by decreasing microbial and enzyme activities in forest and grassland soils in a laboratory experiment. *Sci. Total Envirn.* **2018**, *625*, 1247–1256. [CrossRef]

54. Clough, T.J.; Condron, L.M.; Kammann, C.; Müller, C. A review of biochar and soil nitrogen dynamics. *Agronomy* **2013**, *3*, 275–293. [CrossRef]

55. Jia, X.; Wang, M.; Yuan, W.; Shah, S.; Shi, W.; Meng, X.; Ju, X.; Yang, B. N_2O emission and nitrogen transformation in chicken manure and biochar co-composting. *Trans. Am. Soc. Agric. Biol. Eng.* **2016**, *59*, 1277–1283.

56. Jia, X.; Wang, M.; Yuan, W.; Ju, X.; Yang, B. The influence of biochar addition on chicken manure composting and associated methane and carbon dioxide emissions. *BioResources* **2016**, *11*, 5255–5264. [CrossRef]

57. Lehmann, J.; Rillig, M.C.; Thies, J.; Masiello, C.A.; Hockaday, W.C.; Crowley, D. Biochar effects on soil biota—A review. *Soil Biol. Biochem.* **2011**, *43*, 1812–1836. [CrossRef]

58. Wang, J.; Pan, X.; Liu, Y.; Zhang, X.; Xiong, Z. Effects of biochar amendment in two soils on greenhouse gas emissions and crop production. *Plant Soil* **2012**, *360*, 287–298. [CrossRef]

59. Wu, F.; Jia, Z.; Wang, S.; Chang, S.X.; Startsev, A. Contrasting effects of wheat straw and its biochar on greenhouse gas emissions and enzyme activities in a Chernozemic soil. *Biol. Fertil. Soils* **2013**, *49*, 555–565. [CrossRef]

60. Troy, S.M.; Lawlor, P.G.; O'Flynn, C.J.; Healy, M.G. Impact of biochar addition to soil on greenhouse gas emissions following pig manure application. *Soil Biol. Biochem.* **2013**, *60*, 173–181. [CrossRef]

61. Homagain, K.; Shahi, C.; Luckai, N.; Sharma, M. Life cycle cost and economic assessment of biochar-based bioenergy production and biochar land application in Northwestern Ontario, Canada. *For. Ecosyst.* **2016**, *3*, 21. [CrossRef]

62. Homagain, K.; Shahi, C.; Luckai, N.; Sharma, M. Life cycle environmental impact assessment of biochar-based bioenergy production and utilization in Northwestern Ontario, Canada. *J. For. Res.* **2015**, *26*, 799–809. [CrossRef]

63. Roberts, K.G.; Gloy, B.A.; Joseph, S.; Scott, N.R.; Lehmann, J. Life cycle assessment of biochar systems: Estimating the energetic, economic, and climate change potential. *Environ. Sci. Technol.* **2009**, *44*, 827–833. [CrossRef] [PubMed]

64. Van Zwieten, L.; Kimber, S.W.L.; Morris, S.G.; Singh, B.P.; Grace, P.R.; Scheer, C.; Rust, J.; Downie, A.E.; Cowie, A.L. Pyrolysing poultry litter reduces N_2O and CO_2 fluxes. *Sci. Total Environ.* **2013**, *465*, 279–287. [CrossRef] [PubMed]

65. Stewart, C.E.; Zheng, J.; Botte, J.; Cotrufo, M.F. Co-generated fast pyrolysis biochar mitigates green-house gas emissions and increases carbon sequestration in temperate soils. *GCB Bioenergy* **2013**, *5*, 153–164. [CrossRef]

66. Borken, W.; Muhs, A.; Beese, F. Application of compost in spruce forests: Effects on soil respiration, basal respiration and microbial biomass. *For. Ecol. Manag.* **2002**, *159*, 49–58. [CrossRef]

67. De Urzedo, D.I.; Franco, M.P.; Pitombo, L.M.; do Carmo, J.B. Effects of organic and inorganic fertilizers on greenhouse gas (GHG) emissions in tropical forestry. *For. Ecol. Manag.* **2013**, *310*, 37–44. [CrossRef]

68. Sistani, K.R.; Warren, J.G.; Lovanh, N. Quantification of Greenhouse Gas Emissions from Soil Applied Swine Effluent by Different Methods. In *Livestock Environment VIII, 31 August–4 September 2008, Iguassu Falls, Brazil*; American Society of Agricultural and Biological Engineers: St. Joseph, MI, USA, 2008; p. 26.

69. Rochette, P.; Angers, D.A.; Chantigny, M.H.; Bertrand, N.; Côté, D. Carbon dioxide and nitrous oxide emissions following fall and spring applications of pig slurry to an agricultural soil. *Soil Sci. Soc. Am. J.* **2004**, *68*, 1410–1420. [CrossRef]

70. Rochette, P.; van Bochove, E.; Prévost, D.; Angers, D.A.; Bertrand, N. Soil carbon and nitrogen dynamics following application of pig slurry for the 19th consecutive year II. Nitrous oxide fluxes and mineral nitrogen. *Soil Sci. Soc. Am. J.* **2000**, *64*, 1396–1403. [CrossRef]

71. Rochette, P.; Angers, D.A. Soil carbon and nitrogen dynamics following application of pig slurry for the 19th consecutive year I. Carbon dioxide fluxes and microbial biomass carbon. *Soil Sci. Soc. Am. J.* **2000**, *64*, 1389–1395. [CrossRef]

72. Shah, A.; Lamers, M.; Streck, T. N_2O and CO_2 emissions from South German arable soil after amendment of manures and composts. *Environ. Earth Sci.* **2016**, *75*, 1–12. [CrossRef]

73. Cabrera, M.L.; Chiang, S.C.; Merka, W.C.; Pancorbo, O.C.; Thompson, S.A. Nitrous oxide and carbon dioxide emissions from pelletized and nonpelletized poultry litter incorporated into soil. *Plant Soil* **1994**, *163*, 189–195. [CrossRef]

74. Chadwick, D.; Sommer, S.; Thorman, R.; Fangueiro, D.; Cardenas, L.; Amon, B.; Misselbrook, T. Manure management: Implications for greenhouse gas emissions. *Anim. Feed Sci. Technol.* **2011**, *166*, 514–531. [CrossRef]

75. Huang, G.F.; Wu, Q.T.; Wong, J.W.C.; Nagar, B.B. Transformation of organic matter during co-composting of pig manure with sawdust. *Bioresour. Technol.* **2006**, *97*, 1834–1842. [CrossRef] [PubMed]

76. Zafari, A.; Kianmehr, M.H. Effect of temperature, pressure and moisture content on durability of cattle manure pellet in open-end die method. *J. Agric. Sci.* **2012**, *4*, 203. [CrossRef]

77. Eghball, B.; Ginting, D. *Carbon Sequestration Following Beef Cattle Feedlot Manure, Compost, and Fertilizer Applications*; Nebraska Beef Cattle Reports; University of Nebraska: Lincoln, NE, USA, 2003.

78. Yamane, T. Denitrifying bacterial community in manure compost pellets applied to an Andosol upland field. *Soil Sci. Plant Nutr.* **2013**, *59*, 572–579. [CrossRef]

79. Biala, J.; Rowlings, D.W.; De Rosa, D.; Grace, P. Effects of using raw and composted manures on nitrous oxide emissions: A review. In Proceedings of the XXIX International Horticultural Congress on Horticulture: Sustaining Lives, Livelihoods and Landscapes (IHC2014), Brisbane, Australia, 17–22 August 2014; pp. 425–430.

80. Alemi, H.; Kianmehr, M.; Borghaee, A. Effect of pellet processing of fertilizer on slow-release nitrogen in soil. *Asian J. Plant Sci.* **2010**, *9*, 74. [CrossRef]

81. Nair, P.K.R. The coming of age of agroforestry. *J. Sci. Food Agric.* **2007**, *87*, 1613–1619. [CrossRef]

82. Nair, P.K.R. Agroforestry Systems and Environmental Quality. *J. Environ. Qual.* **2011**, *40*, 784–790. [CrossRef] [PubMed]

forests

Article

Effects of Near Natural Forest Management on Soil Greenhouse Gas Flux in *Pinus massoniana* (Lamb.) and *Cunninghamia lanceolata* (Lamb.) Hook. Plantations

Angang Ming [1,3], **Yujing Yang** [2,3], **Shirong Liu** [2,*], **Hui Wang** [2], **Yuanfa Li** [4], **Hua Li** [1,3], **You Nong** [1,3], **Daoxiong Cai** [1,3], **Hongyan Jia** [1,3], **Yi Tao** [1,3] and **Dongjing Sun** [1,3]

1   Experimental Center of Tropical Forestry, Chinese Academy of Forestry, Pingxiang 532600, China; mingangang0111@163.com (A.M.); lihua782003@163.com (H.L.); imnongyou@163.com (Y.N.); rlzxcdx@126.com (D.C.); rlzxjhy@163.com (H.J.); tyrzjl@126.com (Y.T.); happysdj@163.com (D.S.)
2   Key Laboratory of Forest Ecology and Environment, State Forestry Administration; Institute of Forest Ecology, Environment and Protection, Chinese Academy of Forestry, Beijing 100091, China; yangyujing8809@hotmail.com (Y.Y.); wanghui@caf.ac.cn (H.W.)
3   Guangxi Youyiguan Forest Ecosystem Research Station, Pingxiang 532600, China
4   College of Forestry, Guangxi University, Nanning 530004, China; xianggelilalyf@sina.com
*   Correspondence: liusr@caf.ac.cn; Tel.: +86-10-6288-9311

Received: 29 March 2018; Accepted: 25 April 2018; Published: 26 April 2018

Abstract: Greenhouse gases are the main cause of global warming, and forest soil plays an important role in greenhouse gas flux. Near natural forest management is one of the most promising options for improving the function of forests as carbon sinks. However, its effects on greenhouse gas emissions are not yet clear. It is therefore necessary to characterise the effects of near natural forest management on greenhouse gas emissions and soil carbon management in plantation ecosystems. We analysed the influence of near natural management on the flux of three major greenhouse gases (carbon dioxide (CO_2), methane (CH_4), and nitrous oxide (N_2O)) in *Pinus massoniana* Lamb. and *Cunninghamia lanceolata* (Lamb.) Hook. plantations. The average emission rates of CO_2 and N_2O in the near natural plantations were higher than those in the corresponding unimproved pure plantations of *P. massoniana* and *C. lanceolata*, and the average absorption rate of CH_4 in the pure plantations was lower than that in the near natural plantations. The differences in the CO_2 emission rates between plantations could be explained by differences in the C:N ratio of the fine roots. The differences in the N_2O emission rates could be attributed to differences in soil available N content and the C:N ratio of leaf litter, while the differences in CH_4 uptake rate could be explained by differences in the C:N ratio of leaf litter only. Near natural forest management negatively affected the soil greenhouse gas emissions in *P. massoniana* and *C. lanceolata* plantations. The potential impact of greenhouse gas flux should be considered when selecting tree species for enrichment planting.

Keywords: near natural forest management; *Pinus massoniana* plantation; *Cunninghamia lanceolata* plantation; soil greenhouse gas flux

1. Introduction

Increased emissions of greenhouse gases, dominated by carbon dioxide (CO_2), methane (CH_4), and nitrous oxide (N_2O), are the main cause of global climate change [1]. Most greenhouse gases in the atmosphere are produced and absorbed by soil [2]. Forest soils have the largest carbon pool in terrestrial ecosystems owing to soil respiration processes, mainly root respiration, microbial respiration, and soil animal respiration [3]. N_2O is released from the soil to the atmosphere through microbe-regulated nitrification and denitrification [4], while forest soil usually serves as the absorption sink for atmosphere CH_4 [5]. About 6% of global CH_4 is absorbed through soil processes by methanogenic bacteria [6,7]. The global warming potential of CH_4 and N_2O is 25 and 298 times larger than that of CO_2, respectively, although they are much less abundant than CO_2 in the atmosphere [8]. Therefore, a comprehensive understanding of the rates of greenhouse gas emissions and absorption and their key influencing factors in forest soils is critical to assessing the contribution of forest ecosystems to global climate change [9,10].

Near natural forest management, one of the most promising options for plantation silviculture, has received widespread attention in recent years [11]. Following the principle of near natural forest management, pure plantations are transformed into near natural forests through a series of management strategies, according to the structure and succession of natural forests. The strategies include species introduction, structural adjustment, natural regeneration promotion, and understory protection. Thus, the management of coniferous plantations has a significant impact on the structure, tree species composition, and regeneration of the forests [12,13]. Tree species are considered to alter the soil environment (including soil temperature and moisture), soil physical and chemical properties, and soil biological processes by influencing the composition and quality of the stand root system, canopy, litter, and fine roots [14,15]. As a result, soil greenhouse gas flux is greatly impacted by the composition of tree species. For example, the soil CH_4 flux of *Populus tremula* L., *Picea asperata* Mast. and pine forests in Europe differs significantly [16]. Menyailo and Hungate [17] observed higher CH_4 consumption in aspen, birch and spruce forest soils compared to Scots and Arolla pine forest soils in Siberia. However, average CH_4 uptake rates in mixed and pure beech plantations were about twice as large as that in pure spruce plantations [18]. Soil CO_2 efflux was accelerated after conversion from secondary oak forest to pine plantation in southeastern China [19]. Mature pine plantation soil emits 1.5 and 2.5 times more CO_2 than mature beech and Douglas fir [20]. Studies have also shown significant differences in soil respiration rates among 16 tree species in the tropics, with an emission flux from 2.8 to 6.8 μmol m^{-2} s^{-1} [21].

Although forest soil–atmosphere greenhouse gas exchange in temperate and tropical regions has been studied in depth [5,22–24], little is known about this process in the southern subtropical forests. There is a growing need locally and abroad to reduce greenhouse gas emissions from forests through plantation management. However, few studies have examined the use of plantation management strategies for manipulating soil greenhouse gas flux. Near natural management of coniferous plantations involves the transformation of even-aged pure stands of coniferous species into uneven-aged mixed broad-leaved forests, but it is not well known how this strategy affects the emission and absorption of greenhouse gases. Therefore, a subtropical, near natural *Pinus massoniana* plantation (P(CN)) and an unimproved pure stand of *P. massoniana* (P(CK)), as well as a near natural *Cunninghamia lanceolata* plantation (C(CN)) and an unimproved pure *C. lanceolata* stand (C(CK)) were selected in southern China. The objective of this study was to examine the effects of near natural forest management on soil–atmosphere greenhouse gas exchange and the main factors influencing these processes. The present study provides a theoretical basis for the multi-objective and sustainable management of plantations in southern subtropical regions.

2. Materials and Methods

2.1. Study Site Description

The study site is located in the Experimental Center of Tropical Forestry, Chinese Academy of Forestry (Pingxiang, Guangxi, China). It is one of the forest ecology study stations under the jurisdiction of the State Forestry Administration (22°10′ N, 106°50′ E). The site is within the southwestern region, which has a subtropical monsoon climate, with a semi-humid climate and obvious dry and wet seasons. The annual duration of sunshine is 1200 to 1600 h. The precipitation is abundant, with an annual average precipitation of 1200 to 1500 mm, mainly from April to September each year. The annual evaporation is 1200–1400 mm, the relative humidity is 80–84%, and the average annual temperature is 20.5–21.7 °C. The main types of landforms are low hills and hills. The soil is mainly composed of laterite and red soil based on the Chinese soil classification; this is classified as ferralsols in the World Reference Base for Soil Resources. The soil thickness is generally higher than 80 cm. Subtropical evergreen broad-leaved forests comprise the local vegetation.

There are nearly 20,000 ha of various plantation types in the Experimental Center of Tropical Forestry. *P. massoniana* and *C. lanceolata* are the main coniferous tree species. Native broad-leaved tree species include *Quercus griffithii* (Hook.f. and Thomson ex Miq.), *Erythrophleum fordii* Oliver, *Castanopsis hystrix* Miq., *Mytilaria laosensis* Lecomte., *Betula alnoides* Buch.-Ham. ex D. Don, and *Dalbergia lanceolata* Zipp. ex Span. The main alien tree species are eucalyptus and *Tectona grandis* L.f. Among these species, *E. fordii* and *D. lanceolata* are nitrogen-fixing trees, and *Q. griffithii* is a fast-growing broad-leaved tree species with a strong natural regeneration ability. The near natural management of pure plantations of *P. massoniana* and *C. lanceolata* with *E. fordii* and *Q. griffithii* is widely applied at the center, as it not only meets the need for short-period timbers and precious large-diameter timbers, but also realises the natural regeneration of native broad-leaved species and achieves the goal of near natural management.

2.2. Experimental Design

A single-factor and two-level stochastic block design was used for the present experiment. There were four blocks representing four replicates. Four forest types were set up in each block: near natural *P. massoniana* plantation (P(CN)), unimproved *P. massoniana* plantation (P(CK)), near natural *C. lanceolata* plantation (C(CN)), and unimproved pure *C. lanceolata* plantation (C(CK)). There were thus a total of 16 experimental plots, and the area of each experimental plot was 0.5 ha.

The pure plantations of *P. massoniana* and *C. lanceolata* were established in 1993 after the clear-cutting of *C. lanceolata*, with an initial planting density of 2500 trees ha^{-1}. Felling and afforestation were repeated a total of six times within the first three years after initial afforestation. The release felling was carried out in the seventh year, and the first-increment felling was carried out in the 11th year, retaining a density of 1200 trees ha^{-1}. In 2007, near natural management was carried out, and the main management strategies included reducing the intensity of the intermediate felling of pure stands of *P. massoniana* and *C. lanceolata* forests, while simultaneously preserving natural regeneration (the retention density was 600 trees ha^{-1}). In early 2008, *Q. griffithii* and *E. fordii* were replanted after the intermediate felling of *P. massoniana* and *C. lanceolata*, and the density of the native replanted tree species was 600 trees ha^{-1} (the average density of *Q. griffithii* and *E. fordii* was 300 trees ha^{-1}, respectively). Unevenly-aged mixed broad-leaved forests with a total density of 1200 trees ha^{-1} was formed. During the whole process, pure plantations of *P. massoniana* and *C. lanceolata* were maintained as controls, whose total density was kept at 1200 trees ha^{-1}. At present, the improved plantations have become unevenly-aged mixed stands with multilayer structures. A survey carried out in 2016 showed that the average diameter at breast height (DBH) and average tree height of *Q. griffithii* were 14.7 cm and 15.4 m, respectively, and the average DBH and average tree height of *E. fordii* were 5.2 cm and 6.3 m, respectively. The management processes for the four forests are shown in Table 1.

Table 1. Basic information and management history of the four plantations.

Year	Management	Plantation Type			
		P(CK)	P(CN)	C(CK)	C(CN)
1993	Afforestation	2500 trees ha^{-1}	2500 trees ha^{-1}	2500 trees ha^{-1}	2500 trees ha^{-1}
1993–1995	Tending for new plantations	6 times	6 times	6 times	6 times
2000	Released thinning	1600 trees ha^{-1}	1600 trees ha^{-1}	1600 trees ha^{-1}	1600 trees ha^{-1}
2004	Increment felling	1200 trees ha^{-1}	1200 trees ha^{-1}	1200 trees ha^{-1}	1200 trees ha^{-1}

Table 1. *Cont.*

Year	Management	Plantation Type			
		P(CK)	P(CN)	C(CK)	C(CN)
2007	Intensity thinning	No 1200 trees ha^{-1}	Yes 600 trees ha^{-1}	No 1200 trees ha^{-1}	Yes 600 trees ha^{-1}
2008	Complementary planting	No	Planting *Q. griffithii* and *E. fordii* with 300 trees ha^{-1} respectively	No	Planting *Q. griffithii* and *E. fordii* with 300 trees ha^{-1} respectively
2009	Tending	No	2 times	No	2 times
2016	Average DBH	22.2 ± 1.3 cm for *P. massoniana*	32.2 ± 1.6 cm for *P. massoniana*	17.1 ± 2.1 cm for *C. lanceolata*	22.3 ± 0.8 cm for *C. lanceolata*
2016	Average height	16.7 ± 0.5 m for *P. massoniana*	17.3 ± 0.7 m for *P. massoniana*	17.1 ± 0.4 m for *C. lanceolata*	17.2 ± 0.4 m for *C. lanceolata*

2.3. Measurement and Statistical Analysis

2.3.1. Soil CO_2, N_2O, and CH_4 Measurement

The sampling and analysis of three main greenhouse gases (N_2O, CH_4, and CO_2) in the soils were performed using the static chamber method and gas chromatography [25]. Three static boxes were randomly set in each plot of P(CN), P(CK), C(CN), and C(CK). The static box was 25 cm in diameter and 30 cm in height. A gas extraction valve and a small fan (8 cm in diameter) were installed at the top of the box to facilitate uniform gas mixing during sampling. The bottom of the box was buried in the ground at a depth of 5 cm two or three months before initial sampling [21]. From October 2014 to September 2015, sampling in all four plantations (a total of 16 plots) was completed from 9:00 a.m. to 11:00 a.m. on one day at the end of each month, and the measured values were used to calculate the average daily gas exchange flux [2]. During each sampling period, 100 ml gas samples were taken from static boxes with a medical syringe and timed with a stopwatch. The gas was sampled at 0, 15 and 30 min intervals. Three gas samples at each chamber were collected. The sample was injected into a polyethylene polythene sampling bag, cryopreserved, and sent back to the laboratory for measurement. We analysed gas samples for their N_2O, CH_4, and CO_2 concentrations using a gas chromatograph (Agilent 4890D, Agilent, Santa Clara, CA, USA). The flux of N_2O, CH_4, and CO_2 was calculated using the following formula:

$$F = \rho \times \frac{V}{A} \times \frac{P}{P_0} \times \frac{T_0}{T} \times \frac{dC_1}{dt} \tag{1}$$

where F is the mass change of the gas in the observation box per area and per unit time, ρ is the density of the measured gas in the standard state, V is the gas volume in the box, A is the area covered by the box, P is the atmospheric pressure at the sampling point, T is the absolute temperature at the time of sampling, $dC1/dt$ is the linear slope of gas concentration over time during the sampling, and P_0 and T_0 are the atmospheric pressure and absolute temperature in the standard state, respectively.

2.3.2. Micro-Environmental Data Measurement

Temperature and atmospheric pressure were measured with a thermometer and a barometer at the same time as sampling. The temperature of the soil at a depth of 5 cm was measured with a portable digital thermometer. Soil moisture (volumetric water content) at a depth of 5 cm was measured with an HH2 moisture meter (Delta-T Devices Ltd., Cambridge, UK) and converted into water-filled pore space (*WFPS*) using the following formula:

$$WFPS\ (\%) = \frac{Vol}{1 - \frac{bd}{2.65}} \tag{2}$$

where *bd* is bulk density, *vol* is volumetric water content, and 2.65 is the density of quartz.

2.3.3. Soil and Litterfall Sampling and Measurements

After the fresh and semi-decomposed litter residue at the upper surface of soil was stripped from the woodland near each static box in the four plantations, twelve soil samples at a depth of 0 to 10 cm were randomly collected using a stainless steel soil auger with an inner diameter of 8.7 cm. These samples were placed in mixed sample bags for preservation. The soil samples were then taken back to the laboratory to remove coarse roots, rubble, and other impurities using a 2 mm aperture screen and air dried for physical and chemical analysis.

Six 1 × 1 m leaf litterfall collectors made of nylon gauze (1 mm aperture) were set up randomly in the woodland near each static box in the four plantations. Leaf litterfall was collected once a month, and the leaves, branches, skin, and fruits were picked and sorted by tree species and organ and dried at 65 °C to a constant weight. A total of 12 collections of litterfall samples were prepared over the course of a year.

2.3.4. Fine Root Sampling and Measurements

Fine root biomass was determined by the continuous soil drilling method. Fine roots (diameter < 2 mm) were sampled in the 0–10 cm soil layer using a stainless steel soil auger with a diameter of 8.7 cm for sorting and collection. Twelve soil drillings collected for fine root biomass determination were carefully sorted out at random at the end of each bimonthly period in a sample plot of the four different plantations. In each plantation, the fine root samples were collected six times each year during the experiment. The fine root samples were weighed after drying at 65 °C to a constant weight. The average fine root biomass of the six sampling periods was used as the average fine root biomass [26].

2.3.5. Biogeochemical Properties Analysis of Plant and Soil Samples

Soil bulk density was measured using the volumetric ring during field sampling. Soil pH value was measured using glass electrodes after leaching the soil with 1 mol L^{-1} KCl solution. The organic C contents of the soil, litterfall, and fine root samples were determined by the potassium dichromate external heating method, and total N was determined by the Kjeldahl method. Soil ammonium and nitrate N contents were determined by spectrophotometry. Soil available N was analyzed through quantification of alkali-hydrolysable N in a Conway diffusion unit with Devarda's alloy in the outer chamber and boric acid-indicator solution in the inner chamber [27]. Soil total P was measured by inductively-coupled plasma optical-emission spectrometry (ICP-OES). Soil microbial biomass C and N were determined by the fumigation-extraction method [28].

2.3.6. Statistical Analysis

A one-way analysis of variance (ANOVA) was employed to determine the differences among the annual mean fluxes of soil greenhouse gases, as well as the biogeochemical properties of soil and plant samples in different plantations. Regression models were used to analyse the correlation between soil greenhouse gas flux and soil temperature and soil moisture in the four plantations. Multiple linear regression analyses were used to determine the main factors influencing differences in soil greenhouse gas flux among the four plantations. All of the data in the study followed a normal distribution and satisfied the test of homogeneity of variance. We performed statistical analyses using Windows SPSS 19.0. Statistical significance was determined at a threshold of $p < 0.05$.

3. Results

3.1. Soil Temperature and Moisture

Soil temperature and *WFPS* in the four plantations varied seasonally. The soil was cooler and drier during November 2014 and February 2015, whereas the soil was warmer and more humid from March 2015 to August 2015 (Figure 1). The sampling period in December 2014 was unusual in that it was a short wet period within the cool–dry season. January 2015 and July 2015 could be classified as within the cool–dry season and warm–humid season, respectively.

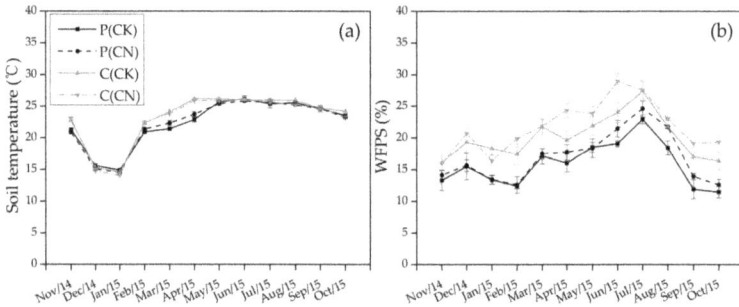

Figure 1. Seasonal patterns of soil temperature (**a**) and soil water-filled pore space (WFPS) (**b**) in the four plantations.

3.2. Seasonal Variation in Soil Greenhouse Gas Flux

The soil CO_2 and N_2O emission rates in the four plantations showed significant seasonal variations. The CO_2 emission rate was highest in July, when it was hot and humid, but lowest in January, during the dry season. All plantations had similar seasonal patterns for N_2O emission and CH_4 uptake (Figure 2).

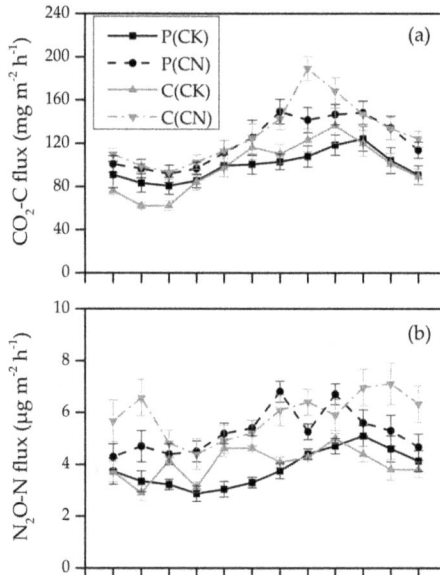

Figure 2. *Cont.*

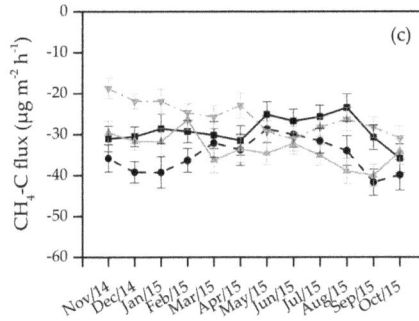

Figure 2. Seasonal patterns of soil CO_2, N_2O, and CH_4 flux in the four plantations.

Soil CO_2 emission rates were positively correlated with soil temperature and soil moisture (Figure 3a,b), but the correlation between CO_2 flux and soil moisture was significant in P(CN) only (Table 2). The N_2O emission was significantly and positively correlated with soil temperature in both P(CK) and C(CK) (Figure 3c). However, no significant correlation was found between soil N_2O flux and soil moisture (Figure 3d and Table 2).

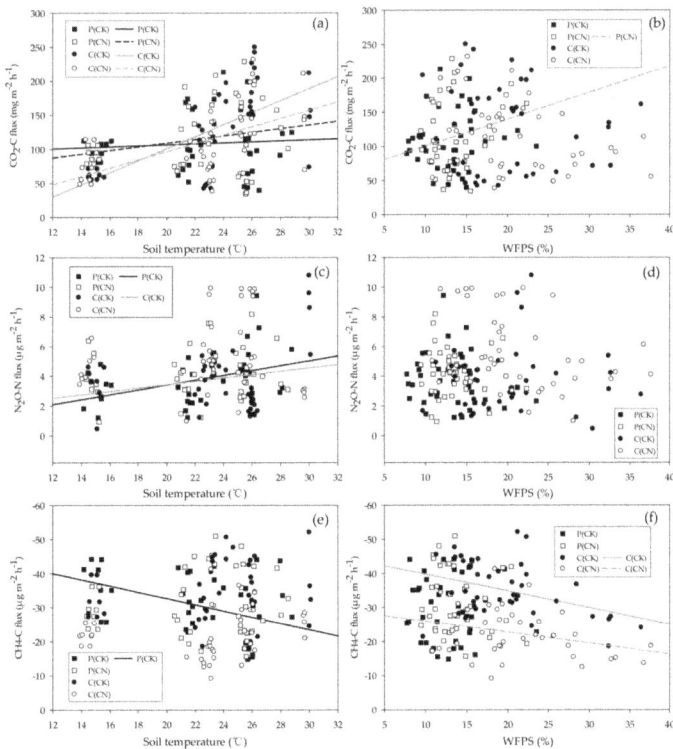

Figure 3. Relationships between soil CO_2 flux and temperature (**a**), CO_2 flux and water-filled pore space (WFPS) (**b**), N_2O flux and temperature (**c**), N_2O flux and WFPS (**d**), CH_4 flux and temperature (**e**), and CH_4 flux and WFPS (**f**) in the four plantations. Significant correlations were shown in solid and dashed lines ($p < 0.05$).

Table 2. Models, coefficients of determination (R^2) and *p*-values of regressions between soil greenhouse gas flux and soil temperature (T) and WFPS (W) in the four plantations. The rows of "T + W" represent the models considering both T and W, while others are those using T and W separately.

Plantation Type	P(CK)	P(CN)	C(CK)	C(CN)
	CO_2-C flux (mg m^{-2} h^{-1})			
T(°C)	$CO_2 = 0.71T + 92.31$ $R^2 = 0.11, p < 0.05$	$CO_2 = 2.67T + 55.83$ $R^2 = 0.15, p < 0.05$	$CO_2 = 9.14T + 80.44$ $R^2 = 0.37, p < 0.001$	$CO_2 = 6.05T + 24.15$ $R^2 = 0.30, p < 0.001$
W(%)	$R^2 = 0.01, p = 0.47$	$CO_2 = 3.92W + 61.71$ $R^2 = 0.13, p < 0.05$	$R^2 = 0.06, p = 0.61$	$R^2 = 0.04, p = 0.28$
T(°C) + W(%)	$R^2 = 0.01, p = 0.80$	$R^2 = 0.09, p = 0.17$	$CO_2 = 9.19T + 0.98W - 103.51$ $R^2 = 0.41, p < 0.001$	$CO_2 = 5.34T - 1.28W + 19.03$ $R^2 = 0.33, p < 0.01$
	N_2O-N flux (µg m^{-2} h^{-1})			
T(°C)	$N_2O = 0.16T + 0.17$ $R^2 = 0.16, p < 0.01$	$R^2 = 0.02, p = 0.43$	$N_2O = 0.11T + 1.19$ $R^2 = 0.16, p < 0.05$	$R^2 = 0.00, p = 0.77$
W(%)	$R^2 = 0.03, p = 0.297$	$R^2 = 0.01, p = 0.54$	$R^2 = 0.01, p = 0.90$	$R^2 = 0.00, p = 0.77$
T(°C) + W(%)	$N_2O = 0.19T - 0.13W + 1.32$ $R^2 = 0.22, p < 0.01$	$R^2 = 0.03, p = 0.54$	$R^2 = 0.06, p = 0.30$	$R^2 = 0.15, p = 0.05$
	CH_4-C flux (µg m^{-2} h^{-1})			
T(°C)	$CH_4 = 0.92T - 51.07$ $R^2 = 0.17, p < 0.01$	$R^2 = 0.01, p = 0.13$	$R^2 = 0.05, p = 0.15$	$R^2 = 0.04, p = 0.25$
W(%)	$R^2 = 0.00, p = 0.998$	$R^2 = 0.01, p = 0.454$	$CH_4 = 0.49W - 44.75$ $R^2 = 0.15, p < 0.01$	$CH_4 = 0.24W - 29.21$ $R^2 = 0.10, p < 0.05$
T(°C) + W(%)	$CH_4 = 0.96T - 0.23W - 48.95$ $R^2 = 0.18, p < 0.05$	$R^2 = 0.03, p = 0.56$	$CH_4 = -0.23T + 0.44W - 38.50$ $R^2 = 0.16, p < 0.05$	$R^2 = 0.09, p = 0.16$

CH_4 flux had a significant correlation with soil temperature in P(CK) only (Figure 3e). In the near natural and pure *C. lanceolata* plantations, soil CH_4 uptake rates decreased with seasonal increases in soil moisture (Figure 3f and Table 2).

When combining soil temperature and moisture in a regression model, significant relations were detected for the CO_2 flux in C(CK) and C(CN), N_2O flux in P(CK), and CH_4 flux in each pure forest (Table 2).

3.3. The Effects of Plantation Type on Soil Greenhouse Gas Flux

Near natural management had significant effects on the annual average emission rate of soil CO_2 and N_2O, and the uptake rate of soil CH_4 (Table 3). The soil CO_2 emission rate in the near natural *P. massoniana* plantation was 17.7% higher than that in the control forest, and the soil CO_2 emission rate of the near natural *C. lanceolata* plantation was 14.5% higher than control. This indicates that the soil CO_2 emission rates for *P. massoniana* and *C. lanceolata* plantations were accelerated by near natural management. Compared with the control forests, the near natural management enhanced the annual average soil N_2O emission rate by 19.4% and 47.4% in the *P. massoniana* and *C. lanceolata* plantation, respectively. Therefore, the soil N_2O emission rates for *P. massoniana* and *C. lanceolata* plantations increased as a result of near natural management.

Table 3. Annual average flux of soil greenhouse gas in the four plantations. Data are shown as means ± standard errors (*n* = 4). Values designated by the different letters within each variable are significant at *p* < 0.05.

Plantation Type	P(CK)	P(CN)	C(CK)	C(CN)
CO_2-C flux (mg m^{-2} h^{-1})	103.3 ± 9.7cd	121.6 ± 4.8ab	112.4 ± 8.9bc	128.7 ± 5.0a
N_2O-N flux (µg m^{-2} h^{-1})	3.6 ± 0.1cd	4.3 ± 0.5b	3.8 ± 0.2bc	5.6 ± 1.1a
CH_4-C flux (µg m^{-2} h^{-1})	−34.7 ± 1.7c	−27.2 ± 1.6b	−34.9 ± 2.8c	−22.4 ± 1.8a

The average soil CH_4 flux was negative for all the four plantations, which indicates that all the forest soils were functioning as CH_4 sinks. The annual average soil CH_4 uptake rate for the near natural plantations was 21.6% and 55.8% lower than the corresponding controls, as for *P. massoniana* and *C. lanceolata*, respectively (Table 3). Therefore, near natural management reduces the soil CH_4 uptake rate of *P. massoniana* and *C. lanceolata* plantations.

3.4. Main Influencing Factors on Soil Greenhouse Gas Flux

Compared with the control, the near natural management of each plantation increased the fine root biomass, soil temperature, pH, and the contents of soil organic C, available N, NH_4^+-N, NO_3^--N, microbial biomass C, and microbial biomass N, while it reduced the C:N of leaf litter and fine roots, as well as soil total P and C:N ($p < 0.05$, Table 4).

Table 4. The biogeochemical properties in the four plantations. Data are shown as means ± standard errors ($n = 4$). Values designated by the different letters within each variable are significant at $p < 0.05$.

Properties	P(CK)	P(CN)	C(CK)	C(CN)
Litterfall quantity (t hm^{-2} r^{-1})	10.23 ± 0.94a	10.84 ± 0.49a	9.02 ± 0.19b	9.54 ± 0.34b
Fine root biomass (t hm^{-2})	0.81 ± 0.07b	1.36 ± 0.22a	0.64 ± 0.26b	1.33 ± 0.28a
C:N of leaf litter	48.07 ± 4.82c	37.49 ± 4.77d	68.13 ± 8.12a	52.70 ± 6.92b
C:N of fine root	57.53 ± 10.7a	39.70 ± 5.70c	55.38 ± 3.30a	45.70 ± 4.40b
Soil porosity (%)	56.80 ± 2.83a	56.04 ± 2.58a	49.05 ± 4.99b	45.17 ± 4.86b
Soil temperature (°C)	22.15 ± 0.12d	22.47 ± 0.17c	22.73 ± 0.04b	23.04 ± 0.03a
Soil WFPS (%)	13.06 ± 0.56b	13.67 ± 0.49b	19.91 ± 1.00a	21.28 ± 1.06a
Soil pH	4.18 ± 0.04d	4.31 ± 0.08c	4.67 ± 0.07b	4.91 ± 0.20a
Soil organic C(g kg^{-1})	25.99 ± 1.32b	29.15 ± 2.42a	17.24 ± 1.85d	21.61 ± 2.58c
Soil total N(g kg^{-1})	2.58 ± 0.04	3.28 ± 0.12	2.29 ± 0.15	3.32 ± 0.13
Soil available N (mg kg^{-1})	94.37 ± 3.94b	103.32 ± 5.62a	77.0 ± 9.07c	96.25 ± 7.27ab
Soil total P (g kg^{-1})	0.28 ± 0.01a	0.25 ± 0.02b	0.24 ± 0.03b	0.21 ± 0.01c
Soil C:N	17.06 ± 0.50a	15.34 ± 0.72c	16.42 ± 0.14b	15.16 ± 0.46c
Soil NH$_4^+$-N content (mg kg^{-1})	20.30 ± 2.07 b	26.67 ± 3.35a	18.44 ± 2.17b	24.56 ± 4.02a
Soil NO$_3^-$-N content (mg kg^{-1})	21.97 ± 1.83b	25.00 ± 2.21a	18.36 ± 2.28b	24.65 ± 4.19a
Soil microbial biomass C (mg kg^{-1})	301.12 ± 24.54b	388.12 ± 11.76a	234.44 ± 29.49c	312.50 ± 32.51b
Soil microbial biomass N (mg kg^{-1})	39.07 ± 6.59bc	53.30 ± 8.11a	36.40 ± 6.45c	46.51 ± 4.21ab

To explain the observed variations in annual average soil greenhouse gas flux among the plantations, the first "stepwise" multiple linear regression model was performed using all the tested biogeochemical properties in the plantations. The model performed on CO_2 emissions indicated that the soil temperature and C:N ratio of the fine roots explained 77.4% of the variation in the soil CO_2 emission rate among the plantations ($R^2 = 0.774$, $p < 0.001$; Table 5). Other independent variables, such as the C:N ratio of leaf litter, soil organic C, soil pH, and soil nitrogen content, were excluded in the model owing to their non-significance or evidence of multicollinearity. The C:N ratio of the fine roots was negatively correlated with the annual average soil CO_2 emission rate, whereas the soil temperature was positively correlated with the annual average soil CO_2 emission rate (Table 5). This indicates that the annual average CO_2 uptake rate increases with an increasing soil temperature and decreasing C:N ratio of the fine roots.

Another multiple linear regression model that examined the variation in the average soil N_2O flux among the four plantations showed that the C:N ratio of leaf litter and soil available N explained 69.3% of the variation in the annual average soil N_2O emission rate ($R^2 = 0.693$, $p < 0.001$; Table 5). The annual average soil N_2O emission rate was negatively correlated with the C:N ratio of leaf litter but positively correlated with soil available N content. This indicates that the annual average N_2O emission rate increases with decreasing C:N ratio in leaf litter and increasing soil available N content.

A final multiple linear regression model showed that the C:N ratio of leaf litter was the only variable that explained a significant proportion (62.4%) of the variation in the annual average soil CH_4 uptake rate among the plantations ($R^2 = 0.624$, $p < 0.001$; Table 5). The annual average soil CH_4 flux was positively correlated with the C:N ratio of leaf litter.

Table 5. Results of multiple linear regression analysis of biogeochemical parameters and annual average soil greenhouse gas flux in the four plantations.

Parameters	Models
CO$_2$-C flux (mg m^{-2} h^{-1}) (Y$_1$)	
C:N ratio of fine root (X$_1$) Soil temperature (°C) (X$_2$)	Y$_1$ = −0.707X$_1$ + 16.2X$_2$ − 217.0, R^2 = 0.774, p < 0.001
N$_2$O-N flux (µg m^{-2} h^{-1}) (Y$_2$)	
C:N ratio of leaf litter (X$_3$) Soil available N (mg kg^{-1}) (X$_4$)	Y$_2$ = −0.044X$_3$ + 0.16X$_4$ + 5.886, R^2 = 0.693, p < 0.001
CH$_4$-C flux (µg m^{-2} h^{-1}) (Y$_3$)	
C:N ratio of leaf litter (X$_5$)	Y$_3$ = 0.343X$_5$ − 6.026, R^2 = 0.624, p < 0.001

4. Discussion

4.1. CO$_2$ Flux and Main Influencing Factors

The present study showed that the seasonal variation in the soil CO$_2$ emission rate in most cases can be attributed to soil temperature rather than soil moisture (Figure 3 and Table 2). Conversely, previous studies have found that soil CO$_2$ emission rates increase with increasing soil moisture and temperature in subtropical forests [2,25]. Therefore, there is no unified understanding of the soil moisture effects on the seasonal variation in soil CO$_2$ flux among different plantations.

Soil CO$_2$ is mainly produced through autotrophic respiration by plant roots and heterotrophic respiration by microorganisms [29]. The spatial variability in soil respiration is due to the differences in soil moisture, bulk density, root biomass, and soil organic matter [30]. The results of the present study indicated that the differences in soil CO$_2$ emission rates among the plantations were caused mainly by the C:N ratio of the fine roots (Table 5). The soil CO$_2$ emission rates of the near natural *P. massoniana* and *C. lanceolata* plantations were significantly higher than those of the control plantations (Table 3). Near natural management alters the composition of tree species, thus influencing the composition and quality of roots and litter, which in turn leads to the differences in CO$_2$ emission rates between the near natural and control forests. This is consistent with the results of previous comparative studies on soil CO$_2$ flux in coniferous pure forests and coniferous and broad-leaved mixed forests [25,31]. The C:N ratio of fine roots plays an important role in regulating microbial activity as an indicator of underground substrate quality, which affects the decomposition of fine roots [32]. The near natural management reduced the C:N ratios of fine root in *P. massoniana* and *C. lanceolata* plantations (Table 4). Therefore, the decomposition rates of fine roots in the near natural plantations can be higher than control, leading to higher soil CO$_2$ emission rates. These results indicate that the higher CO$_2$ emission rates observed in the near natural forest soil can be attributed mainly to the lower C:N ratio and higher decomposition rate. Some studies have also suggested that differences in fine root biomass or the composition and quality of leaf litter due to land use may affect soil respiration [33,34], or that different tree species affect soil respiration through associated differences in leaf litter quantity, chemical properties, and soil environmental conditions [18,21]. However, we found that fine root biomass, litterfall quantity, C:N ratio of leaf litter, and soil environmental conditions were the non-significant variables in our regression model. This indicates that they are not key factors influencing the soil CO$_2$ emission rate in our study area.

4.2. N$_2$O Flux and Main Influencing Factors

The average soil N$_2$O emission rate in our study was 4.3 µg N m^{-2} h^{-1}, which is similar to some other forests [31,35]. However, this is lower than that in tropical rainforests, forests in the northern hemisphere, and those seriously affected by nitrogen deposition [2,36]. This may be attributed to different soil properties. We found no seasonal changes in the N_2O emission rates in the plantations (Figure 2), while soil N_2O emission rates in the near natural *P. massoniana* and *C. lanceolata* plantations were higher than those of control (Table 3). This is in line with previous studies [18,37], and essentially consistent with a study on soil N_2O flux in mixed forests of *C. hystrix* and *P. massoniana* and pure forests of *P. massoniana* in the same study site [25]. The soil N_2O emission rate also differs significantly among vegetation types across Japan [38]. Our present result confirms that tree species composition has significant effects on the soil N_2O emission rate in coniferous plantations.

Soil N_2O emission rates are affected primarily by soil pH [39], soil moisture [40,41], soil carbon and nitrogen pools [41,42], and the C:N ratio of leaf litter [43]. Our present results show that the C:N ratio of leaf litter and soil available N content had the strongest effect on soil N_2O emission rates in the four plantations (Table 5). The N_2O emission rate decreased with increased leaf litter C:N ratio but increased with increased soil available N content. Near natural management enhanced the soil available N content and reduced the C:N ratio of leaf litter (Table 4), thus increasing the soil N_2O emission rates. The increased soil available N content could be largely explained by the introduction of *E. fordii*, which is an N-fixing species. These results were in line with a previous study indicating that the C:N ratio of leaf litter significantly affects the soil nitrification process and nitrogen-containing greenhouse gas flux [43]. The differences in soil N_2O emission rates between near natural and control plantations were therefore due mainly to differences in the C:N ratio of leaf litter and soil available N content. Similarly, the relatively low soil N_2O emissions in the present study compared to other studies may be attributed to low soil N content at the study site.

4.3. CH₄ Flux and Main Influencing Factors

The soil CH_4 flux in the study plantations varied from -22.4 to -34.9, which indicates that the soils are sinks for atmospheric CH_4. This is consistent with previous studies [5,41]. According to the present study, the tree species affects the CH_4 uptake rate (Table 3). The soil CH_4 uptake rate can be higher in broad-leaved forests than in coniferous forests [17,18,20]. However, so far little is known about the soil CH_4 flux in coniferous and broad-leaved mixed forests, particularly in near natural plantations. In this study, the soil CH_4 uptake rate was lower in the near natural plantations than in the control forests (Table 3). This indicates that near natural management reduces the soil CH_4 uptake rate in *P. massoniana* and *C. lanceolata* plantations.

The exchange of CH_4 between the soil and atmosphere is determined by the CH_4 production and consumption processes in the soil. The soil CH_4 production requires a suboxic environment for methanogenic bacteria, whereas CH_4 consumption requires aerobic conditions. Thus, soil aeration and oxygen content are important factors that regulate CH_4 production and consumption [41]. Soil temperature, moisture, pH, substrate availability, and aeration affect the activity and quantity of methanogenic bacteria [44,45], and thus regulate the soil CH_4 flux. However, the near natural management did not affect soil porosity and moisture (Table 4). These were thus not the reason for the differences in CH_4 uptake rates between the forests. Instead, the C:N ratio of the litter can explain the differences in soil CH_4 flux (Table 5). Near natural management significantly reduced the C:N ratio of leaf litter (Table 4), which consumes more oxygen during soil respiration. The hypoxic condition then leads to a production of soil CH_4 that is further released into the atmosphere [44]. Therefore, the net CH_4 absorption in soil decreases at a high rate of soil microbial respiration.

5. Conclusions

Near natural management increased the average soil CO_2 and N_2O emission rates in *P. massoniana* and *C. lanceolata* plantations and reduced the average soil CH_4 absorption rates. The differences in the CO_2 emission rate among plantations can be attributed mainly to the C:N ratio of fine roots, whereas the differences in the N_2O emission rate can be attributed to soil available N content and the C:N ratio of leaf litter. The variation in the CH_4 uptake rate can be attributed only to the C:N ratio of leaf litter. The results of the present study show that near natural management of *P. massoniana* and *C. lanceolata* plantations may increase the emission of greenhouse gases in subtropical China. Therefore, plantation enrichment strategies should take into account potential impacts on greenhouse gas flux. Other research is needed to evaluate the effects of near natural forest management on global climate change.

Author Contributions: A.M. analyzed data and drafted the manuscript. Y.Y. revised the manuscript and participated in collecting the experiment data. S.L. conceived and designed the work. H.W. was involved in planning the study and designing the work. Y.L., D.C., and H.J. contributed to technical advice and refined the ideas of this paper. The remaining authors contributed to carrying out additional analyses and finalizing the paper.

Acknowledgments: We are grateful to Ji Zeng, Zhang Zhao, Zhaoying Li, Lili Li, Wenlian Zhou, Yuan He, Zhongguo Li, Hai Chen and Yanling Huang of Chinese Academy of Forestry's Experimental Center of Tropical Forestry for their assistance in field sampling and data collection. We also gratefully acknowledge the support from the Agriculture College, Guangxi University. This study was funded by the fundamental research funds of CAF (CAFYBB2014QA033), Guangxi forestry science and technology projects (Document of Guangxi forestry department [2016] No.37), Guangxi Natural Science Foundation (2014GXNSFBA118100), International cooperation projects (2015DFA31440) and China's National Natural Science Foundation (31470627).

Conflicts of Interest: The authors declare no conflict of interest.

References

1. IPCC. *Climate Change 2014: Synthesis Report. Contribution of Working Groups I, II, and III to the Fifth Assessment Report of the Intergovernmental Panel on Climate Change*; IPCC: Geneva, Switzerland, 2014.

2. Tang, X.; Liu, S.; Zhou, G.; Zhang, D.; Zhou, C. Soil-atmospheric exchange of CO_2, CH_4, and N_2O in three subtropical forest ecosystems in southern China. *Glob. Chang. Biol.* **2006**, *12*, 546–560. [CrossRef]

3. Lal, R. Forest soils and carbon sequestration. *For. Ecol. Manag.* **2005**, *220*, 242–258. [CrossRef]

4. Butterbach-Bahl, K.; Gasche, R.; Breuer, L.; Papen, H. Fluxes of NO and N_2O from temperate forest soils: Impact of forest type, N deposition and of liming on the NO and N_2O emissions. *Nutr. Cycl. Agroecosyst.* **1997**, *48*, 79–90. [CrossRef]

5. Pitz, S.; Megonigal, J.P. Temperate forest methane sink diminished by tree emissions. *New Phytol.* **2017**, *214*, 1432–1439. [CrossRef] [PubMed]

6. Mer, J.L.; Roger, P. Production, oxidation, emission and consumption of methane by soils: A review. *Eur. J. Soil Biol.* **2001**, *37*, 25–50. [CrossRef]

7. Bodelier, P.L.E.; Laanbroek, H.J. Nitrogen as a regulatory factor of methane oxidation in soils and sediments. *FEMS Microbiol. Ecol.* **2004**, *47*, 265–277. [CrossRef]

8. Solomon, S.; Qin, D.; Manning, M.; Chen, Z. Changes in atmospheric constituents and in radiative forcing. In *Climate Change 2007: The Physical Science Basis. Contribution of Working Group I to the Fourth Assessment Report of the Intergovernmental Panel on Climate Change*; Cambridge University Press: Cambridge, UK, 2007.

9. Butterbach-Bahl, K.; Gasche, R.; Willibald, G.; Papen, H. Exchange of N-gases at the Höglwald Forest— A summary. *Plant Soil* **2002**, *240*, 117–123. [CrossRef]

10. Guo, L.B.; Gifford, R.M. Soil carbon stocks and land use change: A meta analysis. *Glob. Chang. Biol.* **2002**, *8*, 345–360. [CrossRef]

11. Wu, C.; Wei, X.; Mo, Q.; Li, Q.; Li, X.; Shu, C.; Liu, L.; Liu, Y. Effects of stand origin and near-natural restoration on the stock and structural composition of fallen trees in mid-subtropical forests. *Forests* **2015**, *6*, 4439–4450. [CrossRef]

12. Emborg, J.; Christensen, M.; Heilmannclausen, J. The structural dynamics of Suserup Skov, a near-natural temperate deciduous forest in Denmark. *For. Ecol. Manag.* **2000**, *126*, 173–189. [CrossRef]

13. Wang, G.; Liu, F. The influence of gap creation on the regeneration of *Pinus tabuliformis* planted forest and its role in the near-natural cultivation strategy for planted forest management. *For. Ecol. Manag.* **2011**, *262*, 413–423. [CrossRef]

14. Jonard, M.; André, F.; Jonard, F.; Mouton, N.; Procès, P.; Ponette, Q. Soil carbon dioxide efflux in pure and mixed stands of oak and beech. *Ann. For. Sci.* **2007**, *64*, 141–150. [CrossRef]

15. Ullah, S.; Frasier, R.; King, L.; Picotteanderson, N.; Moore, T. Potential fluxes of N_2O and CH_4 from soils of three forest types in Eastern Canada. *Soil Biol. Biochem.* **2008**, *40*, 986–994. [CrossRef]

16. Amanda, M.; Dan, P.; Angela, B.H. Methane and nitrous oxide emissions from mature forest stands in the boreal forest, Saskatchewan, Canada. *For. Ecol. Manag.* **2009**, *258*, 1073–1083.

17. Menyailo, O.V.; Hungate, B.A. Interactive effects of tree species and soil moisture on methane consumption. *Soil Biol. Biochem.* **2003**, *35*, 625–628. [CrossRef]

18. Borken, W.; Beese, F. Methane and nitrous oxide fluxes of soils in pure and mixed stands of European beech and Norway spruce. *Eur. J. Soil Sci.* **2006**, *57*, 617–625. [CrossRef]

19. Shi, Z.; Li, Y.; Wang, S.; Wang, G.; Ruan, H.; He, R.; Tang, Y.; Zhang, Z. Accelerated soil CO_2 efflux after conversion from secondary oak forest to pine plantation in southeastern China. *Ecol. Res.* **2009**, *24*, 1257–1265. [CrossRef]

20. Barrena, I.; Menéndez, S.; Duñabeitia, M.; Merino, P.; Florian Stange, C.; Spott, O.; González-Murua, C.; Estavillo, J.M. Greenhouse gas fluxes (CO_2, N_2O and CH_4) from forest soils in the Basque Country: Comparison of different tree species and growth stages. *For. Ecol. Manag.* **2013**, *310*, 600–611. [CrossRef]

21. Bréchet, L.; Ponton, S.; Roy, J.; Freycon, V.; Coûteaux, M.; Bonal, D.; Epron, D. Do tree species characteristics influence soil respiration in tropical forests? A test based on 16 tree species planted in monospecific plots. *Plant Soil* **2009**, *319*, 235–246. [CrossRef]

22. Leitner, S.; Sae-Tun, O.; Kranzinger, L.; Zechmeister-Boltenstern, S.; Zimmermann, M. Contribution of litter layer to soil greenhouse gas emissions in a temperate beech forest. *Plant Soil* **2016**, *403*, 455–469. [CrossRef]

23. Yamulki, S.; Morison, J.I.L. Annual greenhouse gas fluxes from a temperate deciduous oak forest floor. *Forestry* **2017**, *90*, 541–552. [CrossRef]

24. Daniel, L.; Whendeel, S.; Matteo, D. Temporal dynamics in soil oxygen and greenhouse gases in two humid tropical forests. *Ecosystems* **2011**, *14*, 171–182.

25. Wang, H.; Liu, S.; Wang, J.; Shi, Z.; Lu, L.; Zeng, J.; Ming, A.; Tang, J.; Yu, H. Effects of tree species mixture on soil organic carbon stocks and greenhouse gas fluxes in subtropical plantations in China. *For. Ecol. Manag.* **2013**, *300*, 4–13. [CrossRef]

26. Janssens, I.A.; Sampson, D.A.; Curielyuste, J.; Carrara, A.; Ceulemans, R. The carbon cost of fine root turnover in a Scots pine forest. *For. Ecol. Manag.* **2002**, *168*, 231–240. [CrossRef]

27. Shen, J.; Li, R.; Zhang, F.; Fan, J.; Tang, C.; Rengel, Z. Crop yields, soil fertility and phosphorus fractions in response to long-term fertilization under the rice monoculture system on a calcareous soil. *Field Crop Res.* **2004**, *86*, 225–238. [CrossRef]

28. Huang, X.; Liu, S.; Wang, H.; Hu, Z.; Li, Z.; You, Y. Changes of soil microbial biomass carbon and community composition through mixing nitrogen-fixing species with *Eucalyptus urophylla* in subtropical China. *Soil Biol. Biochem.* **2014**, *73*, 42–48. [CrossRef]

29. Janssens, I.A.; Lankreijer, H.; Matteucci, G.; Kowalski, A.S.; Buchmann, N.; Epron, D.; Pilegaard, K.; Kutsch, W.; Longdoz, B.; Grünwald, T. Productivity overshadows temperature in determining soil and ecosystem respiration across European forests. *Glob. Chang. Biol.* **2001**, *7*, 269–278. [CrossRef]

30. Epron, D.; Bosc, A.; Bonal, D.; Freycon, V. Spatial variation of soil respiration across a topographic gradient in a tropical rain forest in French Guiana. *J. Trop. Ecol.* **2006**, *22*, 565–574. [CrossRef]

31. Livesley, S.J.; Kiese, R.; Miehle, P.; Weston, C.J.; Butterbachbahl, K.; Arndt, S.K. Soil-atmosphere exchange of greenhouse gases in a *Eucalyptus marginata* woodland, a clover-grass pasture, and *Pinus radiata* and *Eucalyptus globulus* plantations. *Glob. Chang. Biol.* **2009**, *15*, 425–440. [CrossRef]

32. Xu, X.; Hirata, E. Decomposition patterns of leaf litter of seven common canopy species in a subtropical forest: N and P dynamics. *Plant Soil* **2005**, *273*, 279–289. [CrossRef]

33. Zhang, Y.; Guo, S.; Liu, Q.; Jiang, J.; Wang, R.; Li, N. Responses of soil respiration to land use conversions in degraded ecosystem of the semi-arid Loess Plateau. *Ecol. Eng.* **2015**, *74*, 196–205. [CrossRef]

34. Hu, X.; Liu, L.; Zhu, B.; Du, E.; Hu, X.; Li, P.; Zhou, Z.; Ji, C.; Zhu, J.; Shen, H. Asynchronous responses of soil carbon dioxide, nitrous oxide emissions and net nitrogen mineralization to enhanced fine root input. *Soil Biol. Biochem.* **2016**, *92*, 67–78. [CrossRef]

35. Rosenkranz, P.; Brüggemann, N. Soil N and C trace gas fluxes and microbial soil N turnover in a sessile oak (*Quercus petraea* (Matt.) Liebl.) forest in Hungary. *Plant Soil* **2006**, *286*, 301–322. [CrossRef]

36. Gundersen, P.; Christiansen, J.R.; Alberti, G.; Brüggemann, N.; Castaldi, S.; Gasche, R.; Kitzler, B.; Klemedtsson, L.; Lobo-do-Vale, R.; Moldan, F.; et al. The response of methane and nitrous oxide fluxes to forest change in Europe. *Biogeosciences* **2012**, *9*, 3999–4012. [CrossRef]

37. Pilegaard, K.; Skiba, U.; Ambus, P.; Beier, C.; Brüggemann, N.; Butterbachbahl, K.; Dick, J.; Dorsey, J.; Duyzer, J.; Gallagher, M. Factors controlling regional differences in forest soil emission of nitrogen oxides (NO and N_2O). *Biogeosciences* **2006**, *3*, 651–661. [CrossRef]

38. Morishita, T.; Sakata, T.; Takahashi, M.; Ishizuka, S.; Mizoguchi, T.; Inagaki, Y.; Terazawa, K.; Sawata, S.; Igarashi, M.; Yasuda, H.; et al. Methane uptake and nitrous oxide emission in Japanese forest soils and their relationship to soil and vegetation types. *Soil Sci. Plant Nutr.* **2007**, *53*, 678–691. [CrossRef]

39. Weslien, P.; Klemedtsson, A.Å.K.; Börjesson, A.G.; Klemedtsson, L. Strong pH influence on N_2O and CH_4 fluxes from forested organic soils. *Eur. J. Soil Sci.* **2009**, *60*, 311–320. [CrossRef]

40. Rowlings, D.W.; Grace, P.R.; Kiese, R.; Weier, K.L. Environmental factors controlling temporal and spatial variability in the soil-atmosphere exchange of CO_2, CH_4 and N_2O from an Australian subtropical rainforest. *Glob. Chang. Biol.* **2012**, *18*, 726–738. [CrossRef]

41. Gütlein, A.; Gerschlauer, F.; Kikoti, I.; Kiese, R. Impacts of climate and land use on N_2O and CH_4 fluxes from tropical ecosystems in the Mt. Kilimanjaro region, Tanzania. *Glob. Chang. Biol.* **2018**, *24*, 1239–1255. [CrossRef] [PubMed]

42. Wen, Y.; Corre, M.D.; Schrell, W.; Veldkamp, E. Gross N_2O emission and gross N_2O uptake in soils under temperate spruce and beech forests. *Soil Biol. Biochem.* **2017**, *112*, 228–236. [CrossRef]

43. Werner, C.; Kiese, R.; Butterbach-Bahl, K. Soil-atmosphere exchange of N_2O, CH_4, and CO_2 and controlling environmental factors for tropical rain forest sites in western Kenya. *J. Geophys. Res.* **2007**, *112*, D3308. [CrossRef]

44. Verchot, L.V.; Davidson, E.A.; Cattânio, J.H.; Ackerman, I.L. Land-use change and biogeochemical controls of methane fluxes in soils of eastern Amazonia. *Ecosystems* **2000**, *3*, 41–56. [CrossRef]

45. Bárcena, T.G.; D'Imperio, L.; Gundersen, P.; Vesterdal, L.; Priemé, A.; Christiansen, J.R. Conversion of cropland to forest increases soil CH_4 oxidation and abundance of CH_4 oxidizing bacteria with stand age. *Appl. Soil Ecol.* **2014**, *79*, 49–58. [CrossRef]

![forests logo] *forests*

MDPI

Article

Seasonal Effects on Microbial Community Structure and Nitrogen Dynamics in Temperate Forest Soil

Tomohiro Yokobe [1,*], Fujio Hyodo [2] and Naoko Tokuchi [3]

[1] Graduate School of Agriculture, Kyoto University, Oiwake-cho, Kitashirakawa, Sakyo-ku, Kyoto 6068502, Japan

[2] Research Core for Interdisciplinary Sciences, Okayama University, 3-1-1 Tsushimanaka, Okayama 7008530, Japan; fhyodo@cc.okayama-u.ac.jp

[3] Field Science Education and Research Center, Kyoto University, Oiwake-cho, Kitashirakawa, Sakyo-ku, Kyoto 6068502, Japan; tokuchi@kais.kyoto-u.ac.jp

* Correspondence: yokobe.tomohiro.65w@kyoto-u.jp; Tel.: +81-75-753-6428

Received: 25 January 2018; Accepted: 13 March 2018; Published: 19 March 2018

Abstract: The soil microbial community and nitrogen (N) dynamics change seasonally due to several factors. The microbial community structure (MCS) can regulate N dynamics. However, there is insufficient information on seasonal changes in MCS and the relationship between MCS and N dynamics. We investigated MCS and N dynamics in forest soils with two different fertilities throughout a year. MCS, measured with phospholipid fatty acid (PLFA) analysis, showed a consistent seasonal trend, regardless of the fertility. Microbial indices (particularly the Saturated-/monounsaturated-PLFA ratio; Sat/mono) indicated a major PLFA shift among seasons, with temperature likely the most important factor. The fungal-/bacterial-PLFA ratio in the dormant season (December–April) was approximately 1.3 times greater than in the growing season (June–November). The trend in N dynamics showed that in summer (June–August), the gross N mineralization potential was greater than immobilization, whereas in winter (December–April), immobilization was dominant. The net mineralization potential in the growing season was approximately 1.6 times higher than in the dormant season. Moreover, a relationship was found between Sat/mono and N transformation potentials. We highlight the microbial sensitivity to seasonal dynamics which can be associated with temperature, as well as carbon and N dynamics.

Keywords: temperature; soil microbial communities; PLFA; seasons; nitrogen dynamics; gross nitrogen transformations

---

1. Introduction

Soil microorganisms play a critical role in the nitrogen (N) dynamics of forest ecosystems, where N often limits primary production [1]. Microbial community structure (MCS) is associated with numerous ecosystem functions [2,3]. For example, fungal versus bacterial dominance can be determined through the C:N stoichiometry of biomass, as fungi are often reported to have a higher C:N ratio than bacteria [4], leading to differences in N usage efficiency during the decomposition of plant litter [5,6]. Also, various extracellular enzyme activities (e.g., labile or recalcitrant C degradation and N release) derived from the MCS can cause differences in the C and N mineralization rates [3,7].

One essential factor affecting MCS is seasonality. For instance, in soils at oak forest and grassland sites in a Mediterranean ecosystem, similar seasonal patterns were observed in soil MCS phospholipid fatty acid (PLFA) levels despite different MCS between two soils [8]. Moreover, in an altitudinal soil transplantation experiment using subalpine grassland soil, a clear shift in the soil MCS was observed between winter and summer, whereas transplanting soils to different altitudes did not affect MCS [9]. Such temporal changes in MCS are derived from a number of factors [10], such as soil moisture [11], tree species (e.g., spatial and temporal substrate inputs, such as leaf and root litter and root exudate [12]), and temperature [13,14].

Among seasonal N dynamics, the pools of water-extractable N species (dissolved organic N, ammonium, and nitrate) and the net ammonification and nitrification rates were higher during summer than during autumn or winter in a northern hardwood forest soil [15] and a temperate beech forest soil [16]. Microorganisms are the driver of these seasonal patterns. Thus, to understand the association between microorganisms and N dynamics, both of these factors must be observed in all seasons. In previous studies, seasonal associations between MCS (e.g., fungi and bacteria) and N dynamics were found in alpine ecosystems [17,18] and temperate forests [16,19]. For instance, Kaiser et al. [16] found that the summer was dominated by microbial N mineralization, measured as a high ammonification rate, whereas winter was the immobilization period, measured as a high N immobilization rate and a large amount of microbial N, likely due to fungal N immobilization in winter. These seasonal patterns are important for N retention in the ecosystem [16], particularly in alpine systems [17]. However, due to the scarcity of studies to date, it is important to confirm the seasonal patterns and associations between microorganisms and N dynamics in various ecosystems.

The aims of our study were to investigate (1) seasonal changes in MCS and N dynamics and (2) the relationships between MCS and N dynamics in a temperate natural forest soil and a nearby plantation forest soil (50-year-old stand). These forests can be represented as N-rich and N-poor systems, respectively, and the importance of N for microbes may differ between them. In N-poor systems, fungi are frequently found to be dominant over bacteria, and a low abundance of available N (e.g., ammonium and nitrate) is often observed [20]. In such a system, the dominant fungi, which have lower biomass turnover rates than bacteria [21], may maintain N in their biomass throughout the year, so no seasonal patterns are apparent. We examined seasonal variation in basic soil properties over the course of a year, including the N pool, microbial biomass, the MCS, and N transformation potentials (net and gross) at two depths, namely the organic layer (O-layer) and the top mineral soil layer (S-layer).

2. Materials and Methods

2.1. Study Sites and Seasonal Soil Sampling

Soil samples were collected in the Tanakami Mountains from a natural forest (NF, N 34°55′, E 135°58′, 510 m a.s.l., 0.10 ha) and a restored forest (RF, N 34°57′, E 135°59′, 250 m a.s.l., 0.68 ha) located in Shiga Prefecture, central Japan (Table 1). This area has a warm temperate climate influenced by the Asiatic monsoon. The mean annual temperature and precipitation were 14.9 °C and 1542 mm, respectively, at the Otsu observation point of the Japan Meteorological Agency (1981–2010, N 34°59′, E 135°54′, 86 m a.s.l., Figure S1), which is near the two sampling sites (middle slope). We expected that seasonal meteorological phenomena would be similar to each other because the locations were 5 km apart.

Land use history, vegetation, and properties of the organic layer and soil are provided in Table 1. The RF area was deforested approximately 1300 years ago due excessive timber harvest, and it remained denuded for a long period. Since the last century, hillside restoration and afforestation projects have been undertaken. In RF, vegetation cover is complete, but the soil nutrient content remains poor [22]. The bedrock is composed of granite at both sites. As both areas are located on steep slopes (20–30°), the soil depths are shallow relative to nearby flat areas. The two soil types differed, i.e., the soil in NF was a Cambisol, whereas the soil in RF was a Regosol due to soil erosion (removal of the A and B horizons) that occurred until recently.

A 20 × 20 m experimental plot was set up at each site. In each plot, five subplots (1.5 × 1.5 m) were established: one subplot was at the intersection of the plot diagonals, and four subplots were at 10 m intervals from the intersection of the diagonals. Seasonal sampling was conducted at each subplot six times a year. The organic layer (O$_{e+a}$: O-layer) and top mineral soil layer (0–10 cm: S-layer) were collected in each subplot. O- and S-layer samples were sieved through a 2- and 4-mm mesh, respectively, and visible roots were removed by hand. The samples were stored at 4 °C, except those for MCS analysis (−20 °C), until further processing.

Table 1. Land use history, vegetation, organic layer, and soil properties of the study sites. Organic layer amount and bulk density are represented as mean values ($n = 5$) ± SE. Total C and N, and C/N are represented as mean values ($n = 30$) ± SE. Letters indicate significant differences between the two sites (Tukey's HSD test, $p < 0.05$).

	Natural Forest (NF)	**Restored Forest (RF)**
Land use history	Natural	Soil erosion over a long period and subsequent reforestation (ca. 100 years ago)
Vegetation	A mature natural forest dominated mainly by Japanese cypress (*Chamaecyparis obtusa* (Siebold & Zucc.) Endl.) and oaks	A semi-mature forest dominated mainly by Japanese cypress (*Chamaecyparis obtusa* (Siebold & Zucc.) Endl.) and oaks
Organic layer amount		
>4 mm (Mg ha^{-1})	3.69 ± 0.33	2.39 ± 0.48
<4 mm (Mg ha^{-1})	11.64 ± 4.81	29.50 ± 2.46
C/N (Organic layer)		
<4 mm	23.8 ± 0.3A	28.3 ± 0.5B
Soil type [1]	Cambisols	Regosols
Bulk density [2] (0–10 cm)		
>2 mm (g cm^{-3})	0.26 ± 0.02	0.26 ± 0.04
<2 mm (g cm^{-3})	0.61 ± 0.04	0.77 ± 0.04
Total C [3] (g C kg^{-1})	53.4 ± 3.2A	17.0 ± 1.3B
Total N [3] (g N kg^{-1})	2.87 ± 0.17A	0.96 ± 0.06B
C/N [3]	18.5 ± 0.2	17.6 ± 0.4

[1] Soil classification by IUSS Working Group WRB [23]. [2] Bulk density was separated into two fractions; the coarse (>2 mm) and fine (<2 mm). [3] Soil layer (0–10 cm; <2 mm).

2.2. Soil Chemical Characteristics

Water content (WC) was determined gravimetrically as water loss. Total C (TC) and N (TN) were measured with an NC analyzer (Sumigraph NC-22A, Sumika Chemical Analysis Service, Ltd., Osaka, Japan) after drying and grinding. Soil pH was measured in water (fresh sample: water, 1:2.5 (w/v)). The N pool and water-extractable organic carbon (WEOC) were extracted with 2 M KCl (fresh soil layer: solution, 1:10 (w/v) or fresh O-layer: solution, 3:50 (w/v)). The extracts were frozen at −20 °C until further analysis. Water-extractable total N (WETN) and WEOC were measured using a TOC/TN analyzer (TOC-L CPH/CPN, Shimadzu, Kyoto, Japan). Ammonium-N (NH_4^+-N) and nitrate-N (NO_3^--N) were also determined using a colorimetric method with a flow injection analyzer (AutoAnalyzer, BL-Tech, Tokyo, Japan). Water-extractable organic N (WEON) was calculated as the difference between WETN and inorganic N (NH_4^+-N + NO_3^--N).

2.3. Microbial Biomass (MB)

MB-C and -N were measured through a fumigation–extraction procedure [24]. Briefly, subsamples were fumigated with chloroform for 24 h at 25 °C, and then extracted in 0.5 M K_2SO_4 (sample: solution, 1:5 (w/v)) along with non-fumigated subsamples. WEOC and WETN of the extracts were measured with a TOC/TN analyzer. MB-C and -N were estimated as the difference in C or N content of extracts from fumigated and non-fumigated subsamples. An extraction coefficient of 0.45 was used for calculating MB-C [25] and N [26].

2.4. Microbial Community Structure (MCS)

In order to assess MCS, the phospholipid fatty acid (PLFA) content of soils was measured [27,28]. Lipids were extracted from freeze-dried samples with an extraction solvent (chloroform: methanol:

0.15 M citric acid buffer, 1:2:0.8, $v/v/v$) and separated into neutral lipids, glycolipids, and phospholipids on a silica acid column. Subsequently, the phospholipids were subjected to mild alkaline methanolysis. Methyl nonadecanoate (19:0) was added as an internal standard. The samples were analyzed using a gas chromatograph equipped with a flame ionization detector (GC-2014, Shimadzu, Kyoto, Japan). Each sample was injected onto a column (DB-5, 30-m length, 0.25-mm i.d., film thickness 0.25 mm; Agilent J & W Scientific, Santa Clara, CA, USA). Peaks were identified by a comparison of retention times with commercial standards (BAMEs, Supelco Bacterial Acid Methyl Esters CP Mix #47080-U, Sigma–Aldrich, Bellefonte, PA, USA). The abundance of individual PLFA markers was determined by comparison to internal standard peak areas in µmol PLFA kg^{-1}. Fatty acid nomenclature was determined following Frostegård et al. [27]. The sum of BAME markers excluding 19:0 was calculated as the abundance of total PLFA. The abundance of each PLFA marker was also determined as mol % relative to the total PLFA abundance. To characterize MCS, we used specific microbial indices closely associated with N dynamics, substrate quality, and temperature (Table 2). The ratios of saturated to monounsaturated PLFAs (Sat/mono) and cyclopropyl to precursors (Cy/pre) were probably associated with the change in MCS and physiological stress or starvation, which cannot be separated [29,30]. Gram-positive and -negative bacteria are important bacteria groups for C and N cycling [31], although their PLFA markers do not coincide completely [29,30]. We used 18:2ω6,9 as fungal-PLFA and the sum of i15:0, a15:0, 15:0, i16:0, i17:0, cy17:0, 17:0, 18:1ω7, and cy19:0 as bacterial-PLFA [27,28].

Table 2. Definition of microbial indices.

Indices (Specific Ratios)	Phospholipid Fatty Acids	Major Association with Increase in PLFA Index
Sat/mono [32,33] Saturated Monounsaturated	14:0 + 15:0 + 16:0 + 17:0 + 18:0 16:1ω7 + 18:1ω7 + 18:1ω9t + 18:1ω9c	High N loading (addition [34,35], deposition [33,36]) Small amount of substrate [32] High temperature [37]
G+/G– [19,32,33] Gram-positive bacteria Gram-negative bacteria	i15:0 + a15:0 + i16:0 + i17:0 16:1ω7 + cy17:0	High N loading (addition [34,38]) Small amount of substrate [32,39] High temperature [37,40]
Cy/pre [32,33] Cyclopropyl Precursor	cy17:0 16:1ω7	High N loading (addition [35], deposition [33]) High temperature [37] High microbial respiration [41]
F/B [27,28] Fungi Bacteria	18:2ω6,9 i15:0 + a15:0 + 15:0 + i16:0 + i17:0 + cy17:0 + 17:0 + 18:1ω7 + cy19:0	Low N loading (addition [35,38], deposition [33,36]) Large amount of substrate [32,39] Low temperature [37,40]

2.5. Gross N Transformation Potential

The gross N transformation potential was estimated using the ^{15}N isotopic dilution method [42]. Briefly, after the addition of ^{15}NH$_4^+$ or ^{15}NO$_3^-$ solution, each subsample was incubated for 2 h or 26 h at 25 °C, and then NH$_4^+$-N and NO$_3^-$-N were extracted and their concentrations measured in the solution, as described above. The ^{15}N atom% of NH$_4^+$ and NO$_3^-$ were also measured in the solutions via the denitrifier method [43] using gas chromatography with a mass spectrometer (GCMS-QP2010 Plus, Shimadzu, Kyoto, Japan) after the conversion of NH$_4^+$ into NO$_3^-$ and finally N$_2$O. The gross N (NH$_4^+$-N and NO$_3^-$-N) production and consumption potentials were calculated according to Davidson et al. [44] and Kuroiwa et al. [42]. NH$_4^+$-N immobilization was calculated as the difference between NH$_4^+$-N consumption and NO$_3^-$-N production, and NO$_3^-$-N consumption was represented as NO$_3^-$-N immobilization. Also, the specific potentials for gross NH$_4^+$-N production and immobilization [45] were defined per unit of MB-N for estimating microbial N processes.

2.6. Net N Transformation Potential

The net transformation potential was estimated with a four-week incubation procedure at 25 °C. First, the samples were incubated in plastic bottles sealed with Parafilm. The initial moisture level was maintained by adding water weekly. After four weeks, the N pool, i.e., NH_4^+-N, NO_3^--N, WEON, and WETN, was analyzed as described above. The net transformation potential was calculated as the difference in the N content of extract between the initial and incubated samples. Rates were expressed in units per day, and the specific potential of net (NH_4^+-N + NO_3^--N) transformation was defined per unit of MB-N for an estimation of microbial N processes.

2.7. Statistical Analyses

All variables measured in each seasonal sampling are expressed as the mean (n = 5 subplots) and standard error (SE). The effect of sampling date on each variable was assessed by a mixed model using a restricted maximum likelihood (REML) estimate for each site and for each depth. The fixed factor was sampling date, while subplot was added as a random factor. Subsequently, if a significant seasonal effect ($p < 0.05$) was found, pairwise comparisons were further analyzed with Tukey's honestly significant differences (HSD) test ($p < 0.05$). Furthermore, the effects of sampling date, sites, and their interactions were assessed by a mixed model using an REML estimate. Fixed factors were sampling date, site, and their interactions, while subplot was added as a random factor. Additionally, the annual mean (n = 30) and SE at both sites are represented for several variables.

To elucidate the seasonal pattern of MCS, principal component analysis (PCA) was performed for PLFA markers (relative abundance; mol %). The mean (n = 5) of each PLFA marker during each season was used. To clarify the seasonal pattern, PLFA markers that were not detected in a season were removed from PCA. Indices of the PC2 (Figure S4) of each site generally reflected the seasonal pattern of MCS. Thus, PC2 scores were adopted for the following statistical analyses as a proxy for seasonal shifts in MCS.

To identify relationships between MCS and variables, Pearson's correlation analysis was performed. Meteorological data were obtained from the Automated Meteorological Data Acquisition System (AMeDAS) [46] Otsu observation station. Temperature, precipitation, and sunshine data used in this analysis were averaged to mean daily values for the 7, 14, 21, or 28 days before the sampling date. These meteorological parameters may directly or indirectly affect MCS and N dynamics.

The effects of microbial indices and several other variables on N transformation potential were analyzed using partial least-square (PLS) regression analysis. PLS regression is used for an estimation of the regression model between one dependent variable and several explanatory variables. PLS regression is similar to multivariate analysis, but it can utilize data with correlated explanatory variables [47,48]. The variable importance of projection (VIP), which indicates the relative importance of each explanatory variable, and standardized coefficients were calculated for each dependent variable. VIP values above 1 represent a significant result [47]. For explanatory variables, we used meteorological phenomena (temperature, precipitation, and sunshine), soil characteristics (WC and pH (H_2O)), substrate availability (WEON and WEOC/WEON), microbial biomass (MB-N, MB-C/N, fungal-PLFA, and bacteria-PLFA), PC2 scores (shown in Figure S4), and the microbial biomarker indices (Sat/mono, Cy/pre, G+/G−, and F/B).

All analyses were conducted using XLSTAT 2017 (Addinsoft, Paris, France). The significance level was set to α = 0.05.

3. Results

3.1. Basic Soil Characteristics

The WC exhibited significant temporal changes, especially in the O-layer (Table S1). Overall, WC was higher in the cool season than in the warm season. Seasonal changes in pH (H_2O) were noted, especially in the O-layer, but an interaction (site × season) effect was also found.

The WEOC also exhibited significant temporal changes, especially in the O-layer (Table S1). Overall, WEOC peaked in February. WEOC/WEON values were higher in the cool than in the warm season (Figure S3).

3.2. N Pool

Seasonal changes in the N pool were significant, especially changes in NF (Figure S2). NH_4^+-N concentrations in the warm season were greater than those in the cool season. NO_3^--N seasonal changes were only significant in NF (O-layer), whereas seasonal changes in WEON were only significant in NF. The inorganic N-to-WEON ratio can be represented as the available N composition index, which had greater values in the warm season than in the cool season.

The NH_4^+-N in both layers was significantly lower in RF than in NF. Similarly, NO_3^--N also tended to be lower in both layers in RF than in NF. Meanwhile, although WEON in the S-layer was significantly lower in RF than in NF, WEON in the O-layer was not significantly different between sites.

3.3. Microbial Biomass (MB) and Fungal and Bacterial PLFA

MB-C and -N significantly changed over time only in NF (Figure S3). In the O-layer, MB-C and -N were higher in February than at other times. MB-C/N significantly changed over time only in RF, while in a comparison between sites, MB-C/N was higher in the O-layer in RF than in NF.

Bacterial PLFA had significant temporal changes in the S-layer in NF (Figure 1). The seasonal trend was similar to MB-C (and weakly similar to MB-N) only in the S-layer. Fungal PLFA significantly changed in the O-layer in RF. Overall, fungal PLFA in the cool season was higher than that in the warm season.

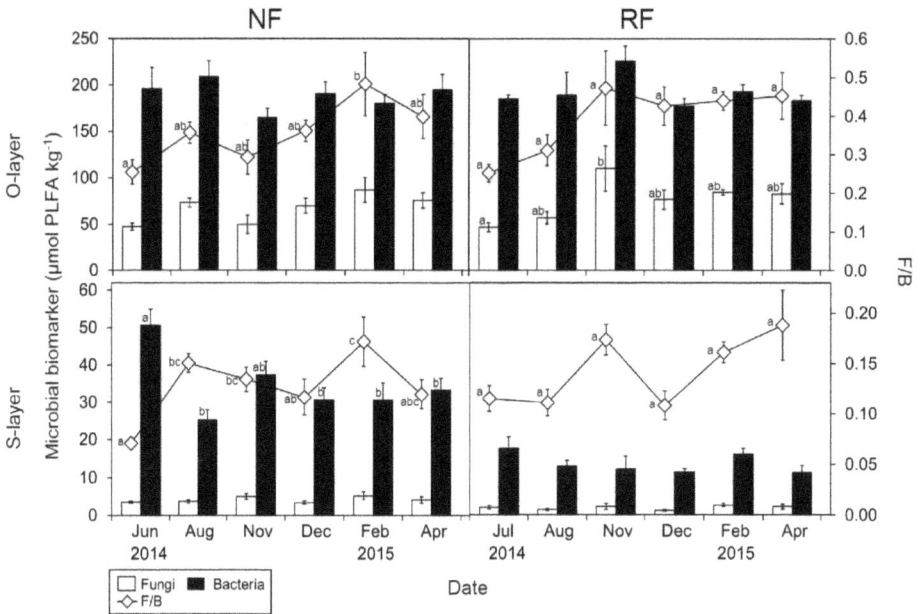

Figure 1. Seasonal changes in fungal and bacterial PLFA, and the F/B ratio of both layers at both sites. Data represent the mean (n = 5) $\pm$ SE. Letters indicate significant differences on a sampling date (Tukey's HSD test, $p < 0.05$).

3.4. Microbial Community Structure (MCS)

PCAs of the PLFA data at each site showed changes in MCS, not only between depths, but also among seasons (Figure S4). PC1 explained 58.0% and 49.8% of the total variance in NF and RF soils, respectively, clearly separating the O-layer and S-layer. Meanwhile, PC2 for NF and RF explained 22.7% and 24.3%, respectively, generally splitting the sampling dates.

In a comparison between NF and RF among seasons, the PCAs at each depth revealed changes in MCS related to both site and season (Figure 2). PC1 in the O- and S-layer explained 42.4% and 42.3%, respectively, while PC2 in NF and RF explained 30.9% and 14.8%, respectively. These PCs did not closely reflect either site or season.

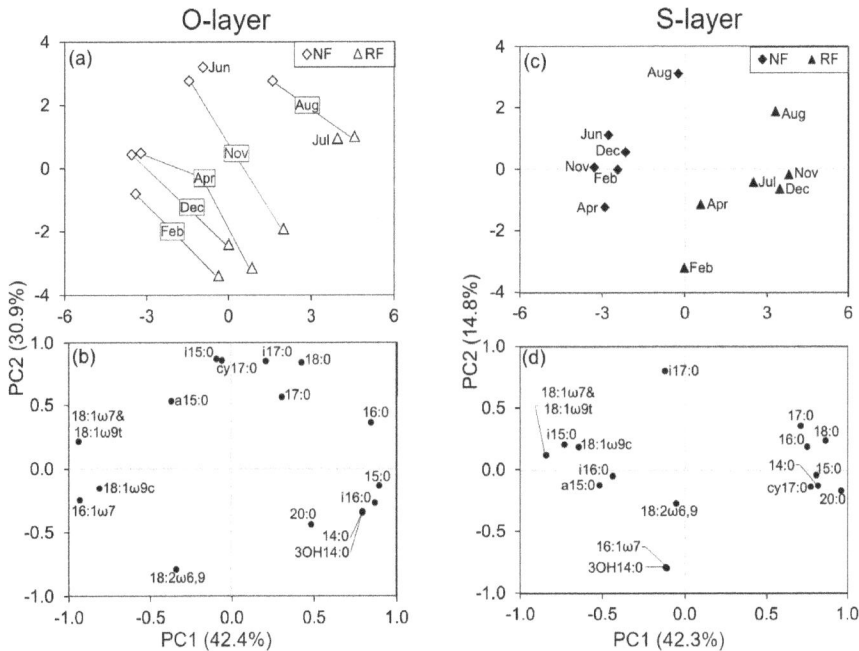

Figure 2. Principal component analysis (PCA) of phospholipid fatty acid (PLFA) data at two depths among seasons (*n* = 6), including the data from both NF and RF (**a,c**) and loading scores for individual PLFAs (**b,d**).

The loading scores of some PLFAs for all PCAs (PC1 and 2) were mostly or partly characterized by seasonal shifts (Figure 2 and Figure S4). To identify which PLFAs were associated with seasonal changes, we assessed seasonal changes in microbial indices (Figure 1 and Figure S5). There were significant temporal changes in all microbial indices. G+/G−, Sat/mono, and Cy/pre were greater in June, July, and August than in December, February, and April. Additionally, the three microbial indices were generally significantly correlated with PC2, as shown in Figure S4, and were also correlated with temperature, especially in the O-layer (Table S2 and Figure 3). In contrast, F/B values were greater in February than in June or July (Figure 1). Comparing NF and RF, Sat/mono was significantly greater at both depths in RF than in NF, but there were no significant differences in the other indices between sites.

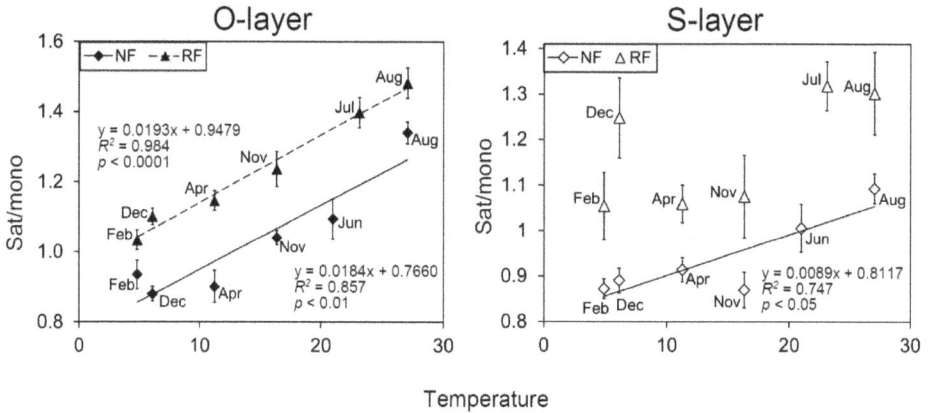

Figure 3. Relationship between the microbial index Sat/mono and temperature (28 day) at each depth, each site, and each season. The data are represented as mean ($n = 5$) $\pm$ SE.

3.5. N Transformation Potential

3.5.1. Gross N Transformation Potential

Seasonal changes in gross NH_4^+-N production and immobilization potential (25 °C) were observed in the O-layer in NF (Figure 4), with the highest values in August. Meanwhile, seasonal changes in the S-layer were less apparent at both sites. As a mean of six measuring seasons, N production potential was approximately equal to immobilization potential, but production potential tended to be greater in the warm than in the cool season.

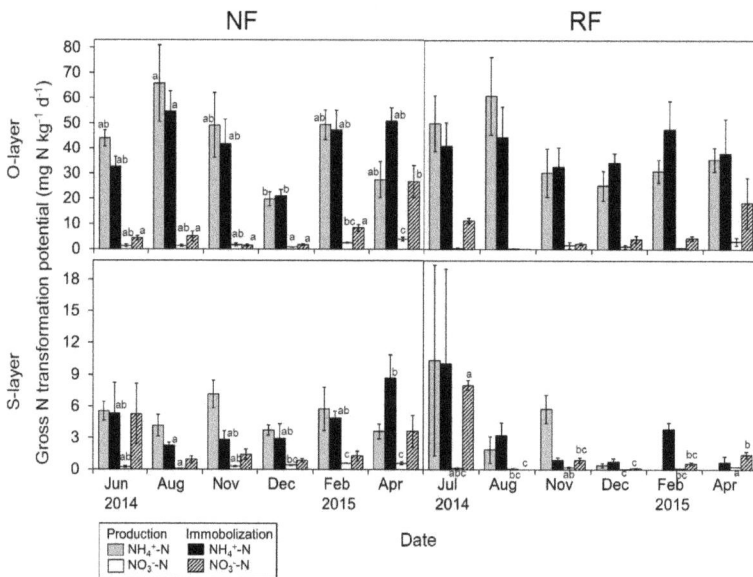

Figure 4. Seasonal changes in gross NH_4^+- and NO_3^--N production potential and gross NH_4^+- and NO_3^--N immobilization potential in both layers at both sites. Data are shown as mean ($n = 5$) $\pm$ SE. Letters indicate significant differences on a sampling date (Tukey's HSD test, $p < 0.05$).

Seasonal changes in gross NO_3^--N production potential were observed in the O-layer in NF. Production potential was higher in April than in other seasons, whereas the immobilization potential was highest in April and June (or July). Immobilization potential tended to be much higher than production potential.

Gross NH_4^+-N production and immobilization potential at both depths exhibited no significant differences between sites. Gross NO_3^--N immobilization potential at both depths also did not differ significantly between sites, whereas gross NO_3^--N production potential in the S-layer was significantly lower in RF than in NF.

3.5.2. Net N Transformation Potential

Seasonal changes in net N transformation potential (25 °C) were clear in the S-layer, but not in the O-layer (Figure 5). Overall, the potential for inorganic N transformation was greater in the warm than in the cool season.

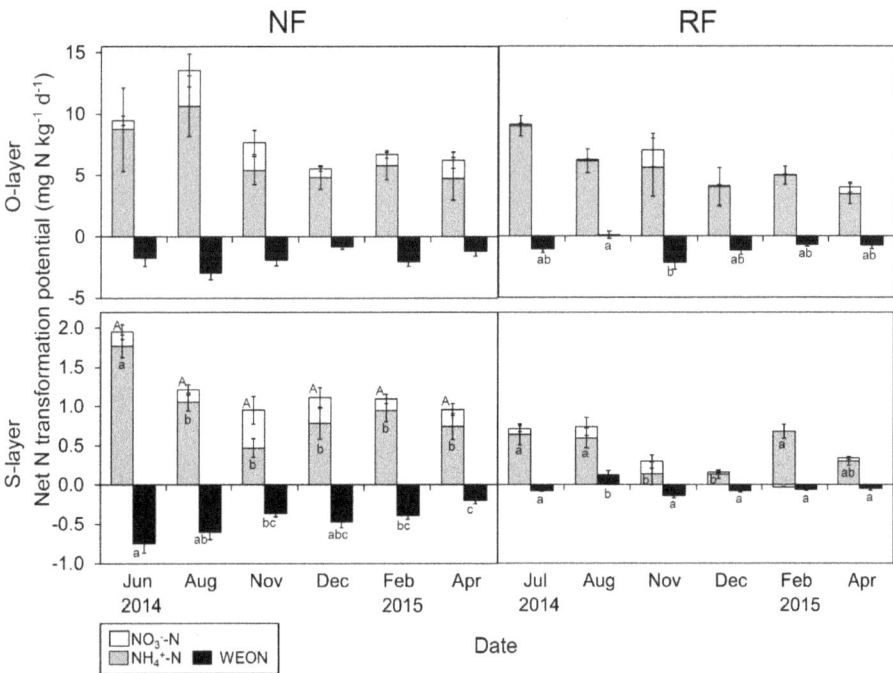

Figure 5. Seasonal changes in net NH_4^+-N, NO_3^--N and water-extractable organic nitrogen (WEON) transformation potential in both layers at both sites. Data are presented as mean ($n = 5$) $\pm$ SE. Letters indicate significant differences on a sampling date (Tukey's HSD test, $p < 0.05$).

The net NH_4^+-N transformation potential in the O-layer did not differ significantly between sites, whereas that in the S-layer was significantly lower in RF than in NF. The net NO_3^--N transformation potential in the O-layer was significantly lower in RF than in NF, whereas that in the S-layer did not show a significant difference between sites. The net WEON transformation potential in both layers was significantly greater in RF than in NF.

3.6. Relationship between N Dynamics and Multi-Variables

In analyzing the seasonal changes in inorganic N, the most important variables from the PLS regression model were temperature, soil environment, substrate availability, and MCS (Table S3). For seasonal changes in gross NH_4^+-N production potential, the major variables from the PLS regression model were temperature, WEON, MB-N bacterial-PLFA, and MCS (Table S4). For the seasonal changes in gross NH_4^+-N immobilization potential, the primary variables from the PLS regression model were WC, substrate availability, and MB-N (Table S5). For seasonal changes in net inorganic N production potential, the major variables from the PLS regression model were temperature, bacterial-PLFA, and MCS (Table S6). In particular, seasonal changes in the specific gross NH_4^+-N and net inorganic N production potential were strongly correlated with Sat/mono (Figure S6).

4. Discussion

4.1. Seasonal Changes in Microbial Community Structure (MCS)

In this study, differences in MCS, as represented by PCAs, were identified between soil depths (Figure S4) and between sites (Figure 2). The difference in the microbial index Sat/mono between sites agreed with McKinley et al. [49] and Bach et al. [50] across ecosystems of differing restoration histories, indicating a relationship between MCS and the amount of organic matter accumulation (total C and N) or nutrient status, e.g., C/N (Table 1; McKinley et al. [49]). Moreover, the difference in F/B between soil depths coincided with the findings of Joergensen and Wichern [51], largely reflecting C/N.

Many field studies have illustrated seasonal shifts in MCS through PLFA patterns [9,52], bacterial communities [53], and fungal communities [54]. The essential factor driving this shift is likely plant C inputs, i.e., leaf and root litter and root exudate [7,12]. First, fresh leaf litter is available mainly in autumn, when it likely changes MCS, e.g., bacterial [55] and fungal communities [54]. In addition, root litter may affect MCS; root exudate is present mainly in summer, when it is essential for specific microbes, such as mycorrhizal fungi [56–59]. Although the relative importance of litter and root exudate for microorganisms is unclear, it is clear that plant C inputs can alter MCS [55,60]. In this study, trends toward lower G+/G− and Sat/mono in winter were identified (Figure S5), and the higher WEOC/WEON in winter seemed to be associated with the large amount of plant litter input in autumn. The responses of the MCS agree with previous studies based on substrate-amended experiments [32,39]. The seasonal C input probably affected MCS at both sites.

Another pivotal driver is likely temperature [61–63]. Our data showed a correlation between MCS and temperature among seasons. These correlations were found regardless of the N fertility. Specifically, the PLFA indices G+/G−, Sat/mono, and Cy/pre were correlated with temperature, consistent with previous microcosm experiments [37]. Dominant G+ compared with G− at high temperatures may be associated with G+ use of relatively recalcitrant soil organic matter (SOM), while dormant G− are found at low temperatures and use labile SOM [31,62,64]. Associations between temperature and Sat/mono have been frequently reported from in vitro [65,66] and microcosm experiments [37], but the association between temperature and Sat/mono has been reported less often in field studies. The shift in Sat/mono is generally considered to maintain microbial membrane fluidity against temperature change [29]. These findings show the importance of temperature for natural microbial communities.

4.2. Seasonal Changes in N Dynamics

Despite the differing N fertility among sites, represented in the variation between WEOC/WEON and MB-C/N (Figure S3), similar seasonal N dynamics were observed. Overall, the gross N mineralization potential tended to be higher than immobilization potential in June–August, while in December–April, immobilization potential was dominant, in agreement with previous studies [16,67]. The net WEON transformation potential tended to be negatively correlated with the net inorganic N transformation potential, especially in NF, suggesting that WEON is important as a source of N mineralization [68]. Also, the potential value was almost negative among seasons, probably due to the disappearance of dissolved organic matter input (e.g., root exudate and compounds from plant litter) in incubation soil.

Contributions to seasonal N dynamics are made by abiotic (e.g., meteorological phenomena and soil environment [15]) and biotic factors (e.g., MB-N [15,16]). Among abiotic factors, temperature seems to be important for the seasonality of N mineralization (Tables S3–S6), although this study did not allow for the precise control of temperature during experiments (i.e., N transformation potential). Precipitation was not a significant determinant of seasonal changes in N dynamics in this study, but it may be critical to short-term or interannual variability [15,69]. In general, precipitation had a less consistent seasonal pattern than temperature (e.g., Figure S1), and thus its effect may be ambiguous. This ambiguity is further confounded by sunshine, which is associated with plant allocation of C to soil [70]. Among biotic factors, the principal direct driver is microbes, with MB-N especially frequently associated with N dynamics [16,71]. However, MB-N only represents the pool size, and thus is insufficient for a full understanding of N dynamics (Tables S3–S6; Bohlen et al. [15]). It is important to note the extent to which microorganisms put available-N into soil, such as N use efficiency (NUE; Mooshammer et al. [6]). In particular, seasonal shifts in the abundance and quality of plant inputs and seasonal differences in MCS likely alter enzyme activities [7], NUE, and subsequent N mineralization.

4.3. Seasonal Changes in the Relationships between MCS and N Dynamics

Our results imply that temperature (and likely plant C inputs) affected MCS, and imply an association between MCS and N dynamics at both sites. Higher F/B in winter is important for N retention in temperate ecosystems [16,72,73], consistent with our results, especially in NF. A trend of greater F/B in the cool season was identified (Figure 1), and MB-N and MB-C/N were also greater in the cool season (Figure S3; Tokuchi et al. [74]). This provides a mechanism for N retention in winter. However, the direct relationship between F/B and N transformation potential was unclear (Tables S4–S6), probably due to the spatial and temporal dependence of different fungal groups (e.g., non-mycorrhizal and mycorrhizal fungi) on different substrates [54,62]. It is likely that greater amounts of plant root exudates lead to an increased biomass of mycorrhizal fungi in summer [75], whereas greater plant leaf and root litter inputs stimulate an increased biomass of saprotrophic fungi in autumn and winter [54], which can cause a functional difference in C and N dynamics.

Microbial indices (Sat/mono, G+/G−, and Cy/pre) explained seasonal changes in N dynamics, except for gross N immobilization potential, better than other variables did. In particular, Sat/mono was positively correlated with the specific N transformation potentials (N transformation potentials normalized by MB-N; Figure S6). Similar relationships have been previously found in several field and laboratory studies using N additions or an N-deposition gradient (Table 2). Here, microbial indices were related to the gross production and net transformation potentials, but not to the gross immobilization potential. Although there was a close association between the production and immobilization (or consumption) of ammonium [76,77], the cause may be the greater control of production (relative to immobilization) through seasonal MCS changes.

Seasonal changes in MCS, specifically in the relative dominance of fungi and bacteria, may be associated with C:N-stoichiometry and N dynamics. In temperate forest ecosystems, plants and microbes are frequently subject to N limitation of growth [1,78], and the extent of this N limitation likely differs with ecosystem age [79,80] and forest stand age [81]. That is, old forests tend to be C limited relative to N, whereas immature forests are more N limited. These differences in the limiting element may lead to different seasonal patterns of MCS and associated differences in N dynamics. However, in this study, similar changes in MCS and N dynamics were found in forest stands of different ages. These findings can be interpreted as indicating similar seasonal changes in the element limitation of soil microbes. Overall, in summer, higher temperatures are likely to cause rapid substrate use along with increasing soil microbial respiration and Cy/pre [41] (possibly explaining our finding of high Cy/pre in summer), higher microbial biomass turnover [82,83], and higher decomposition of SOM [84]. Unless plants can supply enough fresh substrate inputs to exceed the increased decomposition, higher temperatures should convey a lower C/N of the available substrate, which means a stronger demand for C relative to N, i.e., C limitation [85]. In winter, fresh litter inputs with high C/N relative to

SOM arrive in late autumn and likely lead to N limitation [6]. In our results, high WEOC/WEON, representing high substrate quality [68], was also observed in winter, which suggests N limitation. In summary, changes in the limiting elements among seasons based on temperature and plant substrate inputs can influence MCS and N dynamics, regardless of differences in fertility.

4.4. Limitations

In this study, we established two temperate forest sites, but there were no replicated sites with different fertilities. Therefore, we will need to confirm their seasonal effects in systems under different backgrounds, such as with different ratios of C:N:P in the environment [78], which were caused by different parent materials and forest ages. Additionally, we measured the gross and net N transformation potentials of soil among seasons at a constant temperature of 25 °C. However, the results should be interpreted with caution. This method has the advantage of comparability between N dynamics generated by microbial communities among seasons under a constant condition, whereas it has the disadvantage of a difference between laboratory experiments and in situ measurements. The actual N mineralization rates of the cool seasons under field conditions were likely lower than the potentials (25 °C) [86]. Therefore, the difference between warm and cool seasons was probably much greater.

5. Conclusions

We investigated seasonal changes in MCS and N dynamics in temperate forest soils at two sites with differing fertilities over the course of one year. Consistent seasonal trends in MCS were found in both the organic and mineral soil layers, and the specific indicators of PLFA also changed among seasons. These results were mainly attributable to temperature, although we should note that the changes in PLFA were caused not only by the changes in MCS, but also by the microbial physiological response to temperature [30,65,66]. To understand the effects of temperature among seasons, a fully temperature-controlled field study is needed. For N dynamics, summer is likely the N mineralization phase, whereas winter appears to be the immobilization phase. These results imply the importance of gross N production compared with immobilization. Moreover, a relationship between MCS and N dynamics was also found, despite the disclaimer above about PLFA. Plant inputs such as root exudates and fine roots are essential contributors to the seasonal MCS and N dynamics. However, the seasonal allocation of these inputs remains mostly unknown in terms of quantity and spatiotemporal trends. Whether microorganisms can utilize old or fresh substrate (i.e., SOM or plant inputs with different C:N stoichiometries [62]) depends strongly on temperature [64] and the microbial groups present (e.g., G+ and G− [31]). Therefore, to understand the seasonal N dynamics associated with the microbial use of different substrates and with temperature, we recommend PLFA along with stable isotope probing, which can identify the substrate based on $\delta^{13}C$ (e.g., Waldrop and Firestone [62]; Kramer and Gleixner [31]).

Supplementary Materials: Supplementary materials can be found at www.mdpi.com/1999-4907/9/3/153/s1. Table S1: WC, pH, and WEOC in the O-layer and S-layer on each sampling date at both sites, Table S2: Pearson correlation coefficients between microbial community structure (MCS) and other variables among all seasons, Table S3: Variable importance of projection (VIP) and standardized coefficient (SC) of explanatory variables from the PLS regression models for seasonal changes in inorganic N, Table S4: VIP and SC of explanatory variables from the PLS regression models for seasonal changes in gross NH_4^+-N production potential, Table S5: VIP and SC of explanatory variables from the PLS regression models for seasonal changes in gross NH_4^+-N immobilization potential, Table S6: VIP and SC of explanatory variables from the PLS regression models for seasonal changes in net N transformation potential, Figure S1: Seasonal dynamics of sunshine, precipitation, and temperature at Otsu observation station of the Japan Meteorological Agency, Figure S2: Seasonal changes of NH_4^+-N, NO_3^--N, water-extractable organic nitrogen (WEON), and inorganic N/WEON in both layers at both sites, Figure S3: Seasonal changes of microbial biomass (MB)-C, -N, MB-C/N, and WEOC/WEON, Figure S4: Principal component analysis (PCA) of the phospholipid fatty acid (PLFA) data for both sites among all seasons, Figure S5: Seasonal changes in microbial indices determined from PLFA data for both sites and both layers, Figure S6: Relationship between the microbial index (Sat/mono) and specific N transformation potentials at each depth and site among all seasons.

Acknowledgments: We thank Kazuo Isobe for technical support and advice on the ^{15}N isotopic dilution method, the members of Soil Science Laboratory, the University of Tokyo for technical help, and Takahito Yoshioka for useful comments and suggestions. This study was supported by Grants-in-Aid for Scientific Research from the Japanese Society for the Promotion of Science (No. 15H04515).

Author Contributions: Tomohiro Yokobe and Naoko Tokuchi conceived and designed the experiments; Tomohiro Yokobe and Naoko Tokuchi contributed to the field work; Tomohiro Yokobe performed the experiments and analyzed the data; Fujio Hyodo contributed to PLFA analysis; Tomohiro Yokobe wrote the paper; Naoko Tokuchi and Fujio Hyodo reviewed the manuscript and contributed to editing of the manuscript.

Conflicts of Interest: The authors declare no conflict of interest.

References

1. LeBauer, D.S.; Treseder, K.K. Nitrogen limitation of net primary productivity in terrestrial ecosystems is globally distributed. *Ecology* **2008**, *89*, 371–379. [CrossRef] [PubMed]
2. Strickland, M.S.; Lauber, C.; Fierer, N.; Bradford, M.A. Testing the functional significance of microbial community composition. *Ecology* **2009**, *90*, 441–451. [CrossRef] [PubMed]
3. Waldrop, M.P.; Balser, T.C.; Firestone, M.K. Linking microbial community composition to function in a tropical soil. *Soil Biol. Biochem.* **2000**, *32*, 1837–1846. [CrossRef]
4. Strickland, M.S.; Rousk, J. Considering fungal: Bacterial dominance in soils—Methods, controls, and ecosystem implications. *Soil Biol. Biochem.* **2010**, *42*, 1385–1395. [CrossRef]
5. Waring, B.G.; Averill, C.; Hawkes, C.V. Differences in fungal and bacterial physiology alter soil carbon and nitrogen cycling: Insights from meta-analysis and theoretical models. *Ecol. Lett.* **2013**, *16*, 887–894. [CrossRef] [PubMed]
6. Mooshammer, M.; Wanek, W.; Hämmerle, I.; Fuchslueger, L.; Hofhansl, F.; Knoltsch, A.; Schnecker, J.; Takriti, M.; Watzka, M.; Wild, B.; et al. Adjustment of microbial nitrogen use efficiency to carbon: Nitrogen imbalances regulates soil nitrogen cycling. *Nat. Commun.* **2014**, *5*, 3694. [CrossRef] [PubMed]
7. Koranda, M.; Kaiser, C.; Fuchslueger, L.; Kitzler, B.; Sessitsch, A.; Zechmeister-Boltenstern, S.; Richter, A. Seasonal variation in functional properties of microbial communities in beech forest soil. *Soil Biol. Biochem.* **2013**, *60*, 95–104. [CrossRef] [PubMed]
8. Waldrop, M.P.; Firestone, M.K. Seasonal dynamics of microbial community composition and function in oak canopy and open grassland soils. *Microb. Ecol.* **2006**, *52*, 470–479. [CrossRef] [PubMed]
9. Puissant, J.; Cécillon, L.; Mills, R.T.E.; Robroek, B.J.M.; Gavazov, K.; De Danieli, S.; Spiegelberger, T.; Buttler, A.; Brun, J.J. Seasonal influence of climate manipulation on microbial community structure and function in mountain soils. *Soil Biol. Biochem.* **2015**, *80*, 296–305. [CrossRef]
10. Lauber, C.L.; Ramirez, K.S.; Aanderud, Z.; Lennon, J.; Fierer, N. Temporal variability in soil microbial communities across land-use types. *ISME J.* **2013**, *7*, 1641–1650. [CrossRef] [PubMed]
11. Brockett, B.F.T.; Prescott, C.E.; Grayston, S.J. Soil moisture is the major factor influencing microbial community structure and enzyme activities across seven biogeoclimatic zones in western Canada. *Soil Biol. Biochem.* **2012**, *44*, 9–20. [CrossRef]
12. Thoms, C.; Gleixner, G. Seasonal differences in tree species' influence on soil microbial communities. *Soil Biol. Biochem.* **2013**, *66*, 239–248. [CrossRef]
13. Wu, J.; Xiong, J.; Hu, C.; Shi, Y.; Wang, K.; Zhang, D. Temperature sensitivity of soil bacterial community along contrasting warming gradient. *Appl. Soil Ecol.* **2015**, *94*, 40–48. [CrossRef]
14. Treseder, K.K.; Marusenko, Y.; Romero-Olivares, A.L.; Maltz, M.R. Experimental warming alters potential function of the fungal community in boreal forest. *Glob. Chang. Biol.* **2016**, *22*, 3395–3404. [CrossRef] [PubMed]
15. Bohlen, P.J.; Groffman, P.M.; Driscoll, C.T.; Fahey, T.J.; Siccama, T.G. Plant-soil-microbial interactions in a northern hardwood forest. *Ecology* **2001**, *82*, 965–978. [CrossRef]
16. Kaiser, C.; Fuchslueger, L.; Koranda, M.; Gorfer, M.; Claus, F.; Kitzler, B.; Rasche, F.; Strauss, J.; Sessitsch, A.; Zechmeister, S.; et al. Plants control N cycling the seasonal dynamics of microbial in a beech forest soil by belowground C allocation. *Ecology* **2011**, *92*, 1036–1051. [CrossRef] [PubMed]
17. Bardgett, R.D.; Bowman, W.D.; Kaufmann, R.; Schmidt, S.K. A temporal approach to linking aboveground and belowground ecology. *Trends Ecol. Evol.* **2005**, *20*, 634–641. [CrossRef] [PubMed]

18. Schmidt, S.K.; Costello, E.K.; Nemergut, D.R.; Cleveland, C.C.; Reed, S.C.; Meyer, A.F.; Martin, A.M.; Nemergut, R.; Meyer, F.; Reed, S.C. Biogeochemical consequences of rapid microbial turnover and seasonal succession in soil. *Ecology* **2007**, *88*, 1379–1385. [CrossRef] [PubMed]

19. Kaiser, C.; Koranda, M.; Kitzler, B.; Fuchslueger, L.; Schnecker, J.; Schweiger, P.; Rasche, F.; Zechmeister-Boltenstern, S.; Sessitsch, A.; Richter, A. Belowground carbon allocation by trees drives seasonal patterns of extracellular enzyme activities by altering microbial community composition in a beech forest soil. *New Phytol.* **2010**, *187*, 843–858. [CrossRef] [PubMed]

20. Högberg, M.N.; Chen, Y.; Högberg, P. Gross nitrogen mineralisation and fungi-to-bacteria ratios are negatively correlated in boreal forests. *Biol. Fertil. Soils* **2007**, *44*, 363–366. [CrossRef]

21. Rousk, J.; Bååth, E. Growth of saprotrophic fungi and bacteria in soil. *FEMS Microbiol. Ecol.* **2011**, *78*, 17–30. [CrossRef] [PubMed]

22. Hobara, S.; Tokuchi, N.; Ohte, N.; Koba, K.; Katsuyama, M.; Kim, S.-J.; Nakanishi, A. Mechanism of nitrate loss from a forested catchment following a small-scale, natural disturbance. *Can. J. For. Res.* **2001**, *31*, 1326–1335. [CrossRef]

23. IUSS Working Group WRB. *World Reference Base for Soil Resources 2014. International Soil Classification System for Naming Soils and Creating Legends for Soil Maps*; IUSS Working Group WRB: Rome, Italy, 2014; ISBN 9789251083697.

24. Brookes, P.C.; Landman, A.; Pruden, G.; Jenkinson, D.S. Chloroform fumigation and the release of soil nitrogen: A rapid direct extraction method to measure microbial biomass nitrogen in soil. *Soil Biol. Biochem.* **1985**, *17*, 837–842. [CrossRef]

25. Wu, J.; Joergensen, R.G.; Pommerening, B.; Chaussod, R.; Brookes, P.C. Measurement of soil microbial biomass C by fumigation-extraction-an automated procedure. *Soil Biol. Biochem.* **1990**, *22*, 1167–1169. [CrossRef]

26. Jenkinson, D.S. The determination of microbial biomass carbon and nitrogen in soil. In *Advances in Nitrogen Cycling in Agricultural Ecosystems*; Wilson, J.R., Ed.; C.A.B. International: Wallingford, UK, 1988; pp. 368–386. ISBN 085198603X.

27. Frostegård, Å.; Tunlid, A.; Bååth, E. Phospholipid fatty acid composition, biomass, and activity of microbial communities from two soil types experimentally exposed to different heavy metals. *Appl. Environ. Microbiol.* **1993**, *59*, 3605–3617. [PubMed]

28. Bardgett, R.D.; Hobbs, P.J.; Frostegård, Å. Changes in soil fungal: Bacterial biomass ratios following reductions in the intensity of management of an upland grassland. *Biol. Fertil. Soils* **1996**, *22*, 261–264. [CrossRef]

29. Wixon, D.L.; Balser, T.C. Toward conceptual clarity: PLFA in warmed soils. *Soil Biol. Biochem.* **2013**, *57*, 769–774. [CrossRef]

30. Frostegård, Å.; Tunlid, A.; Bååth, E. Use and misuse of PLFA measurements in soils. *Soil Biol. Biochem.* **2011**, *43*, 1621–1625. [CrossRef]

31. Kramer, C.; Gleixner, G. Variable use of plant- and soil-derived carbon by microorganisms in agricultural soils. *Soil Biol. Biochem.* **2006**, *38*, 3267–3278. [CrossRef]

32. Bossio, D.A.; Scow, K.M. Impacts of carbon and flooding on soil microbial communities: Phospholipid fatty acid profiles and substrate utilization patterns. *Microb. Ecol.* **1998**, *35*, 265–278. [CrossRef] [PubMed]

33. Högberg, M.N.; Högbom, L.; Kleja, D.B. Soil microbial community indices as predictors of soil solution chemistry and N leaching in *Picea abies* (L.) Karst. forests in S. Sweden. *Plant Soil* **2013**, *372*, 507–522. [CrossRef]

34. Blaško, R.; Högberg, P.; Bach, L.H.; Högberg, M.N. Relations among soil microbial community composition, nitrogen turnover, and tree growth in N-loaded and previously N-loaded boreal spruce forest. *For. Ecol. Manag.* **2013**, *302*, 319–328. [CrossRef]

35. Högberg, M.N.; Blaško, R.; Bach, L.H.; Hasselquist, N.J.; Egnell, G.; Näsholm, T.; Högberg, P. The return of an experimentally N-saturated boreal forest to an N-limited state: Observations on the soil microbial community structure, biotic N retention capacity and gross N mineralisation. *Plant Soil* **2014**, *381*, 45–60. [CrossRef]

36. Zechmeister-Boltenstern, S.; Michel, K.; Pfeffer, M. Soil microbial community structure in European forests in relation to forest type and atmospheric nitrogen deposition. *Plant Soil* **2011**, *343*, 37–50. [CrossRef]

37. Feng, X.; Simpson, M.J. Temperature and substrate controls on microbial phospholipid fatty acid composition during incubation of grassland soils contrasting in organic matter quality. *Soil Biol. Biochem.* **2009**, *41*, 804–812. [CrossRef]

38. Demoling, F.; Ola Nilsson, L.; Bååth, E. Bacterial and fungal response to nitrogen fertilization in three coniferous forest soils. *Soil Biol. Biochem.* **2008**, *40*, 370–379. [CrossRef]

39. Stevenson, B.A.; Hunter, D.W.F.; Rhodes, P.L. Temporal and seasonal change in microbial community structure of an undisturbed, disturbed, and carbon-amended pasture soil. *Soil Biol. Biochem.* **2014**, *75*, 175–185. [CrossRef]

40. Frey, S.D.; Drijber, R.; Smith, H.; Melillo, J. Microbial biomass, functional capacity, and community structure after 12 years of soil warming. *Soil Biol. Biochem.* **2008**, *40*, 2904–2907. [CrossRef]

41. Schindlbacher, A.; Rodler, A.; Kuffner, M.; Kitzler, B.; Sessitsch, A.; Zechmeister-Boltenstern, S. Experimental warming effects on the microbial community of a temperate mountain forest soil. *Soil Biol. Biochem.* **2011**, *43*, 1417–1425. [CrossRef] [PubMed]

42. Kuroiwa, M.; Koba, K.; Isobe, K.; Tateno, R.; Nakanishi, A.; Inagaki, Y.; Toda, H.; Otsuka, S.; Senoo, K.; Suwa, Y.; et al. Gross nitrification rates in four Japanese forest soils: Heterotrophic versus autotrophic and the regulation factors for the nitrification. *J. For. Res.* **2011**, *16*, 363–373. [CrossRef]

43. Isobe, K.; Suwa, Y.; Ikutani, J.; Kuroiwa, M.; Makita, T.; Takebayashi, Y.; Yoh, M.; Otsuka, S.; Senoo, K.; Ohmori, M.; et al. Analytical techniques for quantifying (15)N/(14)N of nitrate, nitrite, total dissolved nitrogen and ammonium in environmental samples using a gas chromatograph equipped with a quadrupole mass spectrometer. *Microbes Environ.* **2011**, *26*, 46–53. [CrossRef] [PubMed]

44. Davidson, E.A.; Hart, S.C.; Shanks, C.A.; Firestone, M. Measuring gross nitrogen mineralization, and nitrification by 15 N isotopic pool dilution in intact soil cores. *J. Soil Sci.* **1991**, *42*, 335–349. [CrossRef]

45. Corre, M.D.; Beese, F.O.; Brumme, R. Soil nitrogen cycle in high nitrogen deposition forest: Changes under nitrogen saturation and liming. *Ecol. Appl.* **2003**, *13*, 287–298. [CrossRef]

46. Japan Meteorological Agency Statistical Report for Weather in Japan. Available online: http://www.data. jma.go.jp/obd/stats/etrn/index.php (accessed on 18 December 2015).

47. Wold, S.; Sjöström, M.; Eriksson, L. PLS-regression: A basic tool of chemometrics. *Chemom. Intell. Lab. Syst.* **2001**, *58*, 109–130. [CrossRef]

48. Esposito Vinzi, V.; Chin, W.W.; Henseler, J.; Wang, H. *Handbook of Partial Least Squares. Concepts, Methods and Applications*; Springer: Berlin/Heidelberg, Germany, 2010; ISBN 9783540328278.

49. McKinley, V.L.; Peacock, A.D.; White, D.C. Microbial community PLFA and PHB responses to ecosystem restoration in tallgrass prairie soils. *Soil Biol. Biochem.* **2005**, *37*, 1946–1958. [CrossRef]

50. Bach, E.M.; Baer, S.G.; Meyer, C.K.; Six, J. Soil texture affects soil microbial and structural recovery during grassland restoration. *Soil Biol. Biochem.* **2010**, *42*, 2182–2191. [CrossRef]

51. Joergensen, R.G.; Wichern, F. Quantitative assessment of the fungal contribution to microbial tissue in soil. *Soil Biol. Biochem.* **2008**, *40*, 2977–2991. [CrossRef]

52. Bardgett, R.D.; Lovell, R.D.; Hobbs, P.J.; Jarvis, S.C. Seasonal changes in soil microbial communities along a fertility gradient of temperate grasslands. *Soil Biol. Biochem.* **1999**, *31*, 1021–1030. [CrossRef]

53. Lipson, D.A. Relationships between temperature responses and bacterial community structure along seasonal and altitudinal gradients. *FEMS Microbiol. Ecol.* **2007**, *59*, 418–427. [CrossRef] [PubMed]

54. Voříšková, J.; Brabcová, V.; Cajthaml, T.; Baldrian, P. Seasonal dynamics of fungal communities in a temperate oak forest soil. *New Phytol.* **2014**, *201*, 269–278. [CrossRef] [PubMed]

55. Chemidlin Prevost-Boure, N.; Maron, P.-A.; Ranjard, L.; Nowak, V.; Dufrene, E.; Damesin, C.; Soudani, K.; Lata, J.-C. Seasonal dynamics of the bacterial community in forest soils under different quantities of leaf litter. *Appl. Soil Ecol.* **2011**, *47*, 14–23. [CrossRef]

56. Brant, J.B.; Myrold, D.D.; Sulzman, E.W. Root controls on soil microbial community structure in forest soils. *Oecologia* **2006**, *148*, 650–659. [CrossRef] [PubMed]

57. Yarwood, S.A.; Myrold, D.D.; Högberg, M.N. Termination of belowground C allocation by trees alters soil fungal and bacterial communities in a boreal forest. *FEMS Microbiol. Ecol.* **2009**, *70*, 151–162. [CrossRef] [PubMed]

58. De Graaff, M.A.; Classen, A.T.; Castro, H.F.; Schadt, C.W. Labile soil carbon inputs mediate the soil microbial community composition and plant residue decomposition rates. *New Phytol.* **2010**, *188*, 1055–1064. [CrossRef] [PubMed]

59. Eilers, K.G.; Lauber, C.L.; Knight, R.; Fierer, N. Shifts in bacterial community structure associated with inputs of low molecular weight carbon compounds to soil. *Soil Biol. Biochem.* **2010**, *42*, 896–903. [CrossRef]

60. Brant, J.B.; Sulzman, E.W.; Myrold, D.D. Microbial community utilization of added carbon substrates in response to long-term carbon input manipulation. *Soil Biol. Biochem.* **2006**, *38*, 2219–2232. [CrossRef]

61. Zogg, G.P.; Zak, D.R.; Ringelberg, D.B.; White, D.C.; MacDonald, N.W.; Pregitzer, K.S. Compositional and Functional Shifts in Microbial Communities Due to Soil Warming. *Soil Sci. Soc. Am. J.* **1997**, *61*, 475. [CrossRef]

62. Waldrop, M.P.; Firestone, M.K. Altered utilization patterns of young and old soil C by microorganisms caused by temperature shifts and N additions. *Biogeochemistry* **2004**, *67*, 235–248. [CrossRef]

63. Oliverio, A.M.; Bradford, M.A.; Fierer, N. Identifying the microbial taxa that consistently respond to soil warming across time and space. *Glob. Chang. Biol.* **2017**, *23*, 2117–2129. [CrossRef] [PubMed]

64. Biasi, C.; Rusalimova, O.; Meyer, H.; Kaiser, C.; Wanek, W.; Barsukov, P.; Junger, H.; Richter, A. Temperature-dependent shift from labile to recalcitrant carbon sources of arctic heterotrophs. *Rapid Commun. Mass Spectrom.* **2005**, *19*, 1401–1408. [CrossRef] [PubMed]

65. Zhang, Y.-M.; Rock, C.O. Membrane lipid homeostasis in bacteria. *Nat. Rev. Microbiol.* **2008**, *6*, 222–233. [CrossRef] [PubMed]

66. Suutari, M.; Liukkonen, K.; Laakso, S. Temperature adaptation in yeasts: The role of fatty acids. *J. Gen. Microbiol.* **1990**, *136*, 1469–1474. [CrossRef] [PubMed]

67. Jaeger, C.H.; Monson, R.K.; Fisk, M.C.; Schmidt, S.K. Seasonal partitioning of nitrogen by plants and soil microorganisms in an alpine ecosystem. *Ecology* **1999**, *80*, 1883–1891. [CrossRef]

68. Marschner, B.; Kalbitz, K. Controls of bioavailability and biodegradability of dissolved organic matter in soils. *Geoderma* **2003**, *113*, 211–235. [CrossRef]

69. Tiemann, L.K.; Billings, S.A. Changes in variability of soil moisture alter microbial community C and N resource use. *Soil Biol. Biochem.* **2011**, *43*, 1837–1847. [CrossRef]

70. Rowland, L.; Hill, T.C.; Stahl, C.; Siebicke, L.; Burban, B.; Zaragoza-Castells, J.; Ponton, S.; Bonal, D.; Meir, P.; Williams, M. Evidence for strong seasonality in the carbon storage and carbon use efficiency of an Amazonian forest. *Glob. Chang. Biol.* **2014**, *20*, 979–991. [CrossRef] [PubMed]

71. Tahovská, K.; Kaňa, J.; Bárta, J.; Oulehle, F.; Richter, A.; Šantrůčková, H. Microbial N immobilization is of great importance in acidified mountain spruce forest soils. *Soil Biol. Biochem.* **2013**, *59*, 58–71. [CrossRef]

72. De Vries, F.T.; van Groenigen, J.W.; Hoffland, E.; Bloem, J. Nitrogen losses from two grassland soils with different fungal biomass. *Soil Biol. Biochem.* **2011**, *43*, 997–1005. [CrossRef]

73. De Vries, F.T.; Bloem, J.; Quirk, H.; Stevens, C.J.; Bol, R.; Bardgett, R.D. Extensive Management Promotes Plant and Microbial Nitrogen Retention in Temperate Grassland. *PLoS ONE* **2012**, *7*, 1–12. [CrossRef] [PubMed]

74. Tokuchi, N.; Yoneda, S.; Ohte, N.; Usui, N.; Koba, K.; Kuroiwa, M.; Toda, H.; Suwa, Y. Seasonal changes and controlling factors of gross N transformation in an evergreen plantation forest in central Japan. *J. For. Res.* **2014**, *19*, 77–85. [CrossRef]

75. Högberg, P.; Johannisson, C.; Yarwood, S.; Callesen, I.; Näsholm, T.; Myrold, D.D.; Högberg, M.N. Recovery of ectomycorrhiza after "nitorgen saturation" of a coniferous forest. *New Phytol.* **2010**, *189*, 515–525. [CrossRef] [PubMed]

76. Fisk, M.C.; Schmidt, S.K.; Seastedt, T.R. Topographic patterns of above- and belowground production and nitrogen cycling in alpine tundra. *Ecology* **1998**, *79*, 2253–2266. [CrossRef]

77. Christenson, L.M.; Lovett, G.M.; Weathers, K.C.; Arthur, M.A. The Influence of Tree Species, Nitrogen Fertilization, and Soil C to N ratio on Gross Soil Nitrogen Transformations. *Soil Sci. Soc. Am. J.* **2009**, *73*, 638. [CrossRef]

78. Zechmeister-Boltenstern, S.; Keiblinger, K.M.; Mooshammer, M.; Peñuelas, J.; Richter, A.; Sardans, J.; Wanek, W. The application of ecological stoichiometry to plant-microbial-soil organic matter transformations. *Ecol. Monogr.* **2015**, *85*, 133–155. [CrossRef]

79. De Vries, F.T.; Bardgett, R.D. Plant-microbial linkages and ecosystem nitrogen retention: Lessons for sustainable agriculture. *Front. Ecol. Environ.* **2012**, *10*, 425–432. [CrossRef]

80. Blaško, R.; Holm Bach, L.; Yarwood, S.A.; Trumbore, S.E.; Högberg, P.; Högberg, M.N. Shifts in soil microbial community structure, nitrogen cycling and the concomitant declining N availability in ageing primary boreal forest ecosystems. *Soil Biol. Biochem.* **2015**, *91*, 200–211. [CrossRef]

81. Goodale, C.L.; Aber, J.D. The long-term effects of land-use history on nitrogen cycling in northern hardwood forests. *Ecol. Appl.* **2001**, *11*, 253–267. [CrossRef]

82. Joergensen, R.G.; Brookes, P.C.; Jenkinson, D.S. Survival of the soil microbial biomass at elevated temperatures. *Soil Biol. Biochem.* **1990**, *22*, 1129–1136. [CrossRef]

83. Hagerty, S.B.; van Groenigen, K.J.; Allison, S.D.; Hungate, B.A.; Schwartz, E.; Koch, G.W.; Kolka, R.K.; Dijkstra, P. Accelerated microbial turnover but constant growth efficiency with warming in soil. *Nat. Clim. Chang.* **2014**, *4*, 903–906. [CrossRef]

84. Manzoni, S.; Taylor, P.; Richter, A.; Porporato, A.; Ågren, G.I. Environmental and stoichiometric controls on microbial carbon-use efficiency in soils. *New Phytol.* **2012**, *196*, 79–91. [CrossRef] [PubMed]

85. Cookson, W.R.; Osman, M.; Marschner, P.; Abaye, D.A.; Clark, I.; Murphy, D.V.; Stockdale, E.A.; Watson, C.A. Controls on soil nitrogen cycling and microbial community composition across land use and incubation temperature. *Soil Biol. Biochem.* **2007**, *39*, 744–756. [CrossRef]

86. Cookson, W.R.; Cornforth, I.S.; Rowarth, J.S. Winter soil temperature (2–15 °C) effects on nitrogen transformations in clover green manure amended or unamended soils; A laboratory and field study. *Soil Biol. Biochem.* **2002**, *34*, 1401–1415. [CrossRef]

forests

Article

Discriminating between Seasonal and Chemical Variation in Extracellular Enzyme Activities within Two Italian Beech Forests by Means of Multilevel Models

Antonietta Fioretto [1,†], Michele Innangi [1,*,†], Anna De Marco [2], Cristina Menta [3], Stefania Papa [1], Antonella Pellegrino [1] and Amalia Virzo De Santo [2]

[1] Department of Environmental, Biological, and Pharmaceutical Sciences and Technologies, University of Campania "Luigi Vanvitelli", Via Vivaldi 43, 81100 Caserta, Italy; antonietta.fioretto@unicampania.it (A.F.); stefania.papa@unicampania.it (S.P.); dott.a.pellegrino@gmail.com (A.P.)
[2] Department of Biology, University of Naples "Federico II", Complesso Universitario di Monte S. Angelo, Via Cinthia 21, 80126 Naples, Italy; ademarco@unina.it (A.D.M.); virzo@unina.it (A.V.D.S.)
[3] Department of Chemistry, Life Sciences and Environmental Sustainability, University of Parma, Viale delle Scienze 11/A, 43124 Parma, Italy; cristina.menta@unipr.it
* Correspondence: michele.innangi@unina.it; Tel.: +39-0823-274550
† These authors contributed equally to this work.

Received: 10 March 2018; Accepted: 18 April 2018; Published: 19 April 2018

Abstract: Enzymes play a key-role in organic matter dynamics and strong scientific attention has been given to them lately, especially to their response to climate and substrate chemical composition. Accordingly, in this study, we investigated the effects of chemical composition and seasons on extracellular enzyme activities (laccase, peroxidase, cellulase, chitinase, acid phosphomonoesterase, and dehydrogenase) by means of multilevel models within two Italian mountain beech forests. We used chemical variables as the fixed part in the model, season as random variation and layers (decomposition continuum for leaf litter and 0–5, 5–15, 15–30, and 30–40 cm for soil) as nested factors within the two forests. Our results showed that seasonal changes explained a higher amount of variance in enzyme activities compared to substrate chemistry in leaf litter, whereas chemical variation had a stronger impact on soil. Moreover, the effect of seasonality and chemistry was in general larger than the differences between forest sites, soils, and litter layers.

Keywords: seasonal trends; beech forests; soil enzymes; organic matter; multilevel models

1. Introduction

Forests cover about 30% of Earth's surface and are of vital importance for many ecosystem services [1], including the regulation of the global carbon cycle [2]. Forests act as C sinks by storing more than 650 billion tonnes of carbon of which, on average, 11% is found in plant necromass and in the organic horizon, and 45% in the mineral soil [1]. Noticeably, temperate and boreal forests are of great importance as C sinks, containing about 14% and 32% of global forest C pools, respectively [3].

Accumulation of soil organic matter, which is mainly comprised of C, depends on the balance between primary net productivity and detrital decomposition. Several biotic and abiotic factors influence the decomposition processes [4]. In detail, litter decomposition is affected mostly by soil characteristics, nature and abundance of decomposing organisms and their interactions with soil fauna, litter quality and climate [5–8].

Litter decomposition is operated mainly by soil bacteria and fungi by releasing hundreds of different extracellular enzymes into the environment, which break down complex organic compounds into assimilable sources of carbon (C), nitrogen (N), phosphorus (P), and other nutrients. Not every species of microbe produces the same enzymes, and they show different efficiency in substrate use and nutrient demand [9]. In addition, the production of extracellular enzymes is stimulated by substrate supply or reduced nutrient availability [10]. As decomposition goes on, the quality of the organic substance changes due to loss of easily degradable components and, therefore the microbial community that colonizes it changes as well. Thus, a true succession of microbial communities is formed, which tends to be composed predominantly by *K*-strategist species in the advanced phases, decomposers of more recalcitrant components, such as lignin and cellulose [11]. In this view, measuring extracellular enzyme activities could provide functional information on specific aspects and succession of the microbial communities during decomposition [12–14], but it is also useful to assess changes of microbial soil communities in response to environmental variation [15,16]. Kaiser et al. [17] found that changes in enzyme activities during the year suggested a switch of the main substrate to decompose but also a strong relationship between microbial community composition, which responds to environmental changes, and enzyme activities over the seasons. In this view, a relationship between decomposer enzyme activities and temperature and rainfall regimes has been found [18].

Our work focused on change of activity of some enzymes involved in the decomposition process within the organic horizon of forest floor and in the mineral soil, to a depth of 40 cm. We have studied two European beech (*Fagus sylvatica* L.) forest ecosystems under different climatic conditions and on different parent material. We chose beech ecosystems because this species is one of the most important forest trees of Europe, growing in a wide range of site conditions extending from humid to semiarid climates and from alkaline to acidic soils [19]. Beech forest soils contain an extensive carbon stock [20] that is predicted to decrease sharply under climate change scenarios [21]. The two studied beech stands are located on the Italian Apennines, one in the south (Laceno) and the other in the north (Pradaccio). They have been also investigated for litter decomposition dynamics under field and laboratory conditions [22], carbon stock in forest floor and mineral soil [5], and soil fauna communities.

In this study, we investigated the effects of substrate chemical composition and season on extracellular enzyme activities by means of multilevel models. Accordingly, past research demonstrated that enzymes involved in decomposition can respond to substrate availability and can be influenced by seasonality on short timescales, yet there is a need to understand how variation in chemistry vs. temporal is expressed, and how seasonal variation may control soil enzyme activity [23]. Thus, our research is timely and novel, aiming to better understand the role of seasonality versus underlying soil nutrient pools in controlling enzyme activity in short-term ecosystem dynamics [24] The advantage of multilevel models compared to classic statistical approaches is that the hierarchical structure in the data can be specified by considering both the within- and between-group variances, leading to a partial pooling of data across all levels in the hierarchy [25]. We used chemical variables as the fixed part in the model, season as random variation and layers (decomposition continuum for leaf litter and 0–5, 5–15, 15–30, and 30–40 cm for soil) as nested factors within the two forests. As for chemical variables, we measured organic matter, cellulose, acid unhydrolysable residue (AUR, i.e., a proximate content of lignin), and nutrients in leaf litter, while for soil we measured content of organic matter and nitrogen. The extracellular enzyme activities investigated in this study were involved in the main biogeochemical cycles, namely C (laccase, peroxidase, cellulase), N (chitinase), and P (acid phosphomonoesterase), whereas dehydrogenase was chosen as an enzyme proxy of overall biological activity [26]. Given that different factors may affect litter and soil differently, our study aims to evaluate, at high-view level, the differences in controls on microbial enzyme activity between leaf litter and soil.

2. Materials and Methods

2.1. Site Description

The study was conducted in Italy on the Apennines. The northern forest (Pradaccio, N Forest) is located at 1350 m a.s.l. (44°24′ N; 10°01′ E) while the southern forest (Laceno, S Forest) lies at an elevation of 1150 m a.s.l. (40°47′ N; 15°05′ E). Comprehensive tables about forest features and soil characteristics can be found in Innangi et al. (2015) [22] and De Marco et al. (2016) [5]. Both forests have a sub-Mediterranean mountain climate, characterized by temperate and relatively dry summer. However, during the year, temperature is lower and precipitation more abundant in N forest than in S forest. N Forest has a mean temperature of 6.0 °C, averaging on 11.0 °C between April and September and 0.7 °C in the rest of the year. Precipitations have peaks during autumn and spring and a marked reduction between June and August with an overall average rainfall of 2900 mm per year. Snow cover is often very thick and lasting from late November to the beginning of May. Average temperature in S Forest is 13.7 °C between April and September, and 3.7 °C in the rest of the year with an overall mean temperature of 8.7 °C. There was a longer semiarid period (May to September) than in N Forest and an overall average rainfall of 2300 mm per year, lower than in N Forest. Snow cover does not persist long and was commonly present between December and February.

In Figure 1 temperature and precipitation trends in the two forests during the year are shown. Data, provided by Lagdei Metereological Station and Laceno Metereological Station for N and S forests, respectively, are means of 5 years of monitoring (2010–2014, including our observations).

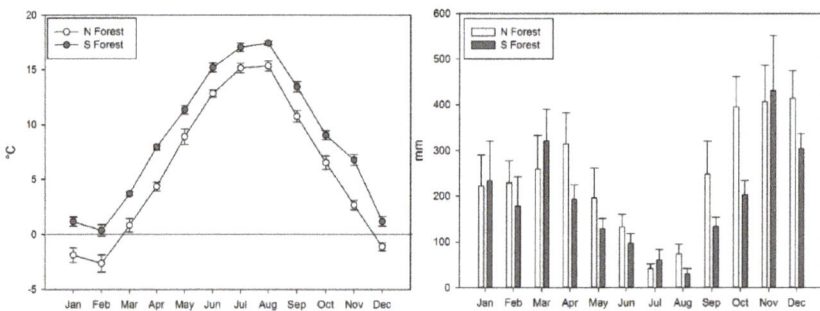

Figure 1. Temperature and precipitation regimes in the two studied forests. Data were collected for five years (2010–2014) at local meteorological stations. Values represent mean ± standard error of the mean.

Pedological surveys done in the past years in the nearby areas acknowledged the soil from Pradaccio as Lithic Haploborolls according to USDA Soil Taxonomy with a loamy-sand to sandy-clay-loam texture [27] while Laceno has been classified as Humic Haplustands according to USDA Soil Taxonomy with a loamy-sand to sandy-loam texture [28].

2.2. Litter and Soil Sampling

In each forest, 6 sampling points were randomly chosen in a 1 ha area. In each of them, litter was sampled by a 20 × 20 cm square steel frame and soil by a steel core sampler with a diameter of 5 cm and a length of 40 cm. Samplings were carried out in October 2011, May and July 2012, henceforward labelled as Autumn, Spring and Summer. Sampling points were marked in the field in order to repeat the sampling in the same spots and avoid introducing biases in the analyses. The samples, enclosed in plastic bags, were kept at 4 °C during transfer to the laboratory. Here, litter and soil were immediately processed.

Following a removal of all non-leaf litter material, leaf litter, which represented the most abundant litter component, was divided into three layers according to the degree of fragmentation/fermentation that can be defined as a decomposition continuum. The discrimination criteria were:

Li: leaf litter where leaves did not show any evident sign of fragmentation and/or chemical alteration. Such layer included mostly newly shed leaves, but also older leaves that have not yet been visibly altered. Leaves in this layer were easily separated from each other;

Lf: leaf litter with evident signs of fragmentation and/or chemical alteration but where the original structure of the leaf itself is still recognizable. Leaves in this layer can be separated from each other; and

Lhf: leaf litter with strong signs of fragmentation and/or chemical alteration, where the original shape of the leaves is hardly recognizable. Leaves in this layer are usually impossible to separate from each other and are sealed together by fungal mycelia.

Subsequently, subsamples for each layer were oven dried at 75 °C for 48 h in triplicate to assess moisture content. Other aliquots were kept at −80 °C for biological analyses and others were air dried and finely ground for chemical analyses.

Soil cores were carefully cut into four layers of 0–5, 5–15, 15–30, and 30–40 cm depth. Subsequently, soil cores were weighed and sieved with a mesh size of 2 mm. Subsamples were oven dried at 105 °C for 48 h in triplicate to assess moisture content. Other aliquots were kept at −80 °C for enzyme analyses and others were air dried and finely ground for chemical analyses.

2.3. Chemical Analyses

For chemical analyses, dry samples were ground by a Fritsch Pulverisette (type 00.502, Oberstein, Germany) equipped with an agate mortar and ball mill. Organic matter was evaluated gravimetrically as the difference between the dry weight and the ashes after incineration in a muffle furnace. Samples of leaf litter were burnt at 550 °C for 4 h into oven dried porcelain capsules [29], while those of mineral soil were burnt at 375 °C for 16 h [30].

Carbon and nitrogen were measured on finely ground dry samples [31] by a CNS elemental analyser (Vario El III, Elementar Analysensysteme GmbH, Hanau, Germany). Approximately 5 and 10 mg were weighted for litter and soil samples, respectively, using a holm oak litter as standards (C = 49.81%, N = 1.86%). As for soil samples, carbonates were removed before analysis [32].

To determine nutrient contents (Ca, P, K, Mg, Mn, and Fe), litter/soil samples were dry-mineralized in a muffle furnace at 480 °C for 16 h. Ashes were rehydrated with 1 mL 1:3 HNO_3 (70%) and 9 mL double-distilled water, filtered and then analytical determinations were performed by inductively coupled plasma atomic emission spectrometry (ICP-AES) without ultrasonic nebulization [33].

Cellulose and acid unhydrolizable residue (AUR) were evaluated according to [34] with modifications [29]. Such method allows for a semi-quantitative determination of AUR, including both native lignin and acid-insoluble lignin-like substances that are formed during decomposition, namely humic substances produced by soil microorganisms [29]. For simplicity, we will refer to this complex of lignin and lignin-like substances as lignin.

2.4. Enzyme Activities

For leaf litter, an extraction of the enzymes in appropriate buffers was performed. The extracts for cellulase and chitinase activities were prepared by suspending 1 g of leaf litter in 10 mL of 0.05 M sodium acetate buffer pH 5.5. For laccase and peroxidase, 0.5 g of fresh litter were suspended in 10 mL of sodium acetate buffer 0.05 M pH 5.0. Extracts for acid phosphomonoesterase were prepared by suspending 0.5 g of fresh litter in 10 mL of modified universal buffer (MUB) pH 6.5. Samples were homogenized using a homogenizer Polytron Heidolph Diax 600 for about 20–30 s while kept in ice to avoid heating. The homogenized samples were then centrifuged for 20 min at 22,000× *g* at 4 °C, the supernatant recovered and used as enzyme extract. For dehydrogenase, no extraction was undertaken, but the colorimetric assay was performed directly on 50 mg homogenized litter/soil in 0.750 mL tris(hydroxymethyl) aminomethane buffer (1 M pH 7.0). Similarly, all enzymatic assays on soil samples were carried out directly on homogenized soil without separating the supernatant.

Carboxymethyl-cellulase (cellulase, EC 3.2.1.4) activity was determined according to Schinner & Von Mersi (1990) [35] with minor modifications [36] using carboxymethyl-cellulose as substrate. *N*-acetylglucosaminidase (henceforward labelled as chitinase, EC 3.2.1.14) and dehydrogenase (EC 1.1.1.x) activities were measured according to Verchot & Borelli (2005) [37] and Von Mersi & Schinner (1991) [38], respectively using 4-Nitrophenyl *N*-acetyl-β-D-glucosaminide and iodonitrotetrazolium chloride as substrates, respectively. Laccase (EC 1.10.3.2) and peroxidase (EC 1.11.1.x) were determined according to Leatham & Stahmann (1981) [39] with modifications [40], while acid (EC 3.1.3.2) phosphomonoesterase was performed according to Eivazi & Tabatabai (1977) [41] (substrates were *o*-Tolidine for laccase and peroxidase and 4-nitrophenyl phosphate bis(cyclohexylammonium) salt for acid phosphomonoesterase). All activities were measured in triplicate on each sample and data expressed as μmol/g dry weight/h. All measurements were done using a Varian Cary 1E UV/Vis spectrophotometer (Santa Clara, CA, USA).

2.5. Statistics

In order to test for variables affecting the variance in each enzyme in leaf litter and soil, we fitted multilevel models (MM), which are also known as random coefficient models and hierarchical linear models [25]. We used chemical data as fixed continuous variables, location and layer (nested within location) as fixed categorical variables and seasons as a random factor. We used grand mean centering for all continuous predictors [42] and, in order to avoid multicollinearity, we tested variance inflation factor (VIF) on each continuous variable, recursively discarding all with VIF ≥ 3 [21]. The procedure kept as chemical variables lignin, Fe, and K in leaf litter dataset and O.M. in soil. In order to ensure normality and homoscedasticity, we applied Box-Cox transformation with optimal λ to each enzyme activity. The transformation applied can be seen in Table S1. Residuals were checked for normality observing normal probability plots of studentized residuals and by fitting generalized additive models (GAMs) to evaluate the presence of non-linear patterns [43]. Multilevel model goodness of fit was expressed as conditional and marginal determination coefficients (R^2_c and R^2_m) in order to quantify unbiased measurements of variance expressed by fixed and fixed + random factors, respectively. All statistical analyses were done in R (v. 3.4.0) using packages 'lme4' and 'sjPlot'.

3. Results

3.1. Leaf Litter

Leaf litter chemical variables can be seen in Table 1. Generally, N Forest was characterized by higher values of O.M. in the leaf litter while S Forest was richer in Ca. Minor differences between forests included Mg, P, K, and Fe, which were higher in S Forest, while Mn and N had higher concentrations in N Forest. With the exception of Ca, all nutrients showed strong differences according to factor Layer, being generally at higher concentration in Lhf compared to Li. During autumn and spring there were higher values of cellulose and C:N, while all other chemical variables, including lignin, tended to increase during summer in both forests.

The trends for all enzyme activities in leaf litter can be found in Figure 2, while results from the MM in leaf litter can be seen in Table 2. Plots for the conditional models of the random effects in leaf litter, including prediction intervals, can be seen in Figure 3. Conditional determination coefficient for laccase was low ($R^2_c = 0.192$), while marginal determination coefficient was higher ($R^2_m = 0.478$), showing that there was a noticeable effect of the random parts in the model. Within the fixed parts, only lignin proved to be significant ($p < 0.050$). Laccase activity was not significantly different between forests, but Lf layer was significantly different from Li in both locations. The random part of the model, expressed by seasonal differences, clearly showed an increase in activity from autumn to summer.

Table 1. Chemical variables in leaf litter. Organic Matter (O.M.) is measured as g × g dry weight^{-1}. Cellulose and lignin are measured as mg × g O.M.$^{-1}$. All macro and micro elements are measured as mg × g dry weight^{-1}. Values are represented as mean ± standard error of the mean.

Season	Location	Layer	O.M.	Cellulose	Lignin	N	Ca	P	K	Mg	Mn	Fe	C:N
Autumn	N Forest	Li	0.95 ± 0.004	526 ± 8	276 ± 9	13.21 ± 0.45	13.15 ± 0.16	0.41 ± 0.01	0.62 ± 0.001	0.85 ± 0.01	0.07 ± 0.001	0.10 ± 0.000	39.2 ± 1.1
		Lf	0.94 ± 0.001	516 ± 12	277 ± 10	16.52 ± 0.71	13.24 ± 0.29	0.45 ± 0.007	0.63 ± 0.001	0.87 ± 0.01	0.088 ± 0.004	0.10 ± 0.000	27.96 ± 1.4
		Lhf	0.91 ± 0.007	482 ± 13	320 ± 5	22.28 ± 0.16	12.61 ± 0.12	0.47 ± 0.004	0.63 ± 0.002	0.91 ± 0.02	0.101 ± 0.003	0.11 ± 0.000	22.0 ± 0.3
	S Forest	Li	0.89 ± 0.005	537 ± 12	275 ± 11	14.06 ± 0.66	16.03 ± 0.13	0.49 ± 0.005	0.63 ± 0.001	0.96 ± 0.03	0.06 ± 0.002	0.10 ± 0.001	35.0 ± 1.2
		Lf	0.89 ± 0.008	510 ± 7	279 ± 10	13.07 ± 0.54	15.84 ± 0.12	0.49 ± 0.005	0.63 ± 0.002	0.94 ± 0.02	0.059 ± 0.002	0.10 ± 0.000	37.1 ± 1.4
		Lhf	0.77 ± 0.017	466 ± 10	313 ± 12	17.69 ± 0.73	15.63 ± 0.32	0.56 ± 0.008	0.64 ± 0.004	1.04 ± 0.004	0.081 ± 0.003	0.12 ± 0.000	22.4 ± 0.45
Spring	N Forest	Li	0.93 ± 0.008	498 ± 11	306 ± 8	18.53 ± 0.32	13.02 ± 0.15	0.44 ± 0.005	0.62 ± 0.001	0.81 ± 0.006	0.074 ± 0.001	0.10 ± 0.000	28.1 ± 0.36
		Lf	0.92 ± 0.010	505 ± 13	312 ± 11	21.35 ± 0.62	12.53 ± 0.23	0.45 ± 0.003	0.62 ± 0.001	0.85 ± 0.01	0.09 ± 0.004	0.10 ± 0.000	23.9 ± 0.8
		Lhf	0.87 ± 0.021	492 ± 13	300 ± 7	19.75 ± 0.94	12.32 ± 0.23	0.46 ± 0.007	0.63 ± 0.003	0.95 ± 0.04	0.098 ± 0.005	0.11 ± 0.000	22.3 ± 0.4
	S Forest	Li	0.89 ± 0.007	503 ± 6	292 ± 2	12.83 ± 0.65	15.79 ± 0.32	0.47 ± 0.006	0.62 ± 0.001	0.86 ± 0.01	0.05 ± 0.003	0.10 ± 0.000	39.4 ± 2.5
		Lf	0.85 ± 0.012	479 ± 18	313 ± 9	15.37 ± 0.67	16.27 ± 0.29	0.51 ± 0.007	0.62 ± 0.001	0.93 ± 0.03	0.06 ± 0.002	0.11 ± 0.000	31.3 ± 1.8
		Lhf	0.75 ± 0.016	450 ± 13	338 ± 4	17.52 ± 0.61	16.38 ± 0.39	0.54 ± 0.013	0.63 ± 0.002	1.00 ± 0.02	0.067 ± 0.003	0.12 ± 0.000	24 ± 0.7
Summer	N Forest	Li	0.954 ± 0.002	426 ± 23	353 ± 20	28.87 ± 1.69	13.87 ± 0.23	0.57 ± 0.011	0.64 ± 0.027	0.92 ± 0.042	0.122 ± 0.007	0.13 ± 0.030	17.6 ± 1.13
		Lf	0.940 ± 0.003	409 ± 20	383 ± 32	37.32 ± 2.43	13.67 ± 0.18	0.60 ± 0.023	0.64 ± 0.022	0.99 ± 0.04	0.150 ± 0.019	0.26 ± 0.070	13.37 ± 0.69
		Lhf	0.906 ± 0.015	398 ± 15	429 ± 38	43.59 ± 2.67	12.55 ± 0.64	0.68 ± 0.017	0.71 ± 0.027	1.14 ± 0.06	0.0163 ± 0.015	0.50 ± 0.150	10.72 ± 0.46
	S Forest	Li	0.932 ± 0.002	426 ± 8	341 ± 25	28.14 ± 1.48	16.33 ± 0.44	0.67 ± 0.014	0.67 ± 0.005	0.95 ± 0.016	0.086 ± 0.006	0.18 ± 0.03	17.7 ± 0.9
		Lf	0.916 ± 0.003	379 ± 11	338 ± 13	31.98 ± 1.12	17.74 ± 0.68	0.75 ± 0.030	0.69 ± 0.015	1.06 ± 0.027	0.111 ± 0.006	0.24 ± 0.04	15.1 ± 0.49
		Lhf	0.830 ± 0.020	387 ± 11	387 ± 25	36.69 ± 0.89	17.11 ± 0.35	0.78 ± 0.014	0.704 ± 0.030	1.28 ± 0.05	0.128 ± 0.003	0.62 ± 0.15	12 ± 0.19

Fixed parts of the model had a weaker effect for peroxidase ($R^2_c = 0.158$), with no chemical variable being significant. There was no significant difference between the two forests, and little to no effect of decomposition layer. Like laccase, random parts improved the model ($R^2_m = 0.283$) with similar activities in spring and summer and lower values in autumn.

Chitinase showed a low effect of the fixed part of the model ($R^2_c = 0.160$). No chemical variable proved to be significant. There was a significant difference between N Forest and S Forest ($p < 0.050$), with the latter expressing lower activity than N Forest. No particular effect of layer was detected. Marginal determination coefficient was sensibly higher ($R^2_m = 0.673$) with a strong seasonal effect ($ICC_{season} = 0.611$). Yet, compared to laccase and peroxidase, the seasonal trend was opposite, with highest activity in autumn and lowest in summer.

As for dehydrogenase, conditional determination coefficient was low ($R^2_c = 0.160$) and no chemical variable was significant. The activity was not different between forests, but in N Forest there was a strong difference between layers compared to S Forest. There was a trend to increase activity from autumn to summer, but this effect was relatively low ($ICC_{season} = 0.284$).

The activity of cellulase also showed a weak effect of the fixed parts of the model ($R^2_c = 0.149$). Accordingly, no chemical variable proved to be significant in the MM. Yet, differences between locations were sharp ($p < 0.001$), with N Forest showing higher activity than S Forest. Little effect of layer could be detected, and only in N Forest. The random part of the model increased the overall explained variance ($R^2_m = 0.566$), with a clear pattern of increasing activity from summer to autumn.

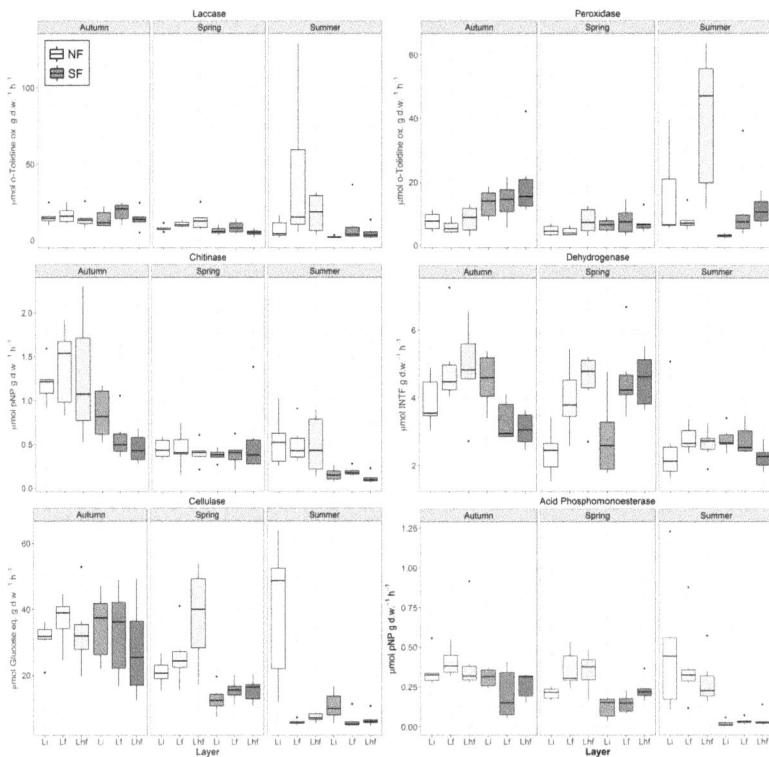

Figure 2. Boxplot representation of extracellular enzyme activities in leaf litter subdivided by season.

Table 2. Results from the MM analysis of Leaf Litter. Dependent variables are reported in columns, while predictors (both fixed and random parts) are reported in rows. Dependent variables have been transformed with Box-Cox transformation as reported in Materials and Methods. Reference levels for the fixed categorical variables have been chosen as N Forest for Location and Li for Layer, respectively. Statistical significance is reported as * ($p < 0.050$), ** ($p < 0.010$), *** ($p < 0.001$).

	Laccase	Peroxidase	Chitinase	Dehydrogenase	Cellulase	Acid Phosph.
	Estimate (CI)	Estimate (CI)	Estimate (CI)	Estimate (CI)	Estimate (CI)	Estimate (CI)
Fixed Parts						
(Intercept)	0.8 (0.76 – 0.84) ***	0.47 (0.40 – 0.54) ***	0.95 (0.89 – 1.02) ***	0.82 (0.78 – 0.86) ***	1.93 (1.72 – 2.14) ***	0.63 (0.49 – 0.76) ***
Lignin	−0.27 (−0.50 – −0.04) *	−0.07 (−0.60 – 0.46)	0.16 (−0.07 – 0.39)	0.15 (−0.07 – 0.37)	−0.83 (−1.72 – 0.06)	0.86 (0.21 – 1.51) *
Fe	0.03 (−0.03 – 0.10)	0.00 (−0.15 – 0.15)	−0.06 (−0.13 – 0.00)	0.03 (−0.03 – 0.09)	−0.17 (−0.42 – 0.07)	−0.24 (−0.42 – −0.06) *
K	0.13 (−0.18 – 0.43)	−0.47 (−1.17 – 0.22)	0.08 (−0.22 – 0.38)	−0.08 (−0.38 – 0.21)	0.50 (−0.66 – 1.66)	0.35 (−0.50 – 1.20)
Location S Forest	0.02 (−0.01 – 0.06)	0.03 (−0.04 – 0.10)	−0.05 (−0.08 – −0.02) *	−0.02 (−0.05 – 0.01)	−0.20 (−0.32 – −0.09) ***	−0.21 (−0.29 – −0.12) ***
Location N Forest: Layer Lf	−0.04 (−0.07 – −0.01) *	0.03 (−0.03 – 0.10)	0.00 (−0.03 – 0.03)	−0.05 (−0.08 – −0.02) **	−0.14 (−0.25 – −0.02) *	0.04 (−0.04 – 0.13)
Location S Forest: Layer Lf	−0.03 (−0.06 – −0.00) *	−0.07 (−0.14 – 0.00)	−0.01 (−0.03 – 0.02)	−0.01 (−0.04 – 0.02)	−0.03 (−0.15 – 0.08)	−0.01 (−0.10 – 0.07)
Location N Forest: Layer Lhf	−0.03 (−0.06 – 0.01)	−0.06 (−0.13 – 0.02)	−0.01 (−0.04 – 0.02)	−0.06 (−0.09 – −0.03) ***	−0.07 (−0.19 – 0.06)	0.00 (−0.08 – 0.09)
Location S Forest: Layer Lhf	0.00 (−0.03 – 0.03)	−0.09 (−0.16 – −0.02) *	−0.02 (−0.05 – 0.01)	−0.01 (−0.04 – 0.03)	−0.01 (−0.13 – 0.11)	0.05 (−0.04 – 0.14)
Random Parts						
σ^2	0.002	0.011	0.002	0.002	0.031	0.016
$\tau_{00,\text{Season}}$	0.001	0.002	0.003	0.001	0.029	0.011
$\text{ICC}_{\text{Season}}$	0.354	0.148	0.611	0.284	0.490	0.406
R^2_c / R^2_m	0.192/0.478	0.158/0.283	0.160/0.673	0.126/0.374	0.149/0.566	0.349/0.613

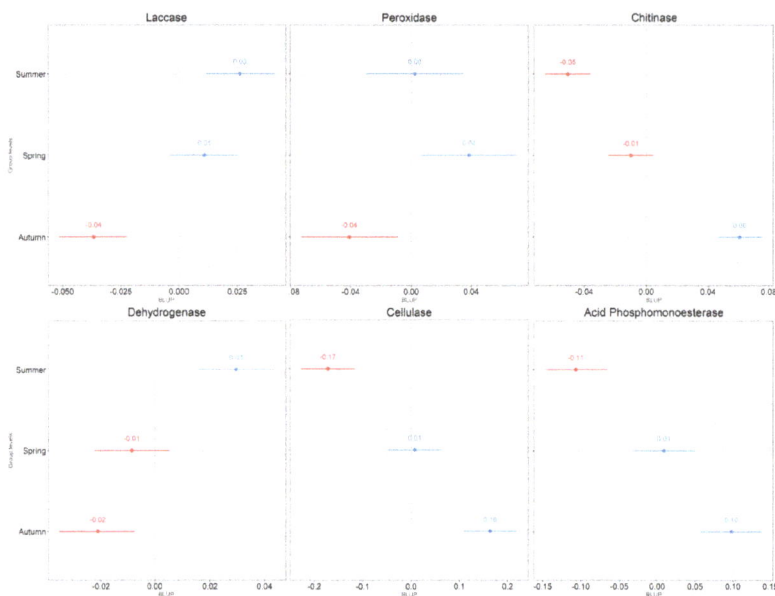

Figure 3. Plots for the conditional models of the random effects in leaf litter, including prediction intervals. Colors indicate a negative value for the conditional model (red) or a positive value (blue). BLUP stands for Best Linear Unbiased Predictor.

Finally, acid phosphomonoesterase showed the highest effect of the fixed parts of the model ($R^2_c = 0.349$). Both lignin and Fe were significant ($p < 0.050$), although with opposite signs. Like cellulase, there was a strong difference between locations ($p < 0.001$) but no relevant layer effect. The seasonal trend was similar to chitinase, although weaker ($ICC_{season} = 0.406$).

3.2. Soil

The chemical variables for soil can be seen in Table 3. The proportion of organic matter, with the exception of topsoil (0–5 cm), was generally higher in S Forest compared to N Forest. S Forest showed a slower decrease of O.M. with depth, whereas in N Forest there was a sharp difference between the topsoil (0–5 cm) and the subsoil (5–40 cm). No particular seasonal effect could be detected in either site. Nitrogen was also different between forests, although there were differences driven by seasons. Topsoil (0–5 cm) of N Forest had higher N content, whereas S Forest showed greater concentrations in the deeper layers of the soil. In contrast to O.M., a general tendency to increase in N was detected from autumn to summer, especially in N Forest. C:N ratio was different between sites, being generally narrower in S Forest. There was an overall similarity of C:N between layers, whereas a sharp seasonal effect was noticed. Although autumn and spring values were largely overlapping, C:N ratio was noticeably lower in summer in both sites.

Results from the MM analysis of the soil can be seen in Table 4, while the trends for enzymes in soil can be seen in Figure 4. Plots for the conditional models of the random effects in soil, including prediction intervals, can be seen in Figure 5. Laccase showed little effect of the random part of the model given that marginal and conditional determination coefficients were similar ($R^2_c/R^2_m = 0.488/544$). Compared to litter, a large amount of variance was derived from the fixed part of the model. In detail, O.M. was significant ($p < 0.050$), whereas there were no differences between forests but a significant effect of layer. Although seasonal effect was not particularly strong, summer was markedly different from both spring and autumn (Figure 5).

Table 3. Chemical variables in studied soil. Organic Matter (O.M.) is measured as $g \times g$ dry weight^{-1}. Total nitrogen (N) is measured as $mg \times g$ dry weight^{-1}. Values are represented as mean ± standard error of the mean.

Season	Autumn							
Location	N Forest				S Forest			
Layer	0–5 cm	5–15 cm	15–30 cm	30–40 cm	5–15 cm	15–30 cm	30–40 cm	0–5 cm
O.M.	0.31 ± 0.06	0.11 ± 0.01	0.10 ± 0.01	0.08 ± 0.01	0.20 ± 0.02	0.16 ± 0.01	0.15 ± 0.01	0.27 ± 0.02
N	6.64 ± 1.01	2.29 ± 0.14	1.42 ± 0.14	1.30 ± 0.10	6.39 ± 0.57	3.97 ± 0.52	3.88 ± 0.15	8.58 ± 0.88
C:N	17.05 ± 0.69	19.37 ± 1.86	21.21 ± 0.44	21.94 ± 0.32	12.27 ± 0.16	12.82 ± 0.27	13.2 ± 0.27	13.11 ± 0.36

Season	Spring							
Location	N Forest				S Forest			
Layer	0–5 cm	5–15 cm	15–30 cm	30–40 cm	5–15 cm	15–30 cm	30–40 cm	0–5 cm
O.M.	0.42 ± 0.08	0.10 ± 0.02	0.09 ± 0.01	0.08 ± 0.01	0.21 ± 0.01	0.16 ± 0.01	0.18 ± 0.00	0.32 ± 0.03
N	10.14 ± 2.06	2.44 ± 0.51	1.53 ± 0.02	1.35 ± 0.12	5.25 ± 0.45	3.8 ± 0.29	4.14 ± 0.35	7.81 ± 0.86
C:N	14.01 ± 0.68	20.82 ± 2.57	21.99 ± 0.42	20.20 ± 0.32	13.24 ± 0.42	12.94 ± 0.19	13.15 ± 0.37	13.20 ± 0.42

Season	Summer							
Location	N Forest				S Forest			
Layer	0–5 cm	5–15 cm	15–30 cm	30–40 cm	5–15 cm	15–30 cm	30–40 cm	0–5 cm
O.M.	0.42 ± 0.1	0.12 ± 0.02	0.10 ± 0.01	0.08 ± 0.01	0.20 ± 0.01	0.20 ± 0.01	0.21 ± 0.01	0.25 ± 0.02
N	21.03 ± 4.21	5.19 ± 0.95	4.95 ± 0.69	4.77 ± 1.11	5.25 ± 0.45	3.8 ± 0.29	4.14 ± 0.35	7.81 ± 0.86
C:N	10.68 ± 0.24	9.68 ± 0.73	6.32 ± 1.39	6.93 ± 1.63	6.39 ± 0.11	6.6 ± 0.21	6.51 ± 0.14	6.99 ± 0.28

Table 4. Results from the MM analysis of studied soil. Dependent variables are reported in columns, while predictors (both fixed and random parts) are reported in rows. Dependent variables have been transformed with Box-Cox transformation as reported in Materials and Methods. Reference levels for the fixed categorical variables have been chosen as N Forest for Location and 0–5 cm for Layer, respectively. Statistical significance is reported as * ($p < 0.050$), ** ($p < 0.010$), *** ($p < 0.001$).

	Laccase	Peroxidase	Chitinase	Dehydrogenase	Cellulase	Acid Phosph.
	Estimate (CI)	Estimate (CI)	Estimate (CI)	Estimate (CI)	Estimate (CI)	Estimate (CI)
Fixed Parts						
(Intercept)	0.69 (0.50 – 0.88) ***	0.16 (0.09 – 0.23) ***	0.95 (0.92 – 0.98) ***	0.81 (0.74 – 0.88) ***	2.37 (1.93 – 2.81) ***	0.62 (0.56 – 0.69) ***
OM	−0.74 (−1.36 – −0.12) *	−0.19 (−0.41 – 0.03)	−0.14 (−0.19 – −0.09) **	0.68 (0.43 – 0.94) ***	−1.07	0.44 (0.28 – 0.61) ***
Location S Forest	0.01 (−0.17 – 0.18)	0.18 (0.11 – 0.24) ***	0.00 (−0.01 – 0.02)	0.09 (0.01 – 0.16) *	−0.43 (−0.59 – −0.27) **	−0.04 (−0.08 – 0.01)
Location N Forest: Layer 5–15 cm	−0.07 (−0.30 – 0.17)	−0.02 (−0.11 – 0.06)	0.00 (−0.02 – 0.02)	−0.01 (−0.11 – 0.09)	0.16 (−0.05 – 0.37)	−0.08 (−0.15 – −0.02) *
Location S Forest: Layer 5–15 cm	0.23 (0.06 – 0.40) *	0.01 (−0.05 – 0.08)	0.01 (−0.00 – 0.03)	−0.1 (−0.17 – −0.03) **	0.15 (−0.01 – 0.30)	0.02 (−0.03 – 0.06)
Location N Forest: Layer 15–30 cm	0.23 (−0.02 – 0.47)	0.13 (0.04 – 0.22) **	0.00 (−0.02 – 0.02)	−0.05 (−0.15 – 0.05)	0.23 (0.01 – 0.44)	−0.06 (−0.12 – 0.01)
Location S Forest: Layer 15–30 cm	0.45 (0.27 – 0.63) ***	0.07 (0.01 – 0.14) *	0.03 (0.02 – 0.05) **	−0.17 (−0.25 – −0.10) ***	0.3 (0.14 – 0.46) *	0.03 (−0.01 – 0.08)
Location N Forest: Layer 30–40 cm	0.41 (0.16 – 0.66) **	0.21 (0.12 – 0.30) ***	0.01 (−0.01 – 0.03)	−0.18 (−0.29 – −0.08) ***	0.43 (0.21 – 0.66) *	−0.1 (−0.17 – −0.04) **
Location S Forest: Layer 30–40 cm	0.46 (0.28 – 0.64) ***	0.06 (−0.00 – 0.13)	0.03 (0.02 – 0.05) **	−0.28 (−0.35 – −0.20) ***	0.38 (0.21 – 0.54) **	0.03 (−0.02 – 0.08)
Random Parts						
σ^2	0.058	0.007	0.000	0.010	0.046	0.004
$\tau_{00,Season}$	0.007	0.001	0.001	0.000	0.133	0.002
ICC_{Season}	0.109	0.135	0.561	0.000	0.742	0.280
R^2_c / R^2_m	0.488 / 0.544	0.569 / 0.627	0.359 / 0.718	0.678 / 0.678	0.397 / 0.844	0.515 / 0.651

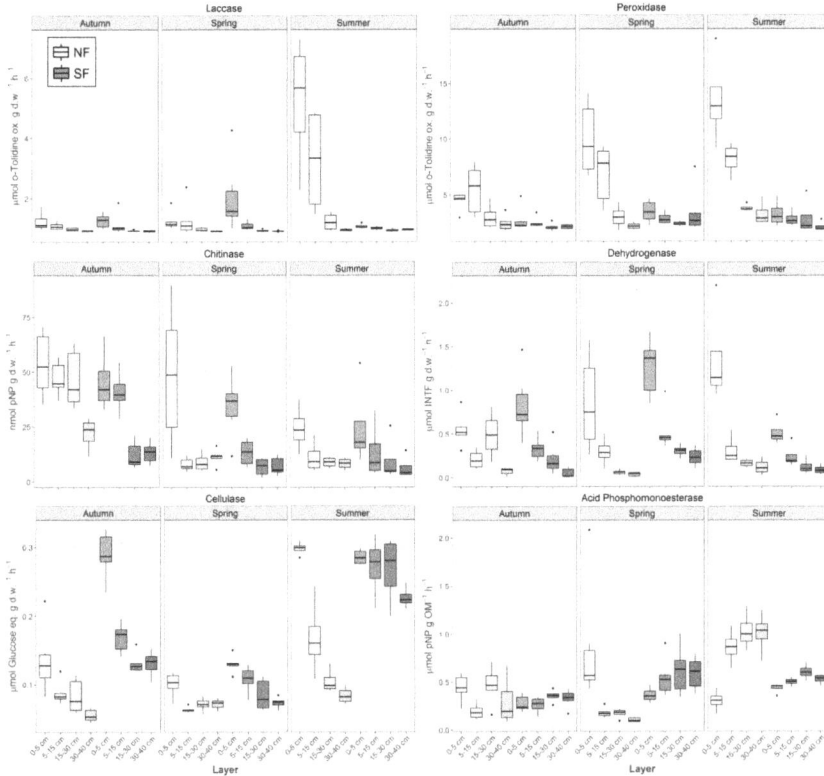

Figure 4. Boxplot representation of extracellular enzyme activities in studied soil subdivided by season.

Fixed parts of the model in peroxidase explained higher amount of variance compared to laccase ($R^2_c = 0.569$), but O.M. was not significant. In contrast to laccase, locations had highly significant differences for peroxidase ($p < 0.001$), with higher activity in N Forest. Differences between layers were sharper in N Forest compared to S Forest. Although the random part moderately contributed to variance ($R^2_m = 0.627$), a weak trend of decreasing activity in autumn was detected when compared to spring and summer.

In contrast with previous enzymes, chitinase had a large difference in conditional and marginal determination coefficients. The proportion of O.M. was highly significant ($p < 0.010$), although fixed parts of the model explained not much variance ($R^2_c = 0.359$). There was no difference between forests and little differences between layers, except for the deeper part of the soil in S Forest. The random part of the model expressed a large amount of variance ($R^2_m = 0.718$) with a clear seasonal effect ($ICC_{season} = 0.561$) with sharp differences in autumn compared to spring and summer.

Dehydrogenase had identical determination coefficients ($R^2_{c/m} = 0.678$), meaning that there was no random effect at all. The proportion of O.M. was highly significant ($p < 0.001$). There was a weak, although significant, difference between forests ($p < 0.050$), with higher activity in S Forest, where also significant differences between layers were found to be more conspicuous than N Forest. As already mentioned, there was no detectable seasonal effect for this enzyme.

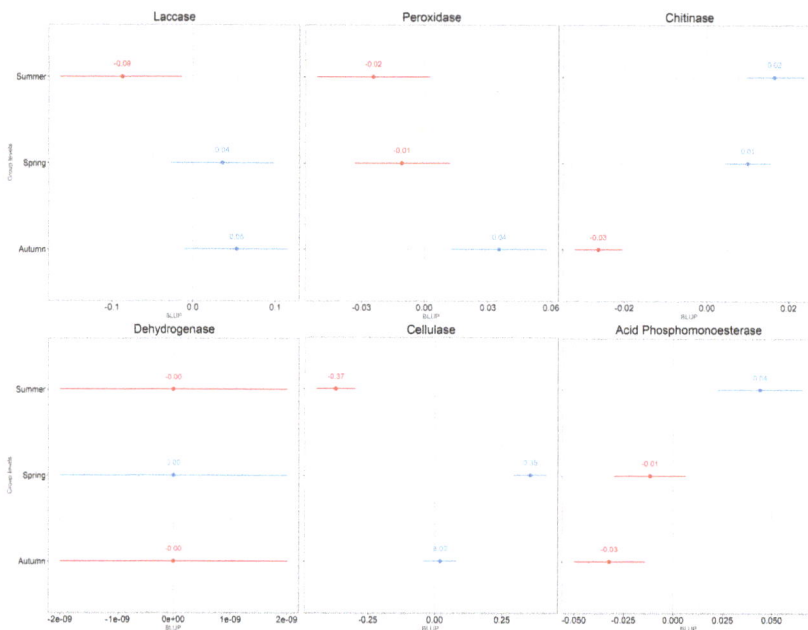

Figure 5. Plots for the conditional models of the random effects in studied soil, including prediction intervals. Colors indicate a negative value for the conditional model (red) or a positive value (blue). BLUP stands for Best Linear Unbiased Predictor.

Like chitinase, a large difference in conditional and marginal determination coefficients was found in cellulase. Fixed parts of the model explained half of the variance ($R^2_c = 0.397$) when compared to random effects ($R^2_m = 0.844$). Again, O.M. proved to be significant ($p < 0.050$) and there was a large difference between forests ($p < 0.010$), with greater activity in S Forest. The deeper layers of the soil, especially in S Forest, showed significant differences. The seasonal effect was broad ($ICC_{season} = 0.742$) with a complex pattern. Although autumn was relatively more similar to summer, the latter season and spring showed sharp differences.

Finally, acid phosphomonoesterase, like laccase and peroxidase, exhibited small differences between fixed and random parts of the model. The proportion of O.M. was highly significant ($p < 0.001$), but no difference was detected between locations. Layers had little effect, being the activity significantly different only in the deepest layer of N Forest. A weak trend of increasing activity from autumn to summer was detected, although seasonal effect was small ($ICC_{season} = 0.280$).

4. Discussion

4.1. Leaf Litter

The trend of organic matter along the decomposition continuum was consistent with the previously observed trend of litter decomposition in the two forests [22] and with the forest floor features in autumn and spring [5]. At the same time, a change in quality was observed as well. Accordingly, there was a decrease of cellulose and an increase of lignin and N (Table 1). Similar trends in the content of cellulose and lignin, mostly for the litter or organic layers, are known from other studies [29,44–47]. The cellulose-rich secondary wall must be at least partially degraded before fungi can attack the lignin. This may explain the increase of ligninolytic enzyme activity with litter decay stage and the significant effect of lignin in MM for laccase (Figure 2 and Table 2). Rihani et al.

(1995) [48] found that, in beech litter, lignin decay rates increased significantly after more than 20% of the cellulose content had disappeared. The small decline of cellulose concentration and the increase of lignin along the decomposition continuum at both sites indicated that even in the most decayed litter, (Lhf) lignin degradation was still at the beginning.

Within the decomposition continuum, the increase of nitrogen suggests that this element was limiting for microbial growth in both forests. Similar trends for N have been already seen during litter decay [4], and similar dynamics in beech litter are known as well [49–51]. The higher N content in summer could be linked to a combined effect of soil fauna activity and higher presence of labile nitrogen compounds, which are subsequently leached away during autumn precipitation [52]. Accordingly, in prairie ecosystems, warming was linked to an increase in labile N, whereas recalcitrant N remained stable [53]. As a consequence of decomposer activities, the C:N ratio decreased along the decomposition continuum, especially in summer, reaching humus-like values in Lhf [54]. This suggested the achievement of a certain stabilization of the organic matter in this leaf litter layer. Our finding is consistent with Michel and Matzner (2002) [55], who hypothesized that low C:N ratios in later stages of decomposition stabilized soil organic matter and could be responsible for increased accumulation of C in forest floors. The sharp decrease of C:N ratio in summer compared to the other seasons could be linked to the subsequent N leaching during autumn. This statement is supported by soluble N data in autumn for the two studied forests, where a decrease from autumn to spring was witnessed [5].

Generally, enzyme activities vary with the qualitative composition of the substrate and are influenced by environmental conditions that act directly on the released enzymes and indirectly by affecting the organisms that produce them [56]. The statistical model applied to leaf litter coming from both forests, showed significant seasonal effects on the activity of all considered enzymes, although effects varied in magnitude between enzymes (Table 2). Activities, such as chitinase, cellulase and, to a lesser degree, acid phosphomonoesterase, were higher in autumn, likely due to higher moisture associated to mild temperature, compared to summer. Other studies showed strong seasonal effects on enzyme activities in Mediterranean ecosystems, with the lowest activity during summer's drought and the winter coldest periods [29,57–59]. Nevertheless, laccase and dehydrogenase showed an opposite trend although, compared to the previous ones, they showed a smaller seasonal effect. Laccase variability was largely due to variation in chemical composition, mostly lignin. The activity of peroxidase, on the contrary, was influenced neither by season nor chemistry, at least for the variables we have measured. Although only laccase showed a weak seasonal trend and there was no difference between locations, another study that used isoelectric focusing on high-stability pH gradients with high resolving power showed strong differences in both laccase and peroxidase isoenzymes, with higher activities of both enzymes in summer and greater diversity of isoenzymes in N Forest [60]. Finally, as for the comparison of the two forests, our model indicated a significant difference only for cellulase, chitinase, and acid phosphomonoesterase.

4.2. Soil

The concentration of organic matter in the soil was shown to be much higher in N Forest in the uppermost layer, but larger in S Forest in the deeper layers. The explanation could be derived to the higher annual litter input of S Forest, due to the milder climate that promotes a longer vegetative season [5]. This allows trees to grow larger leaves and in greater quantity. Thus, in spite of a higher rate of decomposition observed in southern site [22], a greater accumulation of organic matter was evident (Table 3). Quantitative and qualitative changes along soil depth are consistent with current literature [5,28,61–63]. The increased nitrogen content during summer that we have recorded was congruent with reported higher N mineralization and availability in labile-compounds during warmer seasons in temperate ecosystems [64–66].

Along soil depth, enzyme activity varied in relation to the decreasing content of organic matter (Figure 3). The statistical model applied to soil put forward some differences compared to leaf litter: (1) chitinase and, to a lesser degree, acid phosphomonoesterase variation was largely influenced by seasonality but with an opposite trend to leaf litter, being larger in autumn and smaller in summer; (2) laccase and peroxidase were barely influenced by season, but also had opposite trends compared to leaf litter; and (3) the content of organic matter was largely responsible for the variability of enzyme activities, with the exception of peroxidase, as already shown in other beech mountain ecosystems [61].

These findings were congruent with what was observed by Baldrian et al. (2013) [18], given that, along seasons, temperature, humidity and chemical quality have greater changes in litter than soil. The seasonal trend of most enzymes in soil is consistent with previous research in beech forests in Northern Europe, where the activity of chitinase and cellulase was higher during summer [66].

4.3. Conclusions

Admittedly, we cannot make insights beyond a certain resolution, as many factors are not included the modeling analysis and cannot account for the complexity of forest ecosystems. Nevertheless, our work showed that litter is relatively more controlled by seasonality while soil is relatively more controlled by soil properties. Our work contributed to understanding aboveground-belowground interactions and differences in enzyme activities in a short-time perspective, with particular focus on enzymes involved in nutrient cycles that is pivotal to further understand the factors that regulate, for instance, the transfer of plant C to soil and its importance for the microbial community [23]. In conclusion, the main results of our study can be summarized as: (i) seasonal changes, expressed as random variation in our MM, had a stronger effect than chemical variables in leaf litter, whereas chemical variation (organic matter) had a stronger impact on enzyme activities soil; (ii) lignin was an important variable in explaining the variance of laccase and acid phosphomonoesterase in leaf litter, while O.M. was the driving variable for all soil enzymes with the exception of peroxidase; and (iii) the effect of seasonality and chemistry was in general larger than the differences due to both forest origin and layers.

Supplementary Materials: The following are available online at http://www.mdpi.com/2079-4991/9/4/ 219/s1.

Acknowledgments: This research was funded by PRIN 2008, awarded to A. Fioretto (grant No. 2008NMFWYS-001), A. De Marco (grant No. 2008NMFWYS-002) and C. Menta (grant No. 2008NMFWYS-003). We are grateful to the staff of the Guadine Pradaccio Biogenetic Natural Reserve—Parco Nazionale dell'Appennino Tosco-Emiliano and of Regional Natural Reserve of Monti Picentini, for logistical support and technical assistance in the field work. We are thankful to Francesco d'Alessandro, Stefania Pinto and Maria Giordano, along with all the students who contributed to this research throughout the years. We also express appreciation to the two anonymous Reviewers who helped us in improving the quality of this manuscript.

Author Contributions: Antonietta Fioretto, Anna De Marco, and Cristina Menta conceived and designed the elements of this study. All Authors contributed to field samplings in both forests. Laboratory analyses were led by Michele Innangi, Anna De Marco, Cristina Menta, and Antonella Pellegrino. Michele Innangi developed the statistical analyses and created all of the tables and figures presented in this manuscript. The Manuscript was written mainly by Antonietta Fioretto, Michele Innangi, Anna De Marco, Cristina Menta, and Amalia Virzo De Santo, although all Authors read and approved the final version of the manuscript.

Conflicts of Interest: The authors declare no conflict of interest.

References

1. Food and Agriculture Organization (FAO). *Global Forest Resources Assessment 2010*; Food and Agriculture Organization of the United Nations: Rome, Italy, 2010.
2. Lal, R. Soil carbon sequestration to mitigate climate change. *Geoderma* **2004**, *123*, 1–22. [CrossRef]
3. Pan, Y.; Birdsey, R.A.; Fang, J.; Houghton, R.A.; Kauppi, P.E.; Kurz, W.A.; Phillips, O.L.; Shvidenko, A.; Lewis, S.L.; Canadell, J.G.; et al. A large and persistent carbon sink in the world's forests. *Science* **2011**, *333*, 988–993. [CrossRef] [PubMed]
4. Berg, B.; McClaugherty, C. *Plant Litter: Decomposition, Humus Formation, Carbon Sequestration*, 3rd ed.; Springer: Berlin/Heidelberg, Germany, 2014; ISBN 9783642388200.
5. De Marco, A.; Fioretto, A.; Giordano, M.; Innangi, M.; Menta, C.; Papa, S.; Virzo De Santo, A. C stocks in forest floor and mineral soil of two Mediterranean beech forests. *Forests* **2016**, *7*, 181. [CrossRef]

6. Rouifed, S.; Handa, I.T.; David, J.F.; Hättenschwiler, S. The importance of biotic factors in predicting global change effects on decomposition of temperate forest leaf litter. *Oecologia* **2010**, *163*, 247–256. [CrossRef] [PubMed]

7. García-Palacios, P.; Shaw, E.A.; Wall, D.H.; Hättenschwiler, S. Temporal dynamics of biotic and abiotic drivers of litter decomposition. *Ecol. Lett.* **2016**, *19*, 554–563. [CrossRef] [PubMed]

8. Menta, C.; García-Montero, L.G.; Pinto, S.; Conti, F.D.; Baroni, G.; Maresi, M. Does the natural "microcosm" created by *Tuber aestivum* affect soil microarthropods? A new hypothesis based on Collembola in truffle culture. *Appl. Soil Ecol.* **2014**, *84*, 31–37. [CrossRef]

9. Güsewell, S.; Gessner, M.O. N:P ratios influence litter decomposition and colonization by fungi and bacteria in microcosms. *Funct. Ecol.* **2009**, *23*, 211–219. [CrossRef]

10. Brandstätter, C.; Keiblinger, K.; Wanek, W.; Zechmeister-Boltenstern, S. A closeup study of early beech litter decomposition: Potential drivers and microbial interactions on a changing substrate. *Plant Soil* **2013**, *371*, 139–154. [CrossRef] [PubMed]

11. Cotrufo, M.F.; Miller, M.; Zeller, B. Litter decomposition. In *Carbon and Nitrogen Cycling in European Forest Ecosystems*; Springer: Berlin/Heidelberg, Germany, 2000; pp. 276–296.

12. Frankland, J.C. Fungal succession—Unravelling the unpredictable. *Mycol. Res.* **1998**, *102*, 1–15. [CrossRef]

13. Sinsabaugh, R.; Moorhead, D.; Linkins, A. The enzymic basis of plant litter decomposition: Emergence of an ecological process. *Appl. Soil Ecol.* **1994**, *1*, 97–111. [CrossRef]

14. Sinsabaugh, R.L.; Antibus, R.K.; Linkins, A.E. An enzymic approach to the analysis of microbial activity during plant litter decomposition. *Agric. Ecosyst. Environ.* **1991**, *34*, 43–54. [CrossRef]

15. Marx, M.-C.; Wood, M.; Jarvis, S. A microplate fluorimetric assay for the study of enzyme diversity in soils. *Soil Biol. Biochem.* **2001**, *33*, 1633–1640. [CrossRef]

16. Burns, R.G.; DeForest, J.L.; Marxsen, J.; Sinsabaugh, R.L.; Stromberger, M.E.; Wallenstein, M.D.; Weintraub, M.N.; Zoppini, A. Soil enzymes in a changing environment: Current knowledge and future directions. *Soil Biol. Biochem.* **2013**, *58*, 216–234. [CrossRef]

17. Kaiser, C.; Koranda, M.; Kitzler, B.; Fuchslueger, L.; Schnecker, J.; Schweiger, P.; Rasche, F.; Zechmeister-Boltenstern, S.; Sessitsch, A.; Richter, A. Belowground carbon allocation by trees drives seasonal patterns of extracellular enzyme activities by altering microbial community composition in a beech forest soil. *New Phytol.* **2010**, *187*, 843–858. [CrossRef] [PubMed]

18. Baldrian, P.; Šnajdr, J.; Merhautová, V.; Dobiášová, P.; Cajthaml, T.; Valášková, V. Responses of the extracellular enzyme activities in hardwood forest to soil temperature and seasonality and the potential effects of climate change. *Soil Biol. Biochem.* **2013**, *56*, 60–68. [CrossRef]

19. Ellenberg, H.; Leuschner, C. *Vegetation Mitteleuropas Mit Den Alpen in Ökologischer, Dynamischer und Historischer Sicht*; UTB: Stuttgart, Germany, 2010.

20. Gasparini, P.; Tabacchi, G. *L'Inventario Nazionale delle Foreste e dei Serbatoi Forestali di Carbonio INFC 2005*; Secondo Inventario Forestale Nazionale Italiano; Metodi e Risultati. Ministero delle Politiche Agricole, Alimentari e Forestali; Corpo Forestale dello Stato; Edagricole-Il sole 24 Ore: Bologna, Italy, 2011.

21. Innangi, M.; D'Alessandro, F.; Fioretto, A.; Di Febbraro, M. Modeling distribution of Mediterranean beech forests and soil carbon stock under climate change scenarios. *Clim. Res.* **2015**, *66*, 25–36. [CrossRef]

22. Innangi, M.; Schenk, M.K.; d'Alessandro, F.; Pinto, S.; Menta, C.; Papa, S.; Fioretto, A. Field and microcosms decomposition dynamics of European beech leaf litter: Influence of climate, plant material and soil with focus on N and Mn. *Appl. Soil Ecol.* **2015**, *93*, 88–97. [CrossRef]

23. Bardgett, R.D.; Bowman, W.D.; Kaufmann, R.; Schmidt, S.K. A temporal approach to linking aboveground and belowground ecology. *Trends Ecol. Evol.* **2005**, *20*, 634–641. [CrossRef] [PubMed]

24. Knelman, J.E.; Graham, E.B.; Ferrenberg, S.; Lecoeuvre, A.; Labrado, A.; Darcy, J.L.; Nemergut, D.R.; Schmidt, S.K. Rapid shifts in soil nutrients and decomposition enzyme activity in early succession following forest fire. *Forests* **2017**, *8*, 347. [CrossRef]

25. Qian, S.S.; Cuffney, T.F.; Alameddine, I.; McMahon, G.; Reckhow, K.H. On the application of multilevel modeling in environmental and ecological studies. *Ecology* **2010**, *91*, 355–361. [CrossRef] [PubMed]

26. Innangi, M.; Niro, E.; D'Ascoli, R.; Danise, T.; Proietti, P.; Nasini, L.; Regni, L.; Castaldi, S.; Fioretto, A. Effects of olive pomace amendment on soil enzyme activities. *Appl. Soil Ecol.* **2017**, *119*, 242–249. [CrossRef]

27. Leoni, A. *Studio delle Biodiversità Vegetale e del Popolamento a Microartropodi Edafici nella Riserva Naturale "Guadine-Pradaccio"*; Università degli Studi di Parma: Parma, Italy, 2008.

28. Curcio, E.; Danise, T.; Innangi, M.; Alvarez Romero, M.; Coppola, E.; Fioretto, A.; Papa, S. Soil characterization and comparison of organic matter quality and quantity of two stands under different vegetation cover on Monte Faito Soil characterization and comparison of organic matter quality and quantity of two stands under different. *Fresenius Environ. Bull.* **2017**, *26*, 8–18.

29. Fioretto, A.; Di Nardo, C.; Papa, S.; Fuggi, A. Lignin and cellulose degradation and nitrogen dynamics during decomposition of three leaf litter species in a Mediterranean ecosystem. *Soil Biol. Biochem.* **2005**, *37*, 1083–1091. [CrossRef]

30. Pribyl, D.W. A critical review of the conventional SOC to SOM conversion factor. *Geoderma* **2010**, *156*, 75–83. [CrossRef]

31. Song, B.; Niu, S.; Zhang, Z.; Yang, H.; Li, L.; Wan, S. Light and heavy fractions of soil organic matter in response to climate warming and increased precipitation in a temperate steppe. *PLoS ONE* **2012**, *7*, e33217. [CrossRef] [PubMed]

32. Harris, D.; Horwàth, W.R.; Van Kessel, C. Acid fumigation of soils to remove carbonates prior to total organic carbon. *Soil Sci. Soc. Am. J.* **2001**, *65*, 1853–1856. [CrossRef]

33. Bussotti, F.; Prancrazi, M.; Matteucci, G.; Gerosa, G. Leaf morphology and chemistry in *Fagus sylvatica* (beech) trees as affected by site factors and ozone: Results from CONECOFOR permanent monitoring plots in Italy. *Tree Physiol.* **2005**, *25*, 211–219. [CrossRef] [PubMed]

34. Van Soest, P.J.; Wine, R.H. Determination of lignin and cellulose in acid-detergent fiber with permanganate. *J. Assoc. Off. Anal. Chem.* **1964**, *51*, 780–785.

35. Schinner, F.; von Mersi, W. Xylanase-, CM-Cellulase and invertase activity in soil: An improved method. *Soil Biol. Biochem.* **1990**, *22*, 511–515. [CrossRef]

36. Fioretto, A.; Papa, S.; Curcio, E.; Sorrentino, G.; Fuggi, A. Enzyme dynamics on decomposing leaf litter of *Cistus incanus* and *Myrtus communis* in a Mediterranean ecosystem. *Soil Biol. Biochem.* **2000**, *32*, 1847–1855. [CrossRef]

37. Verchot, L.; Borelli, T. Application of para-nitrophenol (pNP) enzyme assays in degraded tropical soils. *Soil Biol. Biochem.* **2005**, *37*, 625–633. [CrossRef]

38. Von Mersi, W.; Schinner, F. An improved and accurate method for determining the dehydrogenase activity of soils with iodonitrotetrazolium chloride. *Biol. Fertil. Soils* **1991**, *11*, 216–220. [CrossRef]

39. Leatham, G.F.; Stahmann, M.A. Studies on the laccase of *Lentinus edodes*: Specificity, localization and association with the development of fruiting bodies. *J. Gen. Microbiol.* **1981**, *125*, 147–157. [CrossRef]

40. Di Nardo, C.; Cinquegrana, A.; Papa, S.; Fuggi, A.; Fioretto, A. Laccase and peroxidase isoenzymes during leaf litter decomposition of *Quercus ilex* in a Mediterranean ecosystem. *Soil Biol. Biochem.* **2004**, *36*, 1539–1544. [CrossRef]

41. Eivazi, F.; Tabatabai, M.A.M. Phosphatases in soils. *Soil Biol. Biochem.* **1977**, *9*, 167–172. [CrossRef]

42. Nicolaus, M.; Tinbergen, J.M.; Ubels, R.; Both, C.; Dingemanse, N.J. Density fluctuations represent a key process maintaining personality variation in a wild passerine bird. *Ecol. Lett.* **2016**, *19*, 478–486. [CrossRef] [PubMed]

43. Green, P.J.; Silverman, B.W. *Nonparametric Regression and Generalized Linear Models: A Roughness Penalty Approach*; Chapman & Hall/CRC Monographs on Statistics & Applied Probability; CRC Press, Taylor & Francis: Boca Raton, FL, USA, 1993; ISBN 9780412300400.

44. Donnelly, P.K.; Entry, J.A.; Crawford, D.L.; Cromack, K. Cellulose and lignin degradation in forest soils: Response to moisture, temperature, and acidity. *Microb. Ecol.* **1990**, *20*, 289–295. [CrossRef] [PubMed]

45. Ververis, C.; Georghiou, K.; Danielidis, D.; Hatzinikolaou, D.G.; Santas, P.; Santas, R.; Corleti, V. Cellulose, hemicelluloses, lignin and ash content of some organic materials and their suitability for use as paper pulp supplements. *Bioresour. Technol.* **2007**, *98*, 296–301. [CrossRef] [PubMed]

46. Sjöberg, G.; Nilsson, S.I.; Persson, T.; Karlsson, P. Degradation of hemicellulose, cellulose and lignin in decomposing spruce needle litter in relation to N. *Soil Biol. Biochem.* **2004**, *36*, 1761–1768. [CrossRef]

47. Romaní, A.M.; Fischer, H.; Mille-Lindblom, C.; Tranvik, L.J. Interactions of bacteria and fungi on decomposing litter: Differential extracellular enzyme activities. *Ecology* **2006**, *87*, 2559–2569. [CrossRef]

48. Rihani, M.; Kiffer, E.; Botton, B. Decomposition of beech leaf litter by microflora and mesofauna. I: In vitro action of white-rot fungi on beech leaves and foliar components. *Eur. J. Soil Biol.* **1995**, *31*, 57–66.

49. Heuck, C.; Spohn, M. Carbon, nitrogen and phosphorus net mineralization in organic horizons of temperate forests: Stoichiometry and relations to organic matter quality. *Biogeochemistry* **2016**, *131*, 1–14. [CrossRef]

50. Albers, D.; Migge, S.; Schaefer, M.; Scheu, S. Decomposition of beech leaves (*Fagus sylvatica*) and spruce needles (*Picea abies*) in pure and mixed stands of beech and spruce. *Soil Biol. Biochem.* **2004**, *36*, 155–164. [CrossRef]

51. Rutigliano, F.A.; Virzo De Santo, A.; Berg, B.; Alfani, A.; Fioretto, A. Lignin decomposition in decaying leaves of *Fagus sylvatica* L. and needles of *Abies alba* Mill. *Soil Biol. Biochem.* **1996**, *28*, 101–106. [CrossRef]

52. MacDonald, J.A.; Dise, N.B.; Matzner, E.; Armbruster, M.; Gundersen, P.; Forsius, M. Nitrogen input together with ecosystem nitrogen enrichment predict nitrate leaching from European forests. *Glob. Chang. Biol.* **2002**, *8*, 1028–1033. [CrossRef]

53. Belay-Tedla, A.; Zhou, X.; Su, B.; Wan, S.; Luo, Y. Labile, recalcitrant, and microbial carbon and nitrogen pools of a tallgrass prairie soil in the US Great Plains subjected to experimental warming and clipping. *Soil Biol. Biochem.* **2009**, *41*, 110–116. [CrossRef]

54. De Nicola, C.; Zanella, A.; Testi, A.; Fanelli, G.; Pignatti, S. Humus forms in a Mediterranean area (Castelporziano Reserve, Rome, Italy): Classification, functioning and organic carbon storage. *Geoderma* **2014**, *235–236*, 90–99. [CrossRef]

55. Michel, K.; Matzner, E. Nitrogen content of forest floor Oa layers affects carbon pathways and nitrogen mineralization. *Soil Biol. Biochem.* **2002**, *34*, 1807–1813. [CrossRef]

56. Nannipieri, P.; Giagnoni, L.; Renella, G. Soil enzymology: Classical and molecular approaches. *Biol. Fertil. Soils* **2012**, *48*, 743–762. [CrossRef]

57. Fioretto, A.; Papa, S.; Pellegrino, A.; Ferrigno, A. Microbial activities in soils of a Mediterranean ecosystem in different successional stages. *Soil Biol. Biochem.* **2009**, *41*, 2061–2068. [CrossRef]

58. Papa, S.; Pellegrino, A.; Fioretto, A. Microbial activity and quality changes during decomposition of *Quercus ilex* leaf litter in three Mediterranean woods. *Appl. Soil Ecol.* **2008**, *40*, 401–410. [CrossRef]

59. Sardans, J.; Peñuelas, J. Drought decreases soil enzyme activity in a Mediterranean *Quercus ilex* L. forest. *Soil Biol. Biochem.* **2005**, *37*, 455–461. [CrossRef]

60. Palumbo, E. Laccase and Peroxidase Enzymes Activities along the Soil Profile in Two Beech Woods: Seasonal Variations and Isoforms Identification. M.Sc. Thesis, University of Campania "Luigi Vanvitelli", Caserta, Italy, 2011.

61. Papa, S.; Pellegrino, A.; Bartoli, G.; Ruosi, R.; Rianna, S.; Fuggi, A.; Fioretto, A. Soil organic matter, nutrient distribution, fungal and microbial biomass and enzyme activities in a forest beech stand on the Apennines of southern Italy. *Plant Biosyst.* **2014**, 1–12. [CrossRef]

62. Guckland, A.; Jacob, M.; Flessa, H.; Thomas, F.M.; Leuschner, C. Acidity, nutrient stocks, and organic-matter content in soils of a temperate deciduous forest with different abundance of European beech (*Fagus sylvatica* L.). *J. Plant Nutr. Soil Sci.* **2009**, *172*, 500–511. [CrossRef]

63. Jobbágy, E.; Jackson, R. The vertical distribution of soil organic carbon and its relation to climate and vegetation. *Ecol. Appl.* **2000**, *10*, 423–436. [CrossRef]

64. Groffman, P.M.; Driscoll, C.D.T.; Fahey, T.J.; Hardy, J.P.; Fitzhugh, R.D.; Tierney, G.L.; Henry, K.S.; Welman, T.A.; Demers, J.D.; Nolan, S. Effects of mild winter freezing on soil nitrogen and carbon dynamics in a northern hardwood forest. *Biogeochemistry* **2001**, *56*, 191–213. [CrossRef]

65. Hogberg, M.N.; Briones, M.J.I.; Keel, S.G.; Metcalfe, D.B.; Campbell, C.; Midwood, A.J.; Thornton, B.; Hurry, V.; Linder, S.; Näsholm, T.; et al. Quantification of effects of season and nitrogen supply on tree below-ground carbon transfer to ectomycorrhizal fungi and other soil organisms in a boreal pine forest. *New Phytol.* **2010**, *187*, 485–493. [CrossRef] [PubMed]

66. Andersson, M.; Kjøller, A.; Struwe, S. Microbial enzyme activities in leaf litter, humus and mineral soil layers of European forests. *Soil Biol. Biochem.* **2004**, *36*, 1527–1537. [CrossRef]

forests

MDPI

Article

Changes in Soil Enzyme Activities and Microbial Biomass after Revegetation in the Three Gorges Reservoir, China

Qingshui Ren, Hong Song, Zhongxun Yuan, Xilu Ni and Changxiao Li *

Key Laboratory of Eco-Environment in the Three Gorges Reservoir Region of the Ministry of Education, School of Life Sciences, Southwest University, Chongqing 400715, China; rqs3468@126.com (Q.R.); m13102302388@163.com (H.S.); yuanzhongxun@email.swu.edu.cn (Z.Y.); nixilu110@163.com (X.N.)
* Correspondence: lichangx@swu.edu.cn; Tel.: +86-23-6825-2365

Received: 28 March 2018; Accepted: 2 May 2018; Published: 4 May 2018

Abstract: Soil enzymes and microbes are central to the decomposition of plant and microbial detritus, and play important roles in carbon, nitrogen, and phosphorus biogeochemistry cycling at the ecosystem level. In the present study, we characterized the soil enzyme activity and microbial biomass in revegetated (with *Taxodium distichum* (L.) Rich. and *Cynodon dactylon* (L.) Pers.) versus unplanted soil in the riparian zone of the Three Gorges Dam Reservoir (TGDR), in order to quantify the effect of revegetation on the edaphic microenvironment after water flooding in situ. After revegetation, the soil physical and chemical properties in revegetated soil showed significant differences to those in unplanted soil. The microbial biomass carbon and phosphorus in soils of *T. distichum* were significantly higher than those in *C. dactylon* and unplanted soils, respectively. The microbial biomass nitrogen in revegetated *T. distichum* and *C. dactylon* soils was significantly increased by 273% and 203%, respectively. The enzyme activities of *T. distichum* and *C. dactylon* soils displayed no significant difference between each other, but exhibited a great increase compared to those of the unplanted soil. Elements ratio (except C/N (S)) did not vary significantly between *T. distichum* and *C. dactylon* soils; meanwhile, a strong community-level elemental homeostasis in the revegetated soils was found. The correlation analyses demonstrated that only microbial biomass carbon and phosphorus had a significantly positive relationship with soil enzyme activities. After revegetation, both soil enzyme activities and microbial biomasses were relatively stable in the *T. distichum* and *C. dactylon* soils, with the wooded soil being more superior. The higher enzyme activities and microbial biomasses demonstrate the C, N, and P cycling and the maintenance of soil quality in the riparian zone of the TGDR.

Keywords: revegetation; microbial biomass; chloroform fumigation extraction; enzyme activities; stoichiometric homeostasis; the Three Gorges Reservoir

1. Introduction

The riparian zone, as an important ecotone, is known to be the key area for improving water quality, controlling flooding, and relieving soil erosion [1]. The riparian zone of the Yangtze River in China, analogous to many other critical interfaces of the river ecosystems in the world, is a unique transition zone in the landscape and plays a crucial role in the health and functioning of the diverse Yangtze River watershed biome as a whole [2]. However, since the building of the Three Gorges Dam (TGD) in the upper reaches of the Yangtze River, the natural flow regime has been disrupted. In this newly formed water level fluctuation zone (WLFZ), the hydrological regime is opposite to the Yangtze River's natural regime, with its peak flows occurring in winter rather than in summer [3]. Plant species adapted to previously terrestrial habitats intensively died out or otherwise were degraded, leading to considerable soil erosion, texture coarsening, habitat loss, biodiversity decline, and environmental pollution in the riparian zone of the TGD region. With the far-reaching and profound challenges exerted on this region [4], reforestation and vegetation reconstruction to restore the newly-formed riparian zone of the TGD region have been thought to be an environmentally friendly alternative [5,6].

Restoration of the degraded riparian zone is a significant management practice [7] that may return an area to a pre-disturbance ecological state, functionally [8] and/or structurally [9]. Since 2000, research has been conducted to determine the suitability of potential plant species for restoring the novel habitat in the WLFZ. Accumulating evidence has demonstrated that species including *Cynodon dactylon* (Bermuda grass) and *Taxodium distichum* (bald cypress) are promising candidates [6,10]. After revegetation, plants such as the abovementioned Bermuda grass and bald cypress will have to experience the hydrological regimes of the WLFZ, i.e., an annual imposed water level fluctuation of 30 meters (from 145 m a.s.l. to 175 m a.s.l.) within the Three Gorges Dam Reservoir (TGDR), and input leaves and plant detritus into the riparian soil. However, riparian zones can effectively remove pollutants and protect water quality through performing functions such as filtration and denitrification. In this regard, soil enzymes can play an essential role in catalyzing reactions necessary for the decomposition of leaves and plant detritus, and also for nutrient cycling, in the WLFZ. It has been reported that any change in soil management and land use is reflected in the soil enzyme activities, and thus soil enzyme activities can anticipate changes in soil quality before necessary further detection through other means of soil analyses [11]. As such, measurements of soil enzyme activities can describe how nutrient cycling is affected by environmental changes after revegetation in the TGD region.

Moreover, microorganisms are the main source of enzymes in soils [12], because they produce extracellular enzymes to acquire energy and resources from complex biomolecules in the soil environment [13]. Parts of the energy and resources are immobilized in their biomass, which thus strongly influences the concentrations of plant available nutrients in the soil [14,15]. Therefore, the soil microbial communities strongly affect the potential of the soil for enzyme-mediated substrate catalysis. Microbial biomass is the labile portion of the organic fraction in soils and serves as both an important source of and sink for plant available nutrients [16]. However, enzyme activities and microbial biomass are closely related to each other because transformations of the important organic elements occur through microorganisms [16].

After re-vegetation in the TGR zone, additional plant leaves, dead roots, and nutrient translocation caused by water fluctuation, may cause a change in the pH and nutrient stocks in the soil [17]. Such changes in nutrient availability in the riparian soil environment would cause substantial variation in soil enzyme activities and microbial biomass in the WLFZ of the TGDR. In particular, the presence of plant roots leads to shifts in the microbial growth strategy, upregulation of enzyme production, and increased microbial respiration [18]. Thus, after revegetation, the in situ monitoring and evaluation of soil enzymes and microbial biomass are both of great significance and necessity for better understanding the dynamic patterns of the riparian ecosystems. Furthermore, understanding the functioning of soils within a riparian zone can be used to evaluate their influence on water quality, and ultimately provide information on the whole ecosystem level environmental quality. Thus, the objective of this work was to monitor and assess microbial biomass and selected enzyme activities relevant to C, N, and P cycling in soils of the WLFZ of the TGDR as affected by re-vegetation. We hypothesized that (1) revegetation of the plants in the riparian zone of the TGDR will enhance soil microbial biomass and enzyme activities as compared to those of the unplanted soil; and (2) the revegetated soils are more homeostatic than unplanted soil.

2. Materials and Methods

2.1. Study Site and Experimental Setup

The study site is located in the Ruxi River basin in Gonghe Village of Shibao Township, Zhong County, Chongqing Municipality (107°32′–108°14′ E, 30°03′–30°35′ N), China. Ruxi River is one of the largest tributaries of the TGDR. The area has a subtropical monsoon climate with four distinctive seasons. It has a mean annual temperature of 19.5 °C and an annual rainfall of 1200 mm. The soil within this region is purple soil (Regosols according to the World Reference Base for Soil Resources), formed in the parent material of calcareous purple sand shale in the subtropical region. The degree of soil development is limited because the rock is not deeply weathered. The soil and water erosion is serious in the less-vegetated riparian zone of the Ruxi River.

Before the inundation, vegetation in this region was dominated by annuals (e.g., *Setaria viridis* (L.) P.Beauv., *Digitaria ciliaris* (Retz.) Koeler, and *Leptochloa chinensis* (L.) Nees), perennials (e.g., *Cynodon dactylon* (L.) Pers., *Hemarthria compressa* (L.f.) R.Br., and *Capillipedium assimile* (Steud.) A. Camus), and woody species (e.g., *Pterocarya stenoptera* C.DC., *Metasequoia glyptostroboides* Hu and W.C.Cheng, and *Alnus cremastogyne* Burk.). However, the reversed flooding seasonality and prolonged flooding duration caused the loss of the former vegetation, and only *S. viridis* and *D. ciliaris* remained [19]. In the study region, revegetation was carried out in March, 2012. Bald cypress and Bermuda grass were planted in the upper portion between 165 m a.s.l. and 175 m a.s.l., with the same slope, solar radiation intensity, and initial soils. The planting density for bald cypress saplings was 1 m × 1 m, and for Bermuda grass was 20 cm × 20 cm, in a total riparian area of 13.3 hm^2. The survival rate of the plants was 100 percent, and they displayed substantial progress in their growth in the following years. In the meantime, there was no sign of anthropogenic disturbance at the experimental site. After a four-year growth, the soil sampling was carried out in June 2016. Growth data on bald cypress saplings and Bermuda grass including tree height, diameter at breast height (DBH), grass stem basal diameter, and crown canopy area (Table 1), were recorded. For the purposes of comparison, we used unplanted plots as controls, henceforth referred to as "CK" treatment.

Table 1. Basic situation of the vegetation (Mean ± SE).

Vegetation	Height (m)	Diameter Breast Height (DBH) (cm)/Stem Basal Diameter (mm)	Canopy (m^2)/Coverage (%)
T. distichum	4.67 ± 0.18	5.21 ± 0.41	2.16 ± 0.60
C. dactylon	0.86 ± 0.02	1.68 ± 0.03	99.17 ± 1.04

The DBH of *T. distichum* was measured, and the stem basal diameter of *C. dactylon* was measured. SE: standard error.

2.2. Soil Sampling and Soil Physicochemical Properties Determination

The water level of the riparian zone in Zhong County in the TGDR is raised from September to October, then maintained at its highest water level (i.e., about 175 m a.s.l.) from November to January. After January, a gradual decrease of the water level will be followed and then maintained at its lowest level (i.e., 145 m a.s.l.) during the months of June through September. Under this circumstance of water-level changes, the soil sampling was conducted in June 2016, when the riparian zone of the TGDR was completely exposed to the air. Nine sampling plots (5 m × 5 m each) were randomly selected in the revegetation area of bald cypress, Bermuda grass, and the unplanted area, between the elevations of 165–175 m a.s.l. When sampling in the field, a cutting ring (10 cm in diameter × 20 cm in height) was used to collect the soil samples, and vegetation was removed from the top of the soil prior to collection.

Within each plot, five 1 m × 1 m soil sampling subplots were made and arranged in a quincunx pattern. Soil samples were collected from the surface 20 cm soil from each of the five 1 m × 1 m sampling subplots, and then used to make one composite sample for that entire plot. Each soil composite sample collected from a particular plot was homogenized and then quartered. All samples were placed on ice and transported to the laboratory immediately. A portion of the soil was stored at 4 °C for biochemistry analysis. The other portion was air-dried separately, then hand-passed through a 2 mm followed by a 0.25 mm sieve, and finally analyzed for the soil nutrients present. Soil pH value was determined using a 1:2.5 ratio of soil to deionized water. Soil organic carbon (SOC) was determined using the potassium dichromate titration method. Total nitrogen (TN) was determined by an element analysis machine (Elementar Vario. EL, Langenselbold, Germany), and available nitrogen (AN) was determined using a micro-diffusion technique after alkaline hydrolysis. Total phosphorus (TP) was determined with acid digestion (HNO_3, HF, and H_2O_2) in a microwave oven and subsequent inductively coupled plasma-optical emission spectrometry measurements (ICP-OES, CIROS, Spectro, Thermo Fisher, NY, USA) [17]. Available phosphorus (AP) was determined using the Mo-Sb colorimetric method. The soil water content (WC) was determined by oven drying method, and soil density was determined by pycnometer method. The soil porosity was calculated from soil density and measured bulk density [20].

2.3. Soil Enzyme and Microbial Biomass Determination

Activities of soil enzymes related to C, N, and P cycling were measured. These enzymes included four hydrolytic enzymes: Catalase activity was determined by measuring the O_2 absorbed by $KMnO_4$ (Chuanjiang Chemical Industry, Chongqing, China) after the addition of H_2O_2 (Chuandong Chemical Industry, Chongqing, China) to the samples [21]. Urease activities were determined in 0.1 M phosphate buffer at pH 7; 1 M urea (Chuandong Chemical Industry, Chongqing, China) and 0.03 M Na-benzoyl-argininamide (BAA) (Fusheng Industry, Shanghai, China) were used as substrates, respectively. A total of 2 mL of buffer and 0.5 mL of substrate were added to 0.5 g of soil sample, and then incubated at 37 °C for 24 h. The activities of urease were determined by the NH_4^+ released [22]. Acid phosphatase, which breaks phosphoester bonds and releases inorganic phosphorus, was measured by p-nitrophenyl phosphate disodium colorimetry, and expressed as mg p-nitrophenol released per gram dry soil over 24 h at 37 °C. The sucrase activity was determined using sucrose as the substrate, and the activity was expressed as mg glucose $g^{-1} \cdot dw\ 24\ h^{-1}$ [23].

Soil microbial biomass C, N and P (SMC, SMN, and SMP) concentrations were determined by the chloroform fumigation extraction method. Specifically, for each soil sample, 25 g (fresh weight) pre-incubated soil was used for chloroform fumigation, and the same weight soils without chloroform fumigation were also conducted for the control. Then, 100 mL 0.5 $mol \cdot L^{-1}$ K_2SO_4 extract was used for the extraction of SMC and SMN [24,25]. Besides, biomass P was measured in 5 g (fresh weight) fumigation soil, with the non-fumigation soil as the control, and 100 mL 0.5 $mol\ L^{-1}$ $NaHCO_3$ extract was used for the extraction of SMP [26]. The following formulas were used to calculate the soil microbial biomass:

$$SMC = EC/kEC \tag{1}$$

$$SMN = EN/kEN \tag{2}$$

$$SMP = EP/kEP \tag{3}$$

EC, EN, and EP were the differences in extractable fractions between fumigated and unfumigated soil. kEC, kEN, and kEP were conversion coefficients, and had values of 0.38, 0.45, and 0.40, respectively [27].

2.4. Statistics Analysis

To determine significant differences between the activities of soil enzymes or edaphic properties after revegetation, one-way analysis of variance (ANOVA) was used, followed by LSD's test. The difference was reported as significant when $p < 0.05$. Redundancy analysis (RDA) with the Monte Carlo permutation test (499 permutations) was performed to determine if soil enzyme activities were correlated with soil properties or microbial biomass. Statistical procedures were conducted using

SPSS version 18.0 (IBM, Armonk, NY, USA), MS Excel (Microsoft, Redmond, WA, USA), and Canoco version 4.5 (Microcomputer Power, Ithaca, NY, USA).

3. Results

3.1. The Edaphic Physicochemical Properties

Soil physicochemical properties are shown in Table 2. Although soil pH values in the planted plots were a little bit lower than the neutral value of 7, indicating a slightly acidic soil, there was no significant difference in soil pH among the three groups. However, bald cypress soils had the highest SOC when compared to that of Bermuda grass and unplanted soils. The planted soils, including both bald cypress and Bermuda grass soils, displayed much higher TN, WC, and porosity than unplanted soil, with a 97% and 58% increase in TN, 41% and 33% increase in WC, and 17% and 28% enhancement in porosity, respectively. In contrast, AN content of planted soils was significantly lower than that of unplanted soil. Moreover, soil density varied among different treatments, with the highest value being found in Bermuda grass soil. The soil density of Bermuda grass was significantly higher than that of bald cypress soil, whereas the soil density of bald cypress showed no obvious difference compared to that of unplanted soil.

Table 2. Comparisons of soil physicochemical properties under different vegetation types (Means ± SE, $n = 3$).

Soils Covered by Vegetation	pH	SOC ($g \cdot kg^{-1}$)	TN ($g \cdot kg^{-1}$)	TP ($g \cdot kg^{-1}$)	AN ($mg \cdot kg^{-1}$)	AP ($mg \cdot kg^{-1}$)	WC (%)	Density ($g \cdot cm^{-3}$)	Porosity (%)
T. distichum	6.84 ± 0.06a	14.76 ± 1.42a	1.32 ± 0.09a	1.14 ± 0.01a	72.31 ± 2.13b	1.04 ± 0.11a	47.7 ± 0.90a	2.48 ± 0.05b	35.25 ± 1.64a
C. dactylon	6.94 ± 0.06a	10.20 ± 2.21b	1.06 ± 0.11a	0.82 ± 0.08a	75.51 ± 2.21b	0.96 ± 0.01a	45.1 ± 1.11a	2.78 ± 0.56a	38.68 ± 7.15a
CK	7.14 ± 0.18a	9.88 ± 0.54b	0.67 ± 0.08b	0.74 ± 0.27a	85.26 ± 1.90a	1.02 ± 0.11a	33.8 ± 1.39b	2.55 ± 0.01ab	30.04 ± 0.35b

Note: CK referred to as unplanted treatment. Different lowercase letters represent significant differences ($p < 0.05$) among different treatments. The same below. SOC: soil organic carbon; TN: total nitrogen; TP: total phosphorus; AN: available nitrogen; AP: available phosphorus; WC: Water content.

3.2. Soil Microbial Biomass

The bald cypress soil exhibited significantly or extremely significantly higher SMC and SMP than the other two soils (Figure 1a,c). Meanwhile, the SMC of Bermuda grass soil was also significantly higher than that of unplanted soil. However, no significant difference was observed for SMP between Bermuda grass and unplanted soils. Compared to the lowest SMN of unplanted soil, the SMN values of revegetated soils were significantly increased by 273% and 203% after revegetation, respectively.

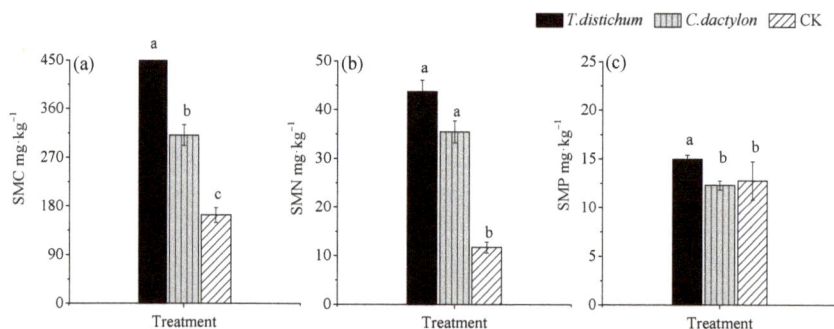

Figure 1. (a) Variations in SMC under different treatments, (b) Variations in SMN under different treatments, (c) Variations in SMP under different treatments CK: unplanted treatment, SMC: Soil microbial biomass C, SMN: Soil microbial biomass N; SMP: Soil microbial biomass P. Different lowercase letters represent significant differences ($p < 0.05$) among different treatments.

3.3. Soil Enzyme Activities

Soil enzyme activities of bald cypress and Bermuda grass soils displayed a similar trend, while showing no significant difference between each other. When compared to the unplanted soil, the activities of all enzymes except phosphatase were strongly augmented after re-vegetation. The activities of catalase, sucrase, and urease of bald cypress soils were significantly increased by 17%, 220%, and 47%, respectively. Likewise, the activities of catalase, sucrase, and urease of Bermuda grass soils were also significantly increased by 9%, 172%, and 65%, respectively.

3.4. Threshold Elemental Ratios of CNP and Stoichiometric Homeostasis

The soil C/N ratio under bald cypress vegetation was significantly higher than that of unplanted soil, but no significant difference in the C/N ratio was observed between the two planted soils or between Bermuda grass and unplanted soils. In contrast, the highest value of microbial C/N ratio occurred in the unplanted soil, which was significantly higher than that of Bermuda grass soil. The ratio of C/P, including both in the soil and in the microbial biomass, displayed a significantly greater value in planted soils than that in unplanted soils. Among the three treatments, the ratios of microbial biomass C, N, and P to soil total C, N, and P correspondingly varied between 2–6%, 1–4%, and 1–2%. The stoichiometric ratios of SMC/SOC and SMN/TN in both bald cypress and Bermuda grass soils were significantly higher than in the unplanted soil, whereas SMP/TP of bald cypress soil showed no obvious difference with that of the unplanted soil.

We analyzed the associations between nutrient elemental ratios in soils and microbial biomass elemental ratios in microbes, and the H value was used to evaluate the strength of soil stoichiometric homeostasis in the present study plots. In the planted soil, the H value was larger than 1, whereas the H value in unplanted soil was around 1.

3.5. Correlations between Soil Variables and Enzyme Activity

Results from the RDA identified that the first two axes could explain 89.8% of the variance caused by soil edaphic properties and microbial biomass. However, only the SMP and SMC significantly positively correlated with soil enzyme activities, which explained 50% ($p = 0.002$, $F = 7.03$) and 16% ($p = 0.048$, $F = 2.7$) of the variances of the total variability of soil enzyme activity, respectively.

4. Discussion

4.1. Edaphic Physicochemical Properties

In our current study, the revegetation with either bald cypress or Bermuda grass modified the soil conditions in the riparian zone of the TGDR (Table 2), which was mainly attributed to the inputs of litter and root exudes. The spatial differences in soil moisture content can partly be ascribed to the differences in throughfall, the vegetative (tree and grass) cover, and water loss through the soil evaporation, since throughfall and evaporation strongly correlate with the gap fraction [28]. In addition, without vegetative covering such as in the group CK, the unplanted soil tends to have a coarser texture and lower porosity, and some soil nutrients might be easily extracted or removed by the water flooding under the dynamic hydrological regime of the TGDR.

4.2. Edaphic Enzyme Activities and Soil Microbial Biomass

Divergence of nutrient availabilities and soil moisture can give feedback to soil microbes' resource requirements, which in turn may influence the microbial biomass [29]. In the present study, the discrepancy of plant characteristics may result in a difference of soil microbial biomass. The leaf type would be a primary factor of microbial biomass accumulation in surface soils due to differences in leaf litter chemistry. The higher SMC in bald cypress soil may be attributed to its higher content of SOC. Meanwhile, in Bermuda grass soil, cellulose represents a crucial proportion of plant detritus that returned to the soil. The inflexible ingredient makes cellulose difficult to decompose in the soil. Additionally, previous studies reported that the SMP is a very important P pool in the organic layer of P-poor soil [30]. In the riparian zone of TGDR, though the P (TP and AP) content displayed no significant difference among the three treatments, the uptake of P was increased by soil microbes in bald cypress soils, which is most likely due to the larger mass of SOC (Table 2) that existed in bald cypress soil through overcompensating for the lower SMP [31]. This result illustrated that the microorganisms in bald cypress soils were capable of immobilizing available P much more efficiently than in the Bermuda grass and unplanted soils, leading to the low P mineralization rates.

Moreover, the more extensive rooting systems and protective canopies of the bald cypress and Bermuda grass will protect the vegetated soil from erosion and maintain more favorable conditions for plant and microbial growth, which were conducive to the accumulation of microbial biomass. In the TGDR region during summer time, the climate was both very hot (the average temperature reaching 37 °C) and dry (the relative humidity index being −0.62), thus promoting extensive evapotranspiration in the studied riparian zone. Consequently, soil impoverishment and nutrient unavailability might occur much more commonly in bare (unplanted) soil compared to that in planted soil which was covered by vegetation, thus leading to the inactive status of the microbes [32]. The active soil microbes under bald cypress and Bermuda grass could build up their new biomass, but, on the contrary, the inactive ones are incapable of accumulating carbon in biomass, even possibly depleting their C by respiring [33]. In such circumstances, the microorganism would also almost completely decrease the production of enzyme, as well as metabolic activities related to decomposition [34], which were consistent with the lower enzyme activities in unplanted soils. However, although the soil microbial biomass under the two re-vegetation schemes in our present study was different, their enzyme activities displayed a similar value. It is also possible that, while the soil microbial community as a whole shifted, the relative abundance of enzyme producers did not [13].

Besides, the soil aggregation could also influence the microbial biomass. A portion of microorganisms live on or closer to the aggregate surface. The increase of microbial biomass as soil aggregate size decreases may depend on the presence of small pore-size microaggregates [35]. This may protect microorganisms from predation by protozoa or from desiccation [36]. Fang et al. [37] also documented that smaller aggregate sizes in the soil have a higher specific surface for microbial cells to attach to, which was of benefit to the build-up of microbial biomass. After the revegetation in the riparian zone of the TGDR, with the root penetrating into the riparian soil, their soil aggregate would be shrunken. However, the unplanted soil maintains its bigger size aggregation without plant effects. The increasing soil aggregation in the CK would prevent the soil enzyme from accessing carbon and nitrogen substrates inside aggregates [38], which was in accordance with the lower nutrient content and enzyme activities in the CK in the present study.

The higher enzyme activities of the soil after revegetation confirmed our first hypothesis, for which one possible reason should be owing to the accumulated microbial biomass. Because the soil enzymes are mainly excreted by microbes to mineralize soil carbon, nitrogen, and phosphorus from SOC, the activities of enzymes could therefore be partially explained by the distribution of soil microbial biomass. Furthermore, soil enzyme activities were also affected by soil water content. In unplanted soil, the water content was significantly lower than that of the planted soils (Table 2), which reduced the diffusion of soil soluble substrates. In such a less hydrologic environment, both the substrate and water become limited, and the soil microorganism down-regulates enzyme production [37,39].

In the riparian zone of the TGDR, as the soil experienced inundation in winter, and strong soil weathering in summer, this resulted in the leaching of nutrient elements of the riparian soil, followed by a decline in the soil nutrients availability [40,41]. According to the lower AN content in planted soil, we could infer that the soil N has a limited availability, and the microbial organism needs to secrete corresponding enzymes to meet their N demand [42]. Therefore, soil urease activities increased significantly in the planted soils. However, the phosphatase activities exhibited no significant difference among the three groups, for which the result was consistent with the undifferentiated P content across the soils of the three groups, further indicating that the P in the riparian soil may not be a limiting nutrient for aboveground productivity in the TGDR region. The former research also pointed out that some enzymes involved in P acquisition seem to be correlated better with the environmental requirement of P [43]. Other studies, however, have reported that the higher N content could stimulate enzyme activities in acidic soil, particularly for phosphatase activity enhancement [44,45]. Under acidic soil conditions, the plant roots would release protons and organic ligands, and thus the rhizosphere would be stimulated and promote the acid phosphatase activity [46]. In our research plots, although the soil TN contents under bald cypress and Bermuda grass were higher than those under unplanted soil, the pH value tended to be neutral, and the moderate pH might weaken the function of soil N or root exudates. Moreover, Cleveland et al. [47] reported that there was similar soil phosphatase activity under both forest and pasture for an oxisol in Costa Rica, and they also found a higher soil microbial biomass under forest than that under pasture. The extremely low activities of catalase throughout unplanted soil profiles suggested that oxygen availability may be a major control factor for catalase, because soil redox status is also an important microbial driver in surface soils [48], particularly in the frequently inundated riparian zone of the TGDR.

4.3. Elemental Ratios in Soils and in Soil Microbial Biomass

Ecological stoichiometry is an important tool to study nutrient cycling and understand the ecosystem dynamics and functioning [49]. Former research showed that, though the soil nutrient ratios varied through the latitude, they did not vary significantly between forests and grasslands [50]. In our present study, the ratios of C/N and C/P in the bald cypress and Bermuda grass soil in the TGDR also revealed no significant difference (Table 3), which is consistent with the finding reported by the previous study [50]. This may be attributed to the fact that the plants are the major sources of soil C and N in the revegetated soils, while because of a lack of litter input into the soil, the ratios of C/N and C/P maintained a relatively lower value in the unplanted soil. Besides, the large amount of needs of N during photosynthesis of the plants from the soil, could lead to a higher ratio of C/N in the revegetated soils as well [50]. In general, biological organisms have an ordered chemical composition, and recent analyses have shown that most plants have relatively constrained element ratios [51]. In the riparian zone of the TGDR, soil microbial C/N ratios varied between 4 and 7 in our present study, which was slightly different to the result obtained from the research conducted in Mongolian grassland, whose soil microbial C/N ratios ranged from 5 to 9 [52]. However, the total soil C/P ratios falling in between 5 to 8 were also different from the result carried out in a hilly area [53]. This kind of divergence in soil C/N and C/P ratios could be mainly generated by the varied environmental habitats and biomes. Furthermore, the ratio of C/N can reflect the overall investment in structural cellular material by soil microbial biomass, for example, the relatively high ratio of C/N representing the high abundance of fungi [51].

Table 3. The element ratio of soil and microbial C, N, P, and stoichiometric homeostasis.

Soils Covered by Vegetation	C/N(S)	C/P(S)	C/N(M)	C/P(M)	SMC/SOC (%)	SMN/TN (%)	SMP/TP (%)	H(C/N)	H(C/P)
T. distichum	6.62 ± 0.08a	7.54 ± 0.68a	10.68 ± 0.19ab	31.08 ± 1.09a	5.25 ± 0.57a	3.12 ± 0.30a	1.32 ± 0.04ab	4.48	2.69
C. dactylon	5.38 ± 0.64ab	7.08 ± 0.85a	8.78 ± 0.06b	25.39 ± 1.64a	4.85 ± 0.41a	3.37 ± 0.26a	1.53 ± 0.14a	434.78	9.61
CK	4.37 ± 0.63b	5.83 ± 0.56b	14.21 ± 1.80a	15.06 ± 1.75b	2.84 ± 0.11b	1.86 ± 0.39b	1.11 ± 0.05b	1.42	1.00

C/N (S), and C/P (S) refer to the ratio of C/N and C/P in the soil; C/N (M) and C/ P (M) refer to the ratio of C/N and C/P in the microbial biomass. H = 1/K. In equation, K is the slope of ln C:N (S) versus ln C:N (M) or slope of ln C:P (S) versus ln C:P (M).

The higher ratios of SMC/SOC and SMN/TN in planted soils reflected the faster nutrient turnover rate after revegetation. The results might be correlated with the higher activities of sucrase and urease (Figure 2), because with the assistance of soil enzymes, the accumulation of elements in the microbes would be accelerated. This finding also indicated that the microbial cells follow the mechanism in terms of element assimilation across C, N, and P, in order to meet the functional demands of microbial progress [54].

Stoichiometric homeostasis can reflect the allocations of different elements in organisms and indicate the stage of microorganism response to variations in soil condition [55] When the stoichiometric composition of the organism does not vary with changes in resource stoichiometry, it is considered to be homeostatic [56]. In order to test the strength of stoichiometric homeostasis, we analyzed the associations between soil nutrient ratios and microbial biomass elemental ratios in the present research. Based on all the soil data, as well as microbial biomass C, N, and P stoichiometric values, a strong community-level elemental homeostasis in both the bald cypress and Bermuda grass soils was detected. Both the H (C/N) and H (C/P) were in close proximity to 1, indicating the unstable condition in unplanted soils in the TGDR region. Thereby, microorganisms could adjust their physiological metabolism to regulate their requirement of C, N, and P resources, thereby acclimating to various habitats [57].

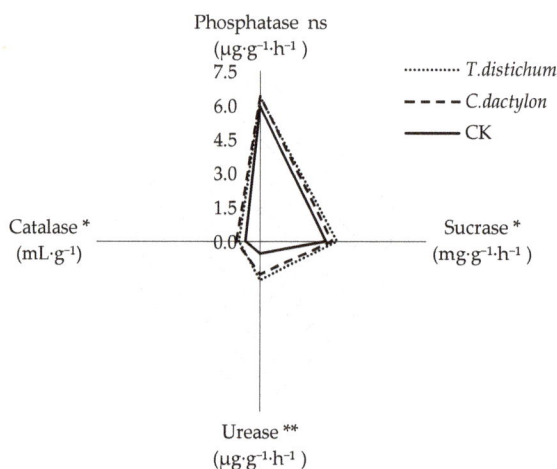

Figure 2. Radar graphs illustrating enzyme activities after revegetation. * and ** refer to significant differences between planted and unplanted soils at $p < 0.05$ and $p < 0.01$, respectively; ns refers to no significant difference.

4.4. Correlations between Enzyme Activities and Soil Parameters

Soil enzyme activities may also reflect the availability of different substrates, as well as microbial energy and nutrient demand, and are often related to C/N ratios of the microbial biomass or of the SOC [58]. In the present study, we found that enzyme activities were more closely correlated to soil microbial biomass than to soil nutrient content, especially to SMC and SMP (Figure 3). This was consistent with the former study conducted by Sari Stark [39], who found that the potential activities of enzymes for nutrient acquisition were not affected when the soil nutrient content increased. In the present study, although the soil physical and chemical parameters did not display direct significant correlations with the enzyme activities, we argued for the potential controls of the physicochemical properties over both soil enzyme activities and microbial biomass. The results may also indicate that for some litter types, such as bald cypress and Bermuda grass, or under some circumstances, such as drying-wetting events, C and P release are more closely coupled [43].

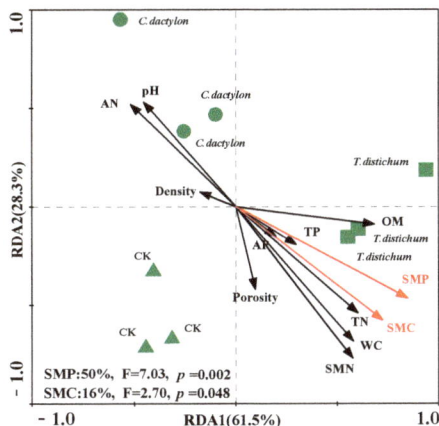

Figure 3. The redundancy analysis (RDA) used to identify the relationship between the soil variables and enzyme activity. The arrows indicate the soil parameters, and the red one indicates significant impacts of the soil parameters on soil enzyme activities ($p < 0.05$). The explained proportion of variability by Monte Carlo permutation is shown at the bottom.

5. Conclusions

Because many soil enzymes and microbial biomass are immediately responsive to soil restoration, they can be used as indices of environmental stability and soil quality for riparian sustainable management. Our results showed that soil enzyme activities and microbial biomass, however, tend to stabilize in the revegetated bald cypress and Bermuda grass soils after revegetation, and the bald cypress showed more preponderance. Therefore, temporal changes in enzyme activities should be accounted for when evaluating the sustainability of revegetation in the TGR region. Besides, the revegetation of bald cypress and Bermuda grass enhanced the ratios of C/N, C/P, SMC/SOC, and SMN/TN, in revegetated soils, indicating a faster nutrient turnover rate after revegetation of the degraded riparian zone of the TGR.

Author Contributions: C.L. and Q.R. conceived of and designed the experiments. Q.R., H.S. and Z.Y. performed the experiments. Q.R. and X.N. analyzed the data. Q.R. wrote the paper. All authors contributed to reviewing the manuscript.

Founding: This research was financially supported by the Chongqing Municipality Key Forestry Research Project [No. Yulinkeyan 2016-8], International Sci-Tech Cooperation Project of Ministry of Science and Technology [No. 2015DFA90900], Forestry Extension Project of China Central Finance [No. Yulinketui 2017-12], China Postdoctoral Science Foundation [2016M602629], and Chongqing Postdoctoral Research Project Special Funding [Xm2016111].

Conflicts of Interest: The authors declare no conflict of interest.

References

1. Clerici, N.; Paracchini, M.L.; Maes, J. Land-cover change dynamics and insights into ecosystem services in European stream riparian zones. *Ecohydrol. Hydrobiol.* **2014**, *14*, 107–120. [CrossRef]
2. Slabbert, E.; Jacobs, S.M.; Jacobs, K. The soil bacterial communities of South African Fynbos riparian ecosystems invaded by Australian Acacia species. *PLoS ONE* **2014**, *9*, e86560. [CrossRef] [PubMed]
3. Lu, Z.J.; Li, L.F.; Jiang, M.X.; Huang, H.D.; Bao, D.C. Can the soil seed bank contribute to revegetation of the drawdown zone in the Three Gorges Reservoir Region? *Plant Ecol.* **2010**, *209*, 153–165. [CrossRef]
4. Yan, T.; Yang, L.; Campbell, C.D. Microbial biomass and metabolic quotient of soils under different land use in the Three Gorges Reservoir area. *Geoderma* **2003**, *115*, 129–138. [CrossRef]

5. Yang, Y.; Li, C. Photosynthesis and growth adaptation of *Pterocarya stenoptera* and *Pinus elliottii* seedlings to submergence and drought. *Photosynthetica* **2016**, *54*, 120–129. [CrossRef]

6. Zhang, Q.; Lou, Z. The environmental changes and mitigation actions in the Three Gorges Reservoir region, China. *Environ. Sci. Policy* **2011**, *14*, 1132–1138. [CrossRef]

7. Veach, A.M.; Dodds, W.K.; Jumpponen, A. Woody plant encroachment, and its removal, impact bacterial and fungal communities across stream and terrestrial habitats in a tallgrass prairie ecosystem. *Fems Microbiol. Ecol.* **2015**, *91*. [CrossRef] [PubMed]

8. Osborne, L.L.; Kovacic, D.A. Riparian vegetated buffer strips in water-quality restoration and stream management. *Freshwater Biol.* **1993**, *29*, 243–258. [CrossRef]

9. Richardson, D.; Holmes, P.; Esler, K.; Galatowitsch, S.; Stromberg, J.; Kirkman, S.; Pysek, P.; Hobbs, R. Riparian vegetation: Degradation, alien plant invasions, and restoration prospects. *Divers. Distrib.* **2007**, *13*, 126–139. [CrossRef]

10. Wang, C.; Li, C.; Wei, H.; Xie, Y.; Han, W. Effects of long-term periodic submergence on photosynthesis and growth of *Taxodium distichum* and *Taxodium ascendens* saplings in the hydro-fluctuation zone of the Three Gorges Reservoir of China. *PLoS ONE* **2016**, *11*, e0162867. [CrossRef] [PubMed]

11. Acosta-Martínez, V.; Cruz, L.; Sotomayor-Ramírez, D.; Pérez-Alegría, L. Enzyme activities as affected by soil properties and land use in a tropical watershed. *Appl. Soil Ecol.* **2007**, *35*, 35–45. [CrossRef]

12. Cui, Y.; Fang, L.; Guo, X.; Wang, X.; Zhang, Y.; Li, P.; Zhang, X. Ecoenzymatic stoichiometry and microbial nutrient limitation in rhizosphere soil in the arid area of the northern Loess Plateau, China. *Soil Biol. Biochem.* **2018**, *116*, 11–21. [CrossRef]

13. Stone, M.M.; DeForest, J.L.; Plante, A.F. Changes in extracellular enzyme activity and microbial community structure with soil depth at the Luquillo Critical Zone Observatory. *Soil Biol. Biochem.* **2014**, *75*, 237–247. [CrossRef]

14. Spohn, M.; Widdig, M. Turnover of carbon and phosphorus in the microbial biomass depending on phosphorus availability. *Soil Biol. Biochem.* **2017**, *113*, 53–59. [CrossRef]

15. Richardson, A.E.; Barea, J.M.; Mcneill, A.M.; Prigent-Combaret, C. Acquisition of phosphorus and nitrogen in the rhizosphere and plant growth promotion by microorganisms. *Plant Soil* **2009**, *321*, 305–339. [CrossRef]

16. Ajwa, H.A.; Dell, C.J.; Rice, C.W. Changes in enzyme activities and microbial biomass of tallgrass prairie soil as related to burning and nitrogen fertilization. *Soil Biol. Biochem.* **1999**, *31*, 769–777. [CrossRef]

17. Tischer, A.; Blagodatskaya, E.; Hamer, U. Microbial community structure and resource availability drive the catalytic efficiency of soil enzymes under land-use change conditions. *Soil Biol. Biochem.* **2015**, *89*, 226–237. [CrossRef]

18. Loeppmann, S.; Semenov, M.; Blagodatskaya, E.; Kuzyakov, Y. Substrate quality affects microbial and enzyme activities in rooted soil. *J. Plant Nut. Soil Sci.* **2016**, *179*, 39–47. [CrossRef]

19. Ye, C.; Cheng, X.; Zhang, Y.; Wang, Z.; Zhang, Q. Soil nitrogen dynamics following short-term revegetation in the water level fluctuation zone of the Three Gorges Reservoir, China. *Ecol. Eng.* **2012**, *38*, 37–44. [CrossRef]

20. Bao, S.D. *Soil Agricultural Chemistry Analysis*, 3rd ed.; China Agriculture Press: Beijing, China, 1999; pp. 30–109, ISBN 9787109066441.

21. Rodríguez-Kábana, R. The effects of crop rotation and fertilization on soil xylanase activity in a soil of the southeastern United States. *Plant Soil* **1982**, *69*, 97–104. [CrossRef]

22. Nannipieri, P.; Ceccanti, B.; Cervelli, S.; Matarese, E. Extraction of phosphatase, urease, proteases, organic carbon, and nitrogen from soil. *Soil Sci. Soc. Am. J.* **1980**, *44*, 1011–1016. [CrossRef]

23. Ye, S.; Yang, Y.; Xin, G.; Wang, Y.; Ruan, L.; Ye, G. Studies of the Italian ryegrass–rice rotation system in southern China: Arbuscular mycorrhizal symbiosis affects soil microorganisms and enzyme activities in the *Lolium mutiflorum* L. rhizosphere. *Appl. Soil Ecol.* **2015**, *90*, 26–34. [CrossRef]

24. Vance, E.D.; Brookes, P.C.; Jenkinson, D.S. An extraction method for measuring soil microbial biomass C. *Soil Biol. Biochem.* **1987**, *19*, 703–707. [CrossRef]

25. Brookes, P.C.; Landman, A.; Pruden, G.; Jenkinson, D.S. Chloroform fumigation and the release of soil nitrogen: A rapid direct extraction method to measure microbial biomass nitrogen in soil. *Soil Biol. Biochem.* **1985**, *17*, 837–842. [CrossRef]

26. Brookes, P.C.; Powlson, D.S.; Jenkinson, D.S. Measurement of microbial biomass phosphorus in soil. *Soil Biol. Biochem.* **1982**, *14*, 319–329. [CrossRef]

27. Wu, J.S.; Lin, Q.M.; Huang, Q.Y.; Xiao, H.A. *Soil Microbial Biomass Determination Method and Its Application*, 1st ed.; Meteorological Press: Beijing, China, 2006; pp. 59–81, ISBN 7-5029-4158-4.

28. Baldrian, P.; Merhautová, V.; Petránková, M.; Cajthaml, T.; Šnajdr, J. Distribution of microbial biomass and activity of extracellular enzymes in a hardwood forest soil reflect soil moisture content. *Appl. Soil Ecol.* **2010**, *46*, 177–182. [CrossRef]

29. Yoshitake, S.; Tabei, N.; Mizuno, Y.; Yoshida, H.; Sekine, Y.; Tatsumura, M.; Koizumi, H. Soil microbial response to experimental warming in cool temperate semi-natural grassland in Japan. *Ecol. Res.* **2015**, *30*, 235–245. [CrossRef]

30. Zavisic, A.; Polle, A. Dynamics of phosphorus nutrition, allocation and growth of young beech (*Fagus sylvatica* L.) trees in P-rich and P-poor forest soil. *Tree Physiol.* **2017**, *38*, 37–51. [CrossRef] [PubMed]

31. Spohn, M.; Zavišić, A.; Nassal, P.; Bergkemper, F.; Schulz, S.; Marhan, S.; Schloter, M.; Kandeler, E.; Polle, A. Temporal variations of phosphorus uptake by soil microbial biomass and young beech trees in two forest soils with contrasting phosphorus stocks. *Soil Biol. Biochem.* **2018**, *117*, 191–202. [CrossRef]

32. Lennon, J.T.; Jones, S.E. Microbial seed banks: The ecological and evolutionary implications of dormancy. *Nat. Rev. Microbiol.* **2011**, *9*, 119–130. [CrossRef] [PubMed]

33. Salazar, A.; Sulman, B.N.; Dukes, J.S. Microbial dormancy promotes microbial biomass and respiration across pulses of drying-wetting stress. *Soil Biol. Biochem.* **2018**, *116*, 237–244. [CrossRef]

34. Blagodatskaya, E.; Kuzyakov, Y. Active microorganisms in soil: Critical review of estimation criteria and approaches. *Soil Biol. Biochem.* **2013**, *67*, 192–211. [CrossRef]

35. Jastrow, J.D.; Amonette, J.E.; Bailey, V.L. Mechanisms controlling soil carbon turnover and their potential application for enhancing carbon sequestration. *Clim. Chang.* **2007**, *80*, 5–23. [CrossRef]

36. Zhang, S.; Li, Q.; Ying, L.; Zhang, X.; Liang, W. Contributions of soil biota to C sequestration varied with aggregate fractions under different tillage systems. *Soil Biol. Biochem.* **2013**, *62*, 147–156. [CrossRef]

37. Fang, X.; Zhou, G.; Li, Y.; Liu, S.; Chu, G.; Xu, Z.; Liu, J. Warming effects on biomass and composition of microbial communities and enzyme activities within soil aggregates in subtropical forest. *Biol. Fert. Soils* **2016**, *52*, 353–365. [CrossRef]

38. Sollins, P.; Homann, P.; Caldwell, B.A. Stabilization and destabilization of soil organic matter: Mechanisms and controls. *Geoderma* **1996**, *74*, 65–105. [CrossRef]

39. Stark, S.; Männistö, M.K.; Eskelinen, A. Nutrient availability and pH jointly constrain microbial extracellular enzyme activities in nutrient-poor tundra soils. *Plant Soil* **2014**, *383*, 373–385. [CrossRef]

40. Bettinahm, S.; Wang, C.P.; Chang, S.C.; Egbert, M. High precipitation causes large fluxes of dissolved organic carbon and nitrogen in a subtropical montane Chamaecyparis forest in Taiwan. *Biogeochemistry* **2010**, *101*, 243–256.

41. Jost, G.; Dirnböck, T.; Grabner, M.T.; Mirtl, M. Nitrogen leaching of two forest ecosystems in a karst watershed. *Water Air Soil Pollut.* **2011**, *218*, 633–649. [CrossRef]

42. Xu, Z.; Yu, G.; Zhang, X.; He, N.; Wang, Q.; Wang, S.; Wang, R.; Zhao, N.; Jia, Y.; Wang, C. Soil enzyme activity and stoichiometry in forest ecosystems along the North-South Transect in eastern China (NSTEC). *Soil Biol. Biochem.* **2017**, *104*, 152–163. [CrossRef]

43. Johnson, D.; Moore, L.; Green, S.; Leith, I.D.; Sheppard, L.J. Direct and indirect effects of ammonia, ammonium and nitrate on phosphatase activity and carbon fluxes from decomposing litter in peatland. *Environ. Pollut.* **2010**, *158*, 3157–3163. [CrossRef] [PubMed]

44. Fatemi, F.R.; Fernandez, I.J.; Simon, K.S.; Dail, D.B. Nitrogen and phosphorus regulation of soil enzyme activities in acid forest soils. *Soil Biol. Biochem.* **2016**, *98*, 171–179. [CrossRef]

45. Gress, S.E.; Nichols, T.D.; Northcraft, C.C.; Peterjohn, W.T. Nutrient limitation in soils exhibiting differing nitrogen availabilities: What lies beyond nitrogen saturation? *Ecology* **2007**, *88*, 119–130. [CrossRef]

46. Lambers, H.; Shane, M.W.; Cramer, M.D.; Pearse, S.J.; Veneklaas, E.J. Root structure and functioning for efficient acquisition of phosphorus: Matching morphological and physiological traits. *Ann. Bot.* **2006**, *98*, 693–713. [CrossRef] [PubMed]

47. Cleveland, C.C.; Townsend, A.R.; Schmidt, S.K.; Constance, B.C. Soil microbial dynamics and biogeochemistry in tropical forests and pastures, Southwestern Costa Rica. *Ecol. Appl.* **2003**, *13*, 314–326. [CrossRef]

48. Hall, S.J.; Silver, W.L. Iron oxidation stimulates organic matter decomposition in humid tropical forest soils. *Glob. Chang. Biol.* **2013**, *19*, 2804–2813. [CrossRef] [PubMed]

49. Zeng, Q.; Liu, Y.; Fang, Y.; Ma, R.; Lal, R.; An, S.; Huang, Y. Impact of vegetation restoration on plants and soil C: N: P stoichiometry on the Yunwu Mountain Reserve of China. *Ecol. Eng.* **2017**, *109*, 92–100. [CrossRef]
50. Cleveland, C.C.; Liptzin, D. C: N: P stoichiometry in soil: Is there a "Redfield ratio" for the microbial biomass? *Biogeochemistry* **2007**, *85*, 235–252. [CrossRef]
51. Reiners, W.A. Complementary models for ecosystems. *Am. Nat.* **1986**, *127*, 59–73. [CrossRef]
52. Li, X.Z.; Qu, Q.H. Soil microbial biomass carbon and nitrogen in Mongolian grassland. *Acta Pedol. Sin.* **2002**, *39*, 91–98.
53. Xue, S.; Liu, G.B.; Dai, Q.H.; Dang, X.H.; Zhou, P. Effect of different vegetation restoration models on soil microbial biomass in eroded hilly Loess Plateau. *J. Nat. Resour.* **2007**, *22*, 20–27. [CrossRef]
54. Xu, X.; Hui, D.; King, A.W.; Song, X.; Thornton, P.E.; Zhang, L. Convergence of microbial assimilations of soil carbon, nitrogen, phosphorus, and sulfur in terrestrial ecosystems. *Sci. Rep.* **2015**, *5*, 17445. [CrossRef] [PubMed]
55. Yu, Q.; Elser, J.J.; He, N.; Wu, H.; Chen, Q.; Zhang, G.; Han, X. Stoichiometric homeostasis of vascular plants in the Inner Mongolia grassland. *Oecologia* **2011**, *166*, 1–10. [CrossRef] [PubMed]
56. Sterner, R.W.; Elser, J.J. *Ecological Stoichiometry: The Biology of Elements from Molecules to the Biosphere*, 1st ed.; Elser Princeton University Press: Princeton, NJ, USA, 2002; pp. 225–226.
57. Tapia-Torres, Y.; Elser, J.J.; Souza, V.; García-Oliva, F. Ecoenzymatic stoichiometry at the extremes: How microbes cope in an ultra-oligotrophic desert soil. *Soil Biol. Biochem.* **2015**, *87*, 34–42. [CrossRef]
58. Sinsabaugh, R.L.; Lauber, C.L.; Weintraub, M.N.; Ahmed, B.; Allison, S.D.; Crenshaw, C.; Contosta, A.R.; Cusack, D.; Frey, S.; Gallo, M.E.; et al. Stoichiometry of soil enzyme activity at global scale. *Ecol. Lett.* **2008**, *11*, 1252–1264. [CrossRef] [PubMed]

forests

MDPI

Article

Distribution Changes of Phosphorus in Soil–Plant Systems of Larch Plantations across the Chronosequence

Fanpeng Zeng [1,2], Xin Chen [1], Bin Huang [1] and Guangyu Chi [1,*]

[1] Key Laboratory of Pollution Ecology and Environmental Engineering, Institute of Applied Ecology, Chinese Academy of Sciences, Shenyang 110016, China; zengfp@iae.ac.cn (F.Z.); chenxin@iae.ac.cn (X.C.); huangbin@iae.ac.cn (B.H.)

[2] University of Chinese Academy of Sciences, Beijing 100049, China

* Correspondence: chigy@iae.ac.cn; Tel.: +86-024-8397-0425

Received: 23 August 2018; Accepted: 11 September 2018; Published: 13 September 2018

Abstract: Phosphorus (P) is one of the most important factors influencing the growth and quality of larch plantations. A systematic knowledge of the dynamic changes of P in soil–plant systems can provide a theoretical basis for the sustainable development of larch plantations. We determined the concentration, biomass, and accumulation of P in five tree components (i.e., leaf, branch, bark, stem, and root), and the concentrations of various soil P fractions of larch plantations in 10-, 25-, and 50-year-old stands in northeast China. Our results showed that the N:P ratio and P concentration in leaves increased with stand age, indicating that the growth of larch plantations might be limited by P in the development of stands. The N:P ratio and P concentration in roots, and P resorption efficiency, increased with stand age, indicating the use efficiency of P could be enhanced in older stands. The concentrations of soil-labile P fractions (Resin-P, $NaHCO_3$-Pi, and $NaHCO_3$-Po) in 25- and 50-year-old stands were significantly lower than those in 10-year-old stands, indicating the availability of soil P decreases with the development of larch plantations.

Keywords: leaf N:P ratio; P resorption efficiency; soil P fractions; P stock; stand age

1. Introduction

Larch (*Larix kaempferi*) is a major plantation species that is widely planted in northeast China because of its high yield and timber quality [1]. Since the 1960s, to meet the great demand for timber, large amounts of secondary forests in northeast China have been replaced by larch plantations [2]. However, due to a lack of efforts to convert larch plantations into mixed forests and the use of improper harvesting and thinning types (whole-tree thinning and harvesting methods) for larch plantations, there has been a decline in soil nutrients in these areas [3–5]. Among these depleted soil nutrients, phosphorus (P) has gradually become a major element affecting the growth of larch plantations with the increase of stand age [6,7].

The dynamic change of P is crucial to assess the growth and function of a forest. Further, it is closely related to the ability of soil to supply P, and the ability of plants to extract available P from the soil and to cycle absorbed P among different plant components [8–10]. Soil P exists in many chemical forms, including labile P and stable P forms in Hedley fractionation [11]. The transformation between different soil P forms is important for the availability of soil P for plants [12]. Although a few studies have reported on the changes of soil P fractions in larch plantations, these studies did not focus on the long-time variation of P throughout the development of larch plantations. Thus, knowledge of the variations in soil P fractions across the chronosequence can help us better understand the supply capacity of soil P with forest development. The concentration and accumulation of nutrients in different components of plants are important factors influencing plant growth [13]. Many studies have reported biomass- and nutrient-allocation strategies in plants [8,13–16]. However, these studies were not based on a destructive sampling method, especially the above- and below-ground components along a chronosequence. Moreover, few studies have focused on the study of the combination of P between soil and plants with the development of pure plantations.

In this study, we examined the changes in soil P fraction concentration and plant P allocation in larch plantations across different age classes (10-, 25-, and 50-year-old) in northeast China. The objectives of the study were to (1) reveal the variation of P distribution in soil–plant systems along an age sequence of a larch plantation, and (2) assess the availability of soil P and the ability of plants to use P across the chronosequence. We hope to provide a theoretical basis for effective P management strategies for the sustainable development of larch plantations.

2. Materials and Methods

2.1. Site Description and Experimental Design

The study was conducted at the Qingyuan Forest CERN (Chinese Ecosystem Research Network), Chinese Academy of Sciences in Liaoning Province, China (41°51′ N, 124°54′ E). The climate of this region belongs to the continental monsoon climate, with humid and rainy summers and cold and dry winters. The annual temperature is 4.7 °C and the minimum monthly and maximum monthly temperatures are −12.1 °C in January and 21.0 °C in July, respectively [17]. Annual rainfall is 700–850 mm, with more than 80% falling from July to August [1].

The study site was firstly occupied by primary mixed broadleaved Korean pine forests until the 1930s and was subsequently subjected to decades of unregulated timber removal. In the early 1950s, the original forests at this study site were completely cleared off by a large fire, and then the forest was replaced by a mixture of naturally regenerating broadleaved native tree species. Since the 1960s, the secondary natural forests at the study site were cleared for larch plantations [2].

Three stands of larch plantations, 10-year-old (young), 25-year-old (half-mature), and 50-year-old (mature) stands, were selected. Each stand was selected on a narrow range of altitudes (525–650 m) and slopes (13–17°) to minimize the differences caused by topographical features. The soil of all stands is typical brown forest soil according to the second edition of the United States Department of Agriculture soil taxonomy, with 25.6% sand, 51.2% silt, and 23.2% clay, on average, and soil depth of 40–50 cm [2]. All of the stands were in their first rotation and were developed by replacing the secondary forests. Therefore, all the stands shared similar geology, microenvironment conditions, and previous land uses, varying only in the age of the plantations. Thus, the preconditions of all the stands were appropriate for our chronosequence study. In each stand, three 20 × 20 m sample plots were laid out with three >10 m buffer zones between them. The diameter at breast height (DBH) and height of all the individual trees in each sample plot were measured. The trees in each stand were divided into 5 DBH classes based on DBH distribution. Five sample trees from each sample plot with different DBHs were selected within each stand. The basic information of the stands is presented in the Table 1.

Table 1. Summary of stand properties of the three age classes of larch plantations.

Stand Properties	Stand Age (Years)		
	10	25	50
Elevation (m)	525 (3)	620 (3)	650 (5)
Slope (°)	13 (0.7)	15 (0.9)	17 (0.6)
Density (trees ha^{-1})	3688 (320)	1966 (220)	960 (130)
DBH (cm)	6.15 (0.1)	15.25 (0.2)	25.08 (0.4)
Tree height (m)	7.20 (0.3)	18.24 (0.1)	24.38 (0.6)
Soil type	Typical brown forest soil	Typical brown forest soil	Typical brown forest soil
Soil bulk density (g cm^{-3})	1.41 (0.05)	1.13 (0.07)	1.23 (0.06)
SOC (%)	2.57 (0.24)	3.00 (0.09)	3.14 (0.39)
Soil total Fe (%)	2.47 (0.13)	2.88 (0.11)	2.32 (0.07)
Soil total Al (%)	6.33 (0.56)	6.34 (0.31)	5.98 (0.54)
Soil pH	5.98 (0.09)	5.79 (0.11)	5.84 (0.06)

Values in the table are the means (standard errors) of the 3 plots per stand $n = 3$. DBH: diameter at breast height; SOC: soil organic carbon.

2.2. Plant Sampling and Analysis

The sample trees of each stand were cut down in August 2015 as leaf biomass reached its peak and nutrient concentrations were stable. The stem of the sample tree was divided into segments with a length of 1 m. Next, the branch was cut down from the stem. All leaves attached to branches were picked off and divided into upper, middle, and lower layers. The roots were dug out completely, and the soil and rocks attached to the roots were cleaned away. The fresh weight of each component was immediately measured with an electronic scale. After weighing, a stem sample with a width of 2 cm was collected from each segment. In the middle of each stem segment, a 10 cm length of bark was stripped as the bark sample. A leaf sample was collected from the three layers. The root sample was collected according to the diameter of the root. At the end of September 2015, when the leaves of the larch plantation were freshly senesced, 10 litter collectors were laid out in each sample plots to obtain the senesced leaf samples. The samples of each component were immediately sent to the laboratory for drying at 65 °C until the weight was unchanged to obtain the moisture content. The biomass of each different component was obtained by adjusting the fresh weight with the respective moisture content. The P accumulation in each component was calculated by multiplying the P concentration by its respective biomass. The total biomass and P accumulation of each individual tree was calculated by summing the different components. The P resorption efficiency (PRE) was calculated as follows: PRE (%) = ((Pg − Ps × MLCF)/Pg) × 100, where Pg and Ps represent P concentrations in green and senesced leaves. The mass-loss correction factor (MLCF) value was 0.745 for conifers [18]. The larch P concentration was determined by H_2SO_4-H_2O_2 digestion, and the amount of P in each component was determined by the Murphy and Riley method [19].

2.3. Soil Sampling and Analysis

Soil bulk density samples were collected by a known volume soil cutting ring for each sample plots, then dried at 105 °C for 12 h, and reweighed to measure soil bulk density. An S-shaped curve (5 sampling plots) was randomly arranged to collect the soil cores at depths of 0–20 cm in each sample plot in July 2015. The 5 soil cores were then mixed into a composite soil sample for each sample plot. Therefore, there were 3 soil samples for each stand. The soil samples were air-dried, ground, and passed through a 0.15 mm sieve for the analysis of soil P fractions.

Soil P fractions were determined by the modified Hedley sequential extraction method [11,20]. Half a gram of each soil sample was weighed into a 50 mL centrifuge tube, and different soil P fractions were sequentially extracted by the following extraction steps: (I) Resin-P: soil was extracted with 30 mL of deionized water and a resin strip. (II) $NaHCO_3$-P: the residue from the first extraction was further extracted with 30 mL of 0.5 M $NaHCO_3$ (adjusted to pH 8.5). One set was oxidized to determine the total $NaHCO_3$-P ($NaHCO_3$-Pt). The other set was used for the determination of $NaHCO_3$ inorganic P

(NaHCO$_3$-Pi). (III) NaOH-P: the residue from the second extraction was then extracted with 30 mL of 0.1 M NaOH. One set was oxidized for the determination of total NaOH-P (NaOH-Pt). The other set was used for the determination of NaOH inorganic P (NaOH-Pi). (IV) HCl-P: the residue from the third extraction was further extracted with 30 mL of 1 M HCl. (V) Residual-P: the residue from the last extraction was overdried at 60 °C and transferred to a conical flask and digested with conc. H$_2$SO$_4$ and HClO$_4$. The amount of Residual-P was determined by the Murphy and Riley method [19]. The amounts of other soil P fractions were determined by the malachite green method [21].

2.4. Statistical Analysis

A one-way ANOVA test was conducted to evaluate the influence of stand age on soil P fractions, larch P concentrations, and N:P ratio. LSDs based on multiple posthoc comparisons ($p < 0.05$) were performed to evaluate the difference between the three stand ages. All statistical analyses were performed using SPSS software package version 23.0.

3. Results

3.1. P Distribution Changes in Tree Components with Stand Age on Larch Plantations

Generally, the concentration of P in different tree components showed a consistent tendency in the three stands; P concentration decreased in the order of leaf > branch > root > bark > stem. The P concentration observed in leaves was approximately 10 times greater than that in the stems. Stand age had a significant influence on the concentration in the leaves, branches, and roots. The concentration of P in leaves decreased with increased stand age. P concentration in the roots in the 50-year-old stand was significantly higher than that in the 10- and 25-year-old stands. Stand age had no significant influence on the P concentration of the bark and stem (Table 2). The green-leaf N:P ratio significantly increased from 13.9 in the 10-year-old stand to 15.4 and 18.4 in the 25- and 50-year-old stands, respectively. The senesced leaf N:P ratio showed consistent tendency with green leaves, ranging from 6.3 to 32.6. In contrast, the root N:P ratio significantly decreased with stand age, ranging from 23.3 to 11.0, while stand age had no significant influence on the N:P ratio of the bark and stem (Table 2).

Table 2. P concentrations and N:P ratio in different larch-plantation components.

Tree Component	Phosphorus (P) Concentration (g kg^{-1})		
	10-Year-Old	**25-Year-Old**	**50-Year-Old**
Green leaf	1.8 (0.04) A	1.6 (0.02) B	1.0 (0.01) C
Root	0.3 (0.01) B	0.3 (0.01) B	0.5 (0.01) A
Bark	0.2 (0.02) A	0.2 (0.01) A	0.2 (0.03) A
Branch	0.5 (0.02) C	0.8 (0.01) A	0.7 (0.05) B
Stem	0.1 (0.01) A	0.1 (0.01) A	0.1 (0.01) A
Senesced leaf	1.3 (0.02) A	0.9 (0.01) B	0.2 (0.01) C
N:P Ratio			
Green Leaf	13.9 (0.22) C	15.4 (0.56) B	18.8 (0.42) A
Root	23.3 (0.32) A	15.8 (0.40) B	11.0 (0.58) C
Bark	12.5 (0.24) A	12.9 (0.12) A	12.7 (0.36) A
Branch	7.5 (0.47) A	4.8 (0.36) C	6.1 (0.67) B
Stem	18.2 (0.46) A	20.0 (0.84) A	18.8 (0.64) A
Senesced leaf	6.3 (0.21) A	8.1 (0.34) B	32.6 (1.21) C

Values in the table are the means (standard errors) of the three plots per stand $n = 3$. The different letters indicate groups with significant differences between different stand ages in each tree component ($p < 0.05$).

On the whole, total tree biomass increased from 9.62 kg tree^{-1} in the 10-year-old stand to 119.79 and 372.38 kg tree^{-1} in the 25- and 50-year-old stands, respectively (Table 3). Individual P accumulation for the 10-, 25-, and 50-year-old stands was 5.64, 35.95, and 62.20 g tree^{-1}, respectively. In increasing order, P was contained in the bark (i.e., 3%, 6%, and 10% in the 10-, 25-, and 50-year-old stands, respectively) and the roots (i.e., 8%, 10%, and 19% in the 10-, 25-, and 50-year-old stands, respectively). The relative contribution of leaves to total P accumulation decreased from 47% in the 10-year-old stand to 17% and 10% in the 25- and 50-year-old stands, respectively. The relative proportion of branches to total P accumulation was 35%, 60%, and 53% in the 10-, 25-, and 50-year-old stands, respectively. The relative share of stems in total P accumulation was 7%, 8%, and 8% in the 10-, 25-, and 50-year-old stands, respectively (Figure 1).

Table 3. Biomass of each tree component for different stand ages in larch plantations.

Tree Component	Biomass (kg tree^{-1})		
	10-Year-Old	**25-Year-Old**	**50-Year-Old**
Leaf	1.50	3.86	5.87
Root	1.69	13.84	52.77
Bark	2.00	21.47	33.23
Branch	0.76	10.85	35.80
Stem	3.67	69.77	244.71
Total tree	9.62	119.79	372.38

The total tree values were obtained as the sum of all components.

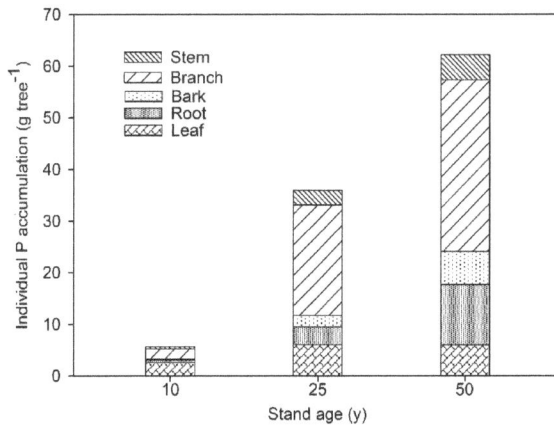

Figure 1. Individual P accumulation of each tree component for different stand ages in larch plantations.

3.2. Variation of Soil P Fractions with Stand Age of Larch Plantations

The concentration of each type of soil P fraction of different stand ages is presented in Figure 2. Generally, Resin-P accounted for around 1% of the soil total P. The proportion of soil P held in the available P form (NaHCO$_3$-P) was decreased from 18.0% in the 10-year-old stand to 10.5% in the 25-year-old stand and 10.0% in the 50-year-old stand. NaOH-P, which is known as a moderately stable P, accounted for the second-largest fraction of the soil total P in each stand, with a fraction of 40.0%, 35.7%, and 32.8% in the 10-, 25-, and 50-year-old stands, respectively. HCl-P accounted for the 12.2%, 11.4%, and 5.3% of the soil total P in the 10-, 25-, and 50-year-old stands, respectively. The percentage of residual P, which belongs to the most stable P pool, increased from 39.6% in the 10-year-old stand to 44.7% in the 25-year-old stand and 53.3% in the 50-year-old stand (Figure 2).

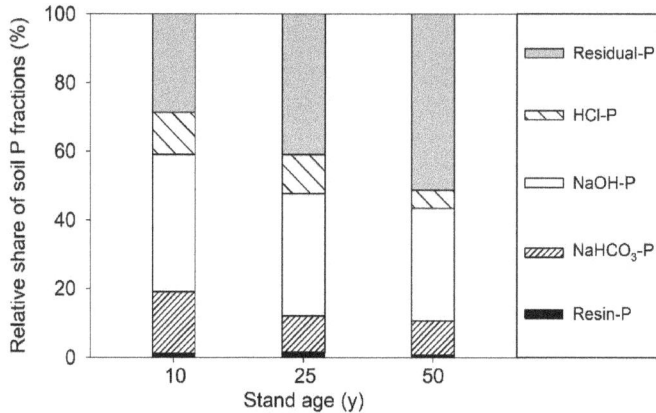

Figure 2. Distribution of soil P fractions for different stand ages on larch plantations. $NaHCO_3$-P is the sum of $NaHCO_3$-Pi and $NaHCO_3$-Po. NaOH-P is the sum of NaOH-Pi and NaOH-Po.

With an increased stand age, the concentrations of soil inorganic P fractions, such as Resin-P, $NaHCO_3$-Pi, and HCl-P, significantly reduced. The highest concentration of each of these inorganic fractions was observed in the 10-year-old stand and was approximately triple of that in the 50-year-old stand. The soil organic P fractions ($NaHCO_3$-Po and NaOH-Po) and the Residual-P in the 25-year-old stand were significantly lower than those in the 50-year-old stand. In contrast, the concentration of NaOH-Pi in the 25-year-old stand was much higher than that in the 50-year-old stand (Table 4).

Table 4. Variations of soil P fractions for different age classes of larch plantations.

Stand Age (years)	Resin-P (mg kg^{-1})	$NaHCO_3$-P (mg kg^{-1})		NaOH-P (mg kg^{-1})		HCl-P (mg kg^{-1})	Residual-P (mg kg^{-1})
		$NaHCO_3$-Pi	$NaHCO_3$-Po	NaOH-Pi	NaOH-Po		
10	4.3 (0.06) A	19.8 (0.23) A	24.0 (1.88) A	29.3 (0.65) B	125.8 (3.97) A	47.2 (0.82) A	153.4 (7.27) A
25	3.7 (0.18) B	11.8 (0.81) B	13.7 (1.35) B	34.9 (1.43) A	52.1 (6.33) C	27.7 (0.45) B	108.9 (5.45) B
50	1.9 (0.08) C	5.2 (0.30) C	21.5 (1.21) A	14.9 (1.52) C	74.3 (8.38) B	14.4 (0.67) C	145.1 (5.90) A

Values in the table are the means (standard errors) of the three plots per stand $n = 3$. Pi: inorganic P; Po: organic P. Different letters indicate groups with significant differences between different stand ages in each soil P fraction ($p < 0.05$).

Soil labile P is usually obtained as the sum of Resin-P, $NaHCO_3$-Pi, and $NaHCO_3$-Po [11]. The highest concentration of soil labile P was observed in the 10-year-old stand, which was 40% higher than that in the 25- and 50-year-old stands. Furthermore, there was no significant difference in soil labile P between the 25- and 50-year-old stands (Figure 3).

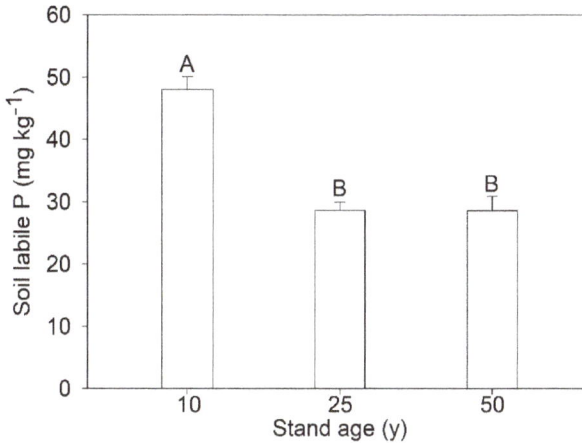

Figure 3. Variations in the concentration of soil labile P (the sum of Resin-P, NaHCO$_3$-Pi, and NaHCO$_3$-Po) with stand age on larch plantation. Thick bars represent the means and thin bars represent standard errors for *n* = 3. Different letters indicate group with significant differences for *p* < 0.05.

3.3. Stock of Soil Labile P and Larch P

The stock of soil labile P at 0–20cm was obtained as: soil labile P stock (kg ha^{-1}) = soil labile P concentration (reported in Figure 3) × soil depth × soil bulk density (reported in Table 1). P stocks in the soil labile P pool for the 10-, 25-, and 50-year-old stands were 135.61, 66.15, and 70.53 kg ha^{-1}, respectively. The stock of larch P was obtained as: larch P stock (kg ha^{-1}) = individual tree P accumulation (reported in Figure 1) × stand density (reported in Table 1). The stock of larch P, which increased with the stand age, varied from 15.87 kg ha^{-1} in the 10-year-old stand to 50.15 and 65.07 kg ha^{-1} in the 25- and 50-year-stands, respectively. The highest P stock in green leaves was observed in the 25-year-old stand, with a value of 11.91 kg ha^{-1}. The P stocks in green leaves at the 10- and 50-year-old stands were 9.79 and 5.47 kg ha^{-1}, respectively. The P stocks in senesced leaves in the 50-year-old stand was 0.93 kg ha^{-1}, which was almost five times lower than that in the 10- and 25-year-old stands. The PRE increased with the stand age and varied from 44.60% to 83.71% (Table 5).

Table 5. The stock of soil labile P and larch P.

Stand Age (years)	Soil Labile P Stock (kg ha^{-1})	Larch P Stock (kg ha^{-1})	Green Leaf P Stock (kg ha^{-1})	Senesced Leaf P Stock (kg ha^{-1})	PRE (%)
10	135.61	15.87	9.79	5.42	44.60
25	66.15	50.15	11.91	5.01	58.01
50	70.53	65.07	5.47	0.93	83.71

The soil labile P stock is the sum of the soil Resin-P, soil NaHCO$_3$-Pi, and soil NaHCO$_3$-Po stocks in 0–20 cm soil. PRE: P resorption efficiency.

4. Discussion

4.1. Variations of Larch N:P Ratio, PRE, P Concentration, and P Accumulation across the Larch-Plantation Chronosequence

N:P stoichiometry is widely used as an indicator of N and P balance and sources of ecosystems [22]. The distribution of nutrient concentrations in different tree components is closely related to nutrient use strategies of trees in certain conditions [23]. In our study, the N:P ratio and P concentration in the roots increased with stand age, while the root N:P ratio and P concentration in the leaves decreased with stand age. These variations of N:P ratio and P concentration might be caused by the different nutrient use strategies at different growth stages of the trees on larch plantations. Some previous studies reported that nutrient use efficiency and retranslocation efficiency in older stands were significantly higher than those in young stands [24–26]. Therefore, older larches tend to decrease P concentration in leaves and resorb more P with forest development. Leaf N:P is usually considered to be a predictor to evaluate the limitation of N and P on the plant [22]. Variations in the larch N:P ratio may indicate that the growth of a larch plantation could be limited by P with the development of stands. The PRE increased with stand age, ranging from 44.60% to 83.71%. The PREs in larches are higher than in other plants. This may be due to the fact that the larch is a kind of deciuous confier that can use its high PRE to make itself less dependent on soil P supply [23,27]. Our findings are consistent with the results of Chen et al., and Yan et al. [6,7], indicating that the larch plantations in northeast China are facing increasing P limitation.

In our study, we observed that the larches increased the relative share of biomass in the remaining component (stem) and decreased the relative share of biomass in the returning components (leaf and branch) with forest development. As for underground biomass allocation, the relative share of the biomass of the roots increased from 12% in the 10-year-old stand to 14% in the 50-year-old stand. Similar findings were also observed in three woody species [28]. Some studies have reported that plants can increase root-biomass allocation and decrease leaf-biomass allocation in response to nutrient limitations [13,29]. However, there is no direct evidence in our study to prove that the different above-ground and below-ground biomass allocation may result from P limitation. The changes in biomass allocation might be mainly caused by stand age.

Our results showed that the relative share of P accumulation in the crown components (the sum of leaf and branch) decreased with stand age. Sardans and Peñuelas [30] found that the allocation of P in leaves has a closer relationship with above-ground growth. The higher portion of P accumulation in the crown components of the 10- and 25-year-old stands may be due to the allocation of more nutrients to leaves and branches in response to rapid growth [31], while the lower relative share of P accumulation in the crown components of the mature stand may result from the trees' nutrient allocation strategies for adapting to soil-fertility conditions [13]. Our results showed the relative contribution of root P concentration and accumulation increased with stand age. Similar findings observed in pines [32] indicated that plants may increase their root allocation to strengthen P acquisition in response to P limitation.

By analyzing the variations of P accumulation in different components, we found that stems constitute less than 10% of the total P accumulation. while P content in leaves, branches, and roots accounts for more than 80% of total P accumulation. Our results may indicate that P loss could be alleviated by leaving the leaf, branch, and root components in the field when larch plantations are thinned and harvested.

4.2. Variations of Soil P Fractions across the Larch-Plantation Chronosequence

In our study, the content of soil labile P was significantly lower in the 25- and 50-year-old stands than that in the 10-year-old stand, while there was no significant difference between the 25- and 50-year-old stands. A possible explanation for the decline of soil labile P in the 25-year-old stand is an improper-harvesting scenario. The whole-tree thinning and harvesting method for larch plantations around the age of 25 years is common in northeast China and can cause a large loss of the amounts of soil nutrients and the depletion of soil P [4]. The variations of soil labile P in the 25- and 50-year-old stand may be the result of increased PRE. According to our results, the PRE in the 50-year-old stand is much higher than that in the 25-year-old stand, indicating that older larches could reduce their dependence on soil P supply. The variations of soil Resin-Pi, $NaHCO_3$-Pi, and $NaHCO_3$-Po fractions during forest development may result from differences in behavior, mobility, and availability [33]. Among these labile P fractions, Resin-P can be absorbed directly by the plant, and $NaHCO_3$-Pi can adhere to the solid phase and become the available P for the plant when the concentration of Resin-P decreases [34]. Some of the moderately stable P pool (especially NaOH-P) can also contribute to the labile P pool under certain conditions [35,36]. In our study, we observed a decrease of NaOH-Po and an increase of NaOH-Pi in the 25-year-old stand. Similar findings were observed in the Luquillo experimental forest [37]. The content variation of NaOH-Po with stand age may be due to the mineralization. The mineralization of organic P forms plays an important role in supplying P nutrients under P deficiency [36]. Although we do not have direct evidence that larches have used the stable P forms in the old stand, the increase of the soil inorganic P fractions and decrease of soil organic P fractions in the 25-year-old stand may help support the increase in mineralization of organic P. In our study, residual P accounted for the largest fraction of the soil total P in each age-class stand, and decreased in the order of 10- > 50- > 25-year-old stand. Higher residual P contents in the youngest plantation (10 years) may be the result of the remainder of the harvest and removal from the prior secondary forest. Our results are in agreement with Walker and Syers' findings, which indicated that soil P is dominated by organic P and occluded P in the late stages of soil development [38].

4.3. Stock of Larch P and Soil Labile P

Soil labile P is considered to be the pool of P most likely to contribute to plant-available P. By measuring the stock of labile P and larch P for the 10–50-year-old stands, we found that the decline of labile P stock is about the same as the increase of larch P stock. This implies that there might be a close balance between the uptake of P from the soil and the increase in P in the biomass. In our study, the stock of larch P increased in the order 10- > 25- > 50-year-old stands, while the stock of soil labile P decreased in the order 10- > 50- > 25-year-old stands. The difference of larch P stock and soil labile P stock with stand age may be caused by the following reasons. First, the peak of larch growth occurs between 10- and 25-years and needs large amounts of P to maintain the plants' rapid growth [39]. Soil labile P is considered to be the pool of P most likely to contribute to the plant-available P [37]. Therefore, the increase of larch P stock and the decline of soil labile P stock mainly occur between the 10- and 25-year-old stands. Second, nutrient resorption is one of the most important nutrient use strategies of plants [23,39,40]. In our study, the highest PRE was observed in the 50-year-old stand, and was around two times higher than that in the 10-year-old stand. Therefore, the 50-year-old stand could resorb more nutrients from senescing leaves and reduce its dependence on the soil labile P supply. Third, the primary source of P was from rock weathering, which occurs at an extremely slow rate [38]. Thus, P recovery could be very difficult when P output exceeds P input for the ecosystem, especially for those in an infertile environment.

5. Conclusions

Leaf N:P ratio increased with stand age, and the content of soil labile P of larch plantations decreased with stand age. This might indicate that larch plantations in northeast China are facing increasing P limitation and the availability of soil P has decreased with the development of larch plantations. Decreased P concentration and the relative share of P accumulation in the leaves, increased P concentration and the relative proportion of P accumulation in the roots, and increased PRE across the chronosequence indicate that larches might improve their use efficiency of P in response to increasingly acute P limitation in older stands. We should pay attention to increasing P limitation across the larch-plantation chronosequence in order to maintain the sustainable development of larch plantations. Nevertheless, the change of P in soil–plant systems is a multifactor effect that depends on soil properties, rhizosphere, microbial activities, plant process, etc. Further research is necessary to better understand these multiple effects and interactions on the growth of larch plantations.

Author Contributions: G.C. and X.C. conceived and designed the experiment; F.Z. performed the experiment and analyzed the data; F.Z and B.H. wrote the manuscript.

funding: This research was funded by grants from the National Natural Science Foundation of China (31470624).

Acknowledgments: We are grateful to Qingyuan Forest CERN, Chinese Academy of Sciences for providing the experimental sites and relevant support. We thank T.Y. and W.J. for the help with field measurements.

Conflicts of Interest: The authors declare no conflict of interest.

References

1. Zhu, J.J.; Liu, Z.G.; Wang, H.X.; Yan, Q.L.; Fang, H.Y.; Hu, L.L.; Yu, L.Z. Effects of site preparation on emergence and early establishment of Laix olgensis in montane regions of northeast China. *New For.* **2008**, *36*, 247–260. [CrossRef]
2. Yang, K.; Zhu, J.J.; Yan, Q.L.; Sun, J.O. Changes in soil P chemistry as affected by conversion of nature secondary forests to larch plantations. *For. Ecol. Manag.* **2010**, *260*, 422–428. [CrossRef]
3. Yang, K.; Shi, W.; Zhu, J.J. The impact of secondary forests conversion into larch plantations on soil chemical and microbiological properties. *Plant Soil* **2013**, *368*, 535–546. [CrossRef]
4. Yan, T.; Zhu, J.J.; Yang, K.; Yu, L.Z.; Zhang, J.X. Nutrient removal under different harvesting scenarios for larch plantations in northeast China: Implications for nutrient conservation and management. *For. Ecol. Manag.* **2017**, *400*, 150–158. [CrossRef]
5. Zheng, X.F.; Yuan, J.; Zhang, T.; Hao, F.; Jose, S.; Zhang, S.X. Soil Degradation and the decline of available nitrogen and phosphorus in soils of the main forest types in the Qinling Mountains of China. *Forests* **2017**, *8*, 460. [CrossRef]
6. Chen, L.X.; Zhang, C.; Duan, W.B. Temporal variations in phosphorus fractions and phosphatase activities in rhizosphere and bulk soil during the development of Larix olgensis plantations. *J. Plant Nutr. Soil Sci.* **2016**, *179*, 67–77. [CrossRef]
7. Yan, T.; Lü, X.T.; Zhu, J.J.; Yang, K.; Yu, L.Z.; Gao, T. Changes in nitrogen and phosphorus cycling suggest a transition to phosphorus limitation with the stand development of larch plantations. *Plant Soil* **2018**, *422*, 385–396. [CrossRef]
8. Slazak, A.; Freese, D.; Matos, E.S.; Hüttl, R.F. Soil organic phosphorus fraction in pine-oak forest stands in Northeastern Germany. *Geoderma* **2010**, *158*, 156–162. [CrossRef]
9. Galván-Tejada, N.C.; Peña-Ramírez, V.; Mora-Palomino, L.; Siebe, C. Soil P fractions in a volcanic soil chronosequence of Central Mexico and their relationship to foliar P in pine trees. *J. Plant Nutr. Soil Sci.* **2014**, *177*, 792–802. [CrossRef]
10. Shiau, Y.J.; Pai, C.W.; Tsai, J.W.; Liu, W.C.; Yam, R.W.; Chang, S.C.; Tang, S.L.; Chiu, C.Y. Characterization of Phosphorus in a Toposequence of Subtropical Perhumid Forest Soils Facing a Subalpine Lake. *Forests* **2018**, *9*, 294. [CrossRef]
11. Hedley, M.J.; Stewart, W.B.; Chauhan, B.S. Changes in inorganic and organic soil phosphorus fractions induced by cultivation practices and by laboratory incubations. *Soil Sci.* **1982**, *46*, 970–976. [CrossRef]
12. Cross, A.F.; Schesinger, W.H. A literature review and evaluation of the Hedley fractionation: Applications to the biogeochemical cycle of soil phosphorus in natural ecosystems. *Geoderma* **1995**, *64*, 197–214. [CrossRef]
13. Zhang, K.; Su, Y.Z.; Yang, R. Biomass and nutrient allocation strategies in a desert ecosystem in the Hexi Corridor, northwest China. *J. Plant Res.* **2017**, *130*, 699–708. [CrossRef] [PubMed]
14. Müller, I.; Schmid, B.; Weiner, J. The effect of nutrient availability on biomass allocation patterns in 27 species of herbaceous plants. *Perspect. Plant Ecol. Evol. Syst.* **2000**, *3*, 115–127. [CrossRef]
15. Glynn, C.; Herms, D.A.; Egawa, M.; Hansen, R.; Mattson, W.J. Effects of nutrient availability on biomass allocation as well as constitutive and rapid induced herbivore resistance in poplar. *Oikios* **2003**, *101*, 385–397. [CrossRef]
16. Bargaza, A.; Noyceb, G.L.; Fulthorpeb, R.; Carlssona, G.; Furzeb, J.R.; Jensena, E.S.; Dhiba, D.; Isaacb, M.E. Species interactions enhance root allocation, microbial diversity and P acquisition in intercropped wheat and soybean under P deficiency. *Appl. Soil Ecol.* **2017**, *120*, 179–188. [CrossRef]
17. Yang, K.; Zhu, J.J. The effects of N and P additions on soil microbial properties in paired stands of temperate secondary forests and adjacent larch plantations in Northeast China. *Soil Biol. Biochem.* **2015**, *90*, 80–86. [CrossRef]
18. Vergutz, L.; Manzoni, S.; Porporato, A.; Novais, R.F.; Jackson, R.B. Global resorption efficiencies and concentrations of carbon and nutrients in leaves of terrestrial plants. *Ecol. Monogr.* **2012**, *82*, 205–220. [CrossRef]

19. Murphy, J.; Riley, J.P. A modified single solution method for determination of phosphate in natural waters. *Anal. Chim. Acta* **1962**, *27*, 31–36. [CrossRef]

20. Tiessen, H.; Moir, J.O. Characterization of available P by sequential extraction. In *Oil Sampling and Methods of Analysis*, 2nd ed.; Carter, M., Gregorich, E.G., Eds.; CRC Press: Boca Raton, FL, USA, 2008; pp. 293–305; ISBN 9781420005271.

21. Ohno, T.; Zibilske, L.M. Determination of low concentrations of phosphorus in soil extracts using malachite green. *Soil Sci. Soc. Am. J.* **1991**, *46*, 892–895. [CrossRef]

22. Koerselman, W.; Meuleman, A.F.M. The vegetation N:P a new tool to detect the nature of nutrient limitation. *J. Appl. Ecol.* **1996**, *33*, 1441–1450. [CrossRef]

23. Kobe, R.K.; Lepczyk, C.A.; Iyer, M. Resorption efficiency decreases with increasing green leaf nutrients in a global data set. *Ecology* **2005**, *86*, 2780–2792. [CrossRef]

24. Gholz, R.F.; Fisher, R.F.; Prichett, W.L. Nutrient Dynamics in Slash Pine Plantation Ecosystems. *Ecology* **1986**, *66*, 647–659. [CrossRef]

25. Hayes, P.; Turner, B.L.; Lambers, H.; Laliberté, E. Foliar nutrient concentrations and resorption efficiency in plants of contrasting nutrient-acquisition strategies along a 2-million-yeardune chronosequence. *J. Ecol.* **2014**, *102*, 396–410. [CrossRef]

26. Viera, M.; Schumacher, M.V.; Bonacina, D.M.; Ramos, L.O.O.; Rodríguez-Soalleiro, R. Biomass and nutrient allocation to aboveground components in fertilized Eucalyptus saligna and E. urograndis plantations. *New For.* **2017**, *48*, 445–462. [CrossRef]

27. Carlyle, J.C.; Malcom, D.C. Larch litter and nitrogen availability in mixed larch spruce stands. 1. Nutrient withdrawal, redistribution, and leaching loss from larch foliage at senescence. *Can. J. For. Res.* **1986**, *16*, 321–326. [CrossRef]

28. Kramer-Walter, K.R.W.; Laughlin, D.C. Root nutrient concentration and biomass allocation are more plastic than morphological traits in response to nutrient limitation. *Plant Soil* **2017**, *416*, 539–550. [CrossRef]

29. Wu, P.F.; Wang, G.Y.; Farooq, T.H.; Li, Q.; Zou, X.H. Low phosphorus and competition affect Chinese fir cutting growth and root organic acid content: Does neighboring root activity aggravate P nutrient deficiency? *J. Soils Sediments* **2017**, *17*, 2775–2785. [CrossRef]

30. Sardans, J.; Peñuelas, J. Tree growth changes with climate and forest type are associated with relative allocation of nutrients, especially phosphorus, to leaves and wood. *Glob. Ecol. Biogeogr.* **2013**, *22*, 494–507. [CrossRef]

31. Ewe, S.M.L.; Sternberg, L.S.L. Growth and gas exchange response of Brazilian pepper (*Schinus terebinthifolius*) and native South Florida species to salinity. *Trees* **2005**, *19*, 119–128. [CrossRef]

32. Peichl, M.; Arain, M.A. Above- and belowground ecosystem biomass and carbon pools in an age-sequence of temperate pine plantation forests. *Agric. For. Meteorol.* **2006**, *140*, 51–63. [CrossRef]

33. Hansen, J.C.; Cade-Menun, B.J.; Strawn, D.G. Phosphorus speciation in manure-amended alkaline soils. *J. Environ. Qual.* **2004**, *33*, 1521–1527. [CrossRef] [PubMed]

34. Frossard, E.; Condron, L.M.; Oberson, A.; Sinaj, S.; Fardeau, J.C. Processes governing phosphorus availability in temperate soils. *J. Environ. Qual.* **2000**, *29*, 15–23. [CrossRef]

35. Cumming, J.R.; Weinstein, L.H. Utilization of $AlPO_4$ as a phosphorus source by ectomycorrhizal Pinus rigida Mill. seedlings. *New Phytol.* **1990**, *116*, 99–106. [CrossRef]

36. Turner, B.L.; Condron, L.M.; Richardson, S.J.; Peltzer, D.A.; Allison, V.J. Soil organic phosphorus transformations during pedogenesis. *Ecosystems* **2007**, *10*, 1166–1181. [CrossRef]

37. Frizano, J.; Johnson, A.H.; Vann, D.R.; Scatena, F.N. Soil phosphorus fractionation during forest development on landslide scars in the Luquillo Mountains, Puerto Rico. *Biotropica* **2002**, *34*, 17–26. [CrossRef]

38. Walker, T.W.; Syers, J.K. The fate of phosphorus during pedogenesis. *Geoderma* **1976**, *15*, 1–19. [CrossRef]

39. Aerts, R. Nutrient resorption from senescing leaves of perennials: Are there general patterns? *J. Ecol.* **1996**, *84*, 597–608. [CrossRef]

40. Brant, A.N.; Chen, H.Y.H. Patterns and mechanisms of nutrient resorption in plants. *Crit. Rev. Plant Sci.* **2015**, *34*, 471–486. [CrossRef]

Article

Soil Nitrogen Responses to Soil Core Transplanting Along an Altitudinal Gradient in an Eastern Tibetan Forest

Li Zhang [1,2], **Ao Wang** [3], **Fuzhong Wu** [1,2], **Zhenfeng Xu** [1,2], **Bo Tan** [1,2], **Yang Liu** [1,2], **Yulian Yang** [1,2,4], **Lianghua Chen** [1,2] and **Wanqin Yang** [1,2,*]

[1]   Long-term Research Station of Alpine Forest Ecosystems, Institute of Ecology & Forestry, Sichuan Agricultural University, Chengdu 611130, China; zhangli19830116@hotmail.com (L.Z.); wufzchina@163.com (F.W.); sicauxzf@163.com (Z.X.); bobotan1984@163.com (B.T.); sicauliuyang@163.com (Y.L.); yangyulian2015@163.com (Y.Y.); sicauchenlh@126.com (L.C.)

[2]   Collaborative Innovation Center of Ecological Security in the Upper Reaches of Yangtze River, Chengdu 611130, China

[3]   Institute of Product Quality Inspection and Testing of Zunyi, Guizhou 563000, China; wangaocn@gmail.com

[4]   Ecological Security and Protection Key Laboratory of Sichuan Province, Mianyang Normal University, Mianyang 621000, China

*   Correspondence: scyangwq@163.com; Tel.: +86-28-8629-1112

Received: 8 March 2018; Accepted: 27 April 2018; Published: 2 May 2018

Abstract: To understand the differential effects of altitudinal gradient on soil inorganic nitrogen concentration and associated ammonia-oxidizingbacteria (AOB) and archaea (AOA), intact soil cores from a primary coniferous forest were in situ incubated in an alpine forest at a 3582-m altitude (A1) and transplanted to subalpine forests at a 3298-m altitude (A2) and 3023-m altitude (A3) on the eastern Tibetan Plateau. Transplant cooled the soil temperature of A2 but warmed the A3 soil temperature. Both AOA and AOB were found at the three altitudes. Compared to A1, A2 had greater AOA and AOB abundance, but A3 showed lower AOA abundance in organic soil. The AOA abundance was negatively correlated with ammonium concentration at all three altitudes, but AOB showed the reverse trend. Our results suggested that the soil nitrogen process responded differentially to soil core transplanting at different altitudes.

Keywords: alpine forest; ammonia-oxidizing bacteria; ammonia-oxidizing archaea; ammonium; nitrate

1. Introduction

Ongoing climate change, characterized by warming winters, snow cover decline and extreme weather events, is changing the processes of terrestrial ecosystems in cold biomes. Until now, direct soil warming and snow removal experiments along latitudinal and altitudinal gradients have been widely used to understand the effects of climate warming on soil processes [1–4]. However, different soil processes have been observed during cold winters in many areas, and increasing air temperatures in the winter have led to soil cooling [5–7]. Therefore, direct soil warming experiments cannot fully reflect the impact of climate warming on soil processes in cold regions. In the alpine-gorge area, the duration and depth of seasonal snow cover, seasonal freeze-thaw cycles, and temperature vary along the altitudinal gradient within a small range [8], which provides an ideal platform for investigating the effects of warming, snow cover decline, and seasonal freeze-thaw cycles on soil processes.

As a main limiting factor for plant growth and net primary productivity, soil nitrogen availability and its responses to global environmental changes are crucial in terms of understanding how an ecosystem will be affected by climate change [9,10]. The changes of environmental factors, such as temperature, water, and soil freezing, may consequently affect plant, soil and microbial processes and nutrient losses, and thus influence soil nitrogen process [6,11]. In high altitude and latitude regions, different elevations can lead to a continuous change in environmental factors, such as temperature, precipitation, and freeze-thaw characteristics [12–14], and the consequence of these factors may affect the soil nitrogen dynamics. To fully understand the nitrogen cycle under climate change scenarios, it is necessary to investigate the changes in soil nitrogen pools at different altitudes.

Ammonia oxidation is the first and rate-limiting step of nitrification, which plays a crucial role in the global nitrogen cycle [15]. Along with ammonia-oxidizing bacteria (AOB), ammonia-oxidizing archaea (AOA) have also been detected in extreme environments, such as deep marine areas, hot springs, and soils [16–19]. The active expression of AOA and AOB genes also has been detected in alpine areas [4,20–22]. In cold biomes, the seasonal soil freeze-thaw cycles create extreme conditions for soil microorganisms [23]. The dramatic temperature fluctuations lead to seasonal variations in the abundance and structure of AOA and AOB communities in alpine areas [4,23,24]. Although previous studies have documented that warming decreases the abundance of ammonia oxidizing bacteria and archaea under nitrogen fertilization [24] and the temperature influences the ammonia oxidizer population [25], the responses of AOA and AOB abundance to altitudinal gradient in both organic and mineral soils remain poorly understood.

The Tibetan Plateau is one of the most sensitive areas to global climate change [26]; it has experienced pronounced warming in recent decades that is expected to increase by 2.6–5.2 °C by 2100 [27]. Alpine forests in the upper reaches of the Yangtze River and the eastern Tibetan Plateau play important and irreplaceable roles in conserving water and soil, harboring biodiversity, sequestering atmospheric carbon dioxide, and indicating climate change [28]. As the area is affected by low temperatures and frequent geological disasters, the forest soils are characterized by a thick organic soil and a thin mineral soil [29]. However, how the inorganic nitrogen concentration and ammonia-oxidizing microbial community in both the organic and mineral soils respond to different altitudes remains unknown.

In this study, an altitudinal gradient experiment in combination with soil core transplanting was conducted to investigate the changes of soil nitrogen processes and the related ammonia-oxidizing microbial community (AOA and AOB) at different altitudes. We hypothesized that soil core transplanting might (1) increase the soil temperature at the two lower altitudes and (2) enhance the soil inorganic nitrogen concentration and related microbial abundance.

2. Materials and Methods

2.1. Site Description

This study was conducted at the Long-term Research Station of Alpine Forest Ecosystems in the Miyaluo Nature Reserve (102°53′–102°57′ E, 31°14′–31°19′ N, 2458–4619 m a.s.l.), which is located in Li County, western Sichuan, China (Figure 1). This is a transitional area situated between the Tibetan Plateau and the Sichuan Basin. The mean annual temperature is approximately 3 °C, with maximum and minimum temperatures of 23 °C (July) and −18 °C (January), respectively. The annual precipitation is approximately 850 mm. The forests consist of conifers and natural mixed hardwoods depending on the altitude and are mainly dominated by Minjiang fir (*Abies faxoniana* Rehd. et Wils.), Dragon spruce (*Picea purpurea* Mast.), and Red birch (*Betula albosinensis* Burk.). The forest soils are classified as Cambisols [30]. Seasonal soil freezing and thawing are observed in this area [31].

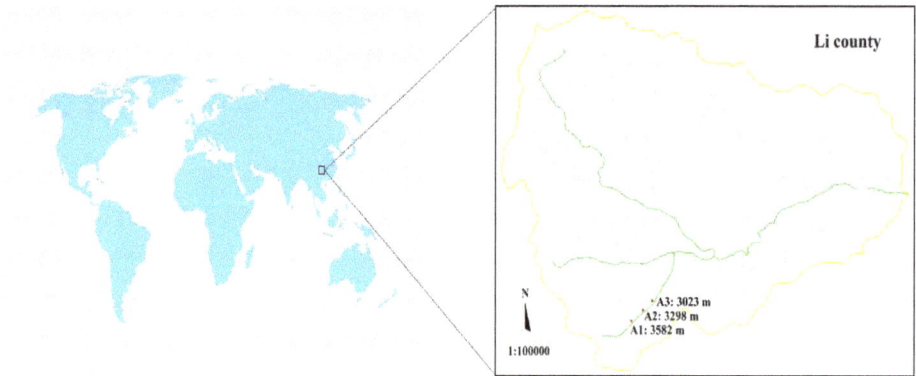

Figure 1. Location of experimental sites in this study.

2.2. Temperature Monitoring

Air temperature at 2 m height in the study site and soil temperatures (5 and 20 cm depths) were recorded at three locations using buried Thermochron iButton DS1923-F5 Recorders (Maxim/Dallas Semiconductor Corp, Sunnyvale, CA, USA) every hour between May 2010 and April 2011.

2.3. Soil Incubation

A 3 × 3 m sampling plot was randomly selected in a representative primary conifer alpine forest dominated by Minjiang fir at 3582 m. The basic properties of the organic and mineral soils are shown in Table 1. After clearing plants and fresh litter from the ground, Polyvinyl Chloride (PVC) cylinders (20 cm in length, 5 cm in diameter) were inserted into the soil to take undisturbed soil cores; forty-five soil cores were taken from this plot in May 2010. These soil cores were divided into three groups (fifteen cores in each group) and incubated in the 3582-m (A1, in situ incubate), 3298-m (A2) and 3023-m (A3) altitude sites.

Table 1. The basic properties of the soil organic layer and mineral soil layer in the eastern Tibetan forest.

	Organic Carbon $(g \cdot kg^{-1})$	Total Nitrogen $(g \cdot kg^{-1})$	NH_4^+ $(mg \cdot kg^{-1})$	NO_3^- $(mg \cdot kg^{-1})$	Bulk Density $(g \cdot cm^{-3})$	pH
Organic soil	138.56 ± 4.04	7.28 ± 0.07	18.73 ± 0.36	140.75 ± 2.73	1.09 ± 0.05	5.6 ± 0.3
Mineral soil	25.03 ± 0.88	1.69 ± 0.03	11.32 ± 1.35	14.04 ± 1.45	1.2 ± 0.03	5.3 ± 0.2

2.4. Sample Collection

One incubated soil core was retrieved from each sampling plot during the early growing season (EGS, 24 May to 12 August 2010), late growing season (LGS, 12 August to 17 October 2010), onset of the freezing period (OFP, 17 October to 23 December 2010), freezing period (FP, 23 December 2010 to 3 March 2011), and thawing period (TP, 3 March to 19 April 2011). Soil samples of the organic and mineral soils were collected in each plot [32]. All soil samples were hand-sorted to remove gravel and coarse roots. The samples were stored in freezer boxes, transported to the laboratory within 24 h, and stored at −20 °C. Soils were sieved through a 2-mm mesh before chemical analysis.

2.5. Inorganic Nitrogen Concentration

Ammonium (NH_4^+) and nitrate (NO_3^-) concentrations in the extract were measured using indophenol-blue and phenol disulfonic acid colorimetry. A 10-g soil sample from each soil core was taken, to which 50 mL 2 M KCl at room temperature was added. The mixture of soil and extractant was shaken for 1 h. After shaking, the soil suspension was filtered (Whatman filter paper, 12.5 cm in diameter). Soil solutions were kept frozen prior to analysis for ammonium and nitrate using a TU-1901 Analyzer (Beijing Purkinje General Instrument Co. Ltd., Beijing, China).

2.6. DNA Extraction

DNA was extracted from 0.8 g to 1.0 g (fresh weight) of soil using the E.Z.N.A.® Soil DNA Kit (Omega Bio Inc., Norcross, GA, USA). The extracted DNA was checked on 1% agarose gel, and the concentration was determined using a Nanodrop® ND-1000 UV-Vis spectrophotometer (Nano-Drop Technologies, Wilmington, DE, USA).

2.7. Quantification of amoA Genes by Real-Time PCR

The *amoA* gene cloning and the method to create standard curves were as previously described [21]. The primer pairs Arch-amoAF/Arch-amoAR [33] and amoA-1F/amoA-2R [34] were used for real-time polymerase chain reaction (PCR) quantification of the archaeal and bacterial *amoA* genes, respectively. Real-time PCR was performed using the CFX96 System (Bio-Rad Laboratories Inc., Hercules, CA, USA) in 25 μL reactions containing 12.5 μL of SYBR® Premix Ex TaqTM (TaKaRa Biotechnology Co. Ltd., Dalian, China), 0.4 mg·mL^{-1} of bovine serum albumin, 200 nmol·L^{-1} of each AOA primer or 400 nmol·L^{-1} of each AOB primer, and 1 μL of DNA (1–10 ng) as the template. Three analytical replicates were performed for each soil sample. Amplifications were carried out as follows: 95 °C for 1 min followed by 40 cycles of 10 s at 95 °C, 25 s at 63 °C for AOA or 57 °C for AOB, and 1 min at 72 °C. The plates were read at 72 °C after each cycle. The product specificity was confirmed using a melting curve analysis (65–95 °C, 0.5 °C per read with a hold time of 5 s) at the end of each PCR run.

2.8. Statistical Analyses

All statistical tests were performed using the software Statistical Package for the Social Sciences (SPSS Inc., IBM, Armonk, NY, USA) version 16.0. Data were subjected to one-way analysis of variance, and significant differences between treatment means for each variable were compared by the LSD post hoc test at $p < 0.05$. The relationships between the abundances of AOA and AOB, ammonium and nitrate concentration at each altitude were tested by Pearson correlation analyses.

3. Results

3.1. Air and Soil Temperatures

The temperature dynamics of air and soil at the three altitudes from May 2010 to April 2011 are shown in Figure 2. Air temperature increased with a decrease in altitude; compared with A1, A2 and A3 increased by 1.39 °C and 2.64 °C, respectively. However, the soil annual temperature displayed different characteristics at different altitudes. Generally, compared to A1, A3 increased by 1.03 °C and 1.08 °C, but A2 decreased by 0.26 °C and 0.25 °C in the organic and mineral soils, respectively.

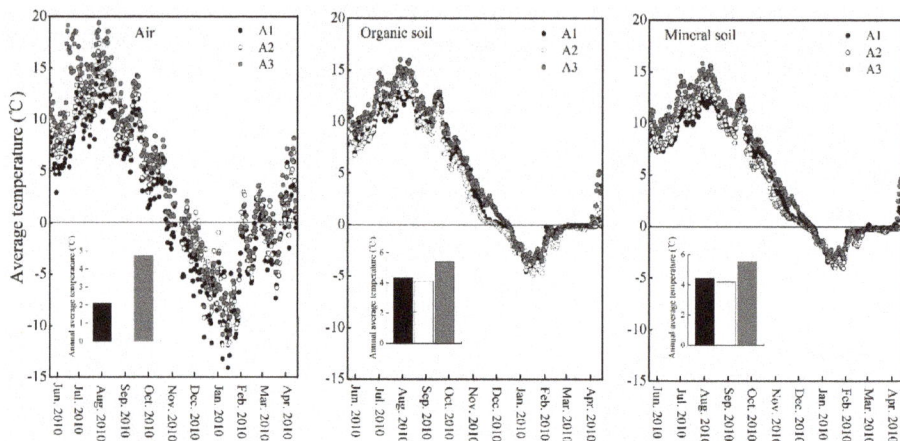

Figure 2. Daily and annual air and soil average temperatures at different altitudes. A1: 3582-m altitude; A2: 3298-m altitude; A3: 3023-m altitude. The inserts represent the mean values at the A1 (black), A2 (open) and A3 (gray) sites.

3.2. Inorganic Nitrogen Concentration

Transplanting soil significantly affected the soil inorganic nitrogen (ammonium and nitrate) concentration in both soil layers. With respect to organic soil, A3 had the highest ammonium and nitrate concentration during most of the sampling period except for the ammonium concentration during the onset of the freezing and thawing periods (Figure 3a,c). However, with respect to mineral soil, A3 showed the lowest ammonium concentration during the onset of the freezing period and during the freezing period (Figure 3a,c). In both soil layers, A2 had the highest ammonium and nitrate concentrations during the early growing season, but a lower ammonium concentration during the onset of the freezing period and the thawing period was found in the organic soil (Figure 3c). At all altitudes, the change in inorganic N concentration in the soil organic layer was more sensitive than that in the mineral soil (Figure 3c).

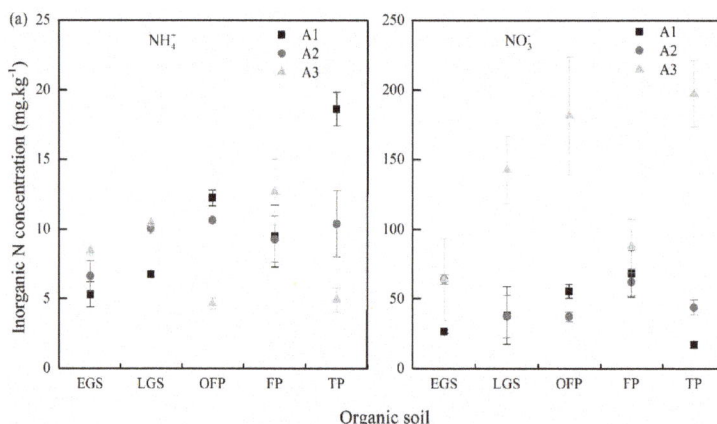

Figure 3. *Cont.*

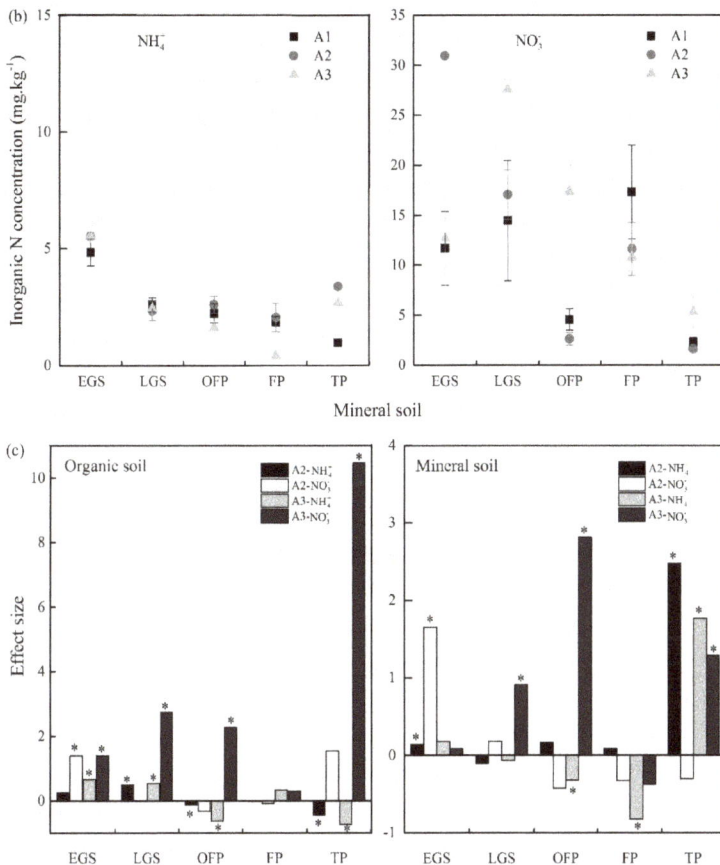

Figure 3. Ammonium (NH_4^+) and nitrate (NO_3^-) concentrations in (**a**) organic and (**b**) mineral soils and (**c**) the effect size of soil core transplanting in different periods. The effect size is the average difference in the ammonium and nitrate concentration at A2 or A3 with respect to the original site (A1), A2 = (A2 − A1)/A1 and A3 = (A3 − A1)/A1. EGS: early growing season, LGS: late growing season, OFP: onset of freezing period, FP: freezing period, TP: thawing period. A1: 3582-m altitude; A2: 3298-m altitude; A3: 3023-m altitude. * indicates a significant difference ($p < 0.05$) between different altitudes.

3.3. Abundance of amoA Genes

Both archaeal *amoA* genes (8.15×10^4 to 1.50×10^9 g^{-1} soil) and bacterial *amoA* genes (1.11×10^5 to 1.20×10^8 g^{-1} soil) were detected in the soil at the three altitudes (Figure 4). Compared to A1, A2 had greater AOA and AOB abundance at most sampling times, except for the AOA abundance during the early growing season in the organic soil and the AOB abundance during the freezing stage in the mineral soil (Figure 4c). A3 had lower AOA abundance but greater AOB abundance in organic soil compared to A1. Regarding mineral soil, A3 had lower AOA and AOB abundance during the early growing season and at the onset of the freezing period, but greater abundance was observed during other sampling periods. The abundance of AOB and AOA showed significant correlations with the ammonium and nitrate concentration, respectively, at 3023 m altitude (Table 2).

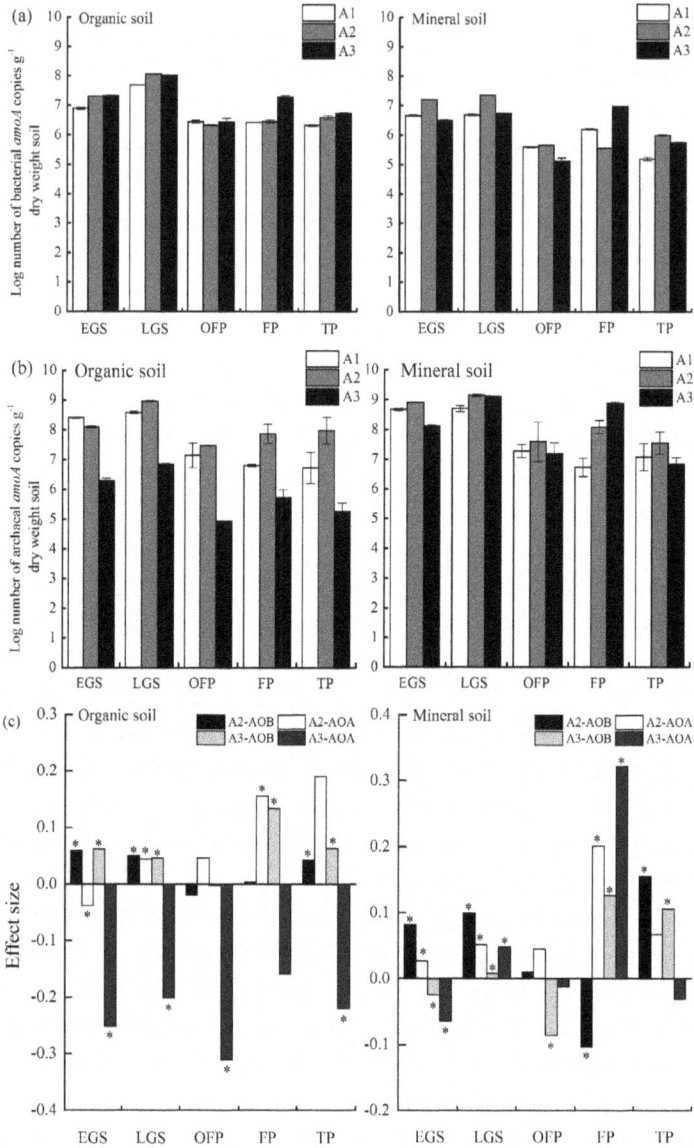

Figure 4. (**a**) Bacterial and (**b**) archaeal *amoA* gene copy numbers at different altitudes (mean ± SD, *n* = 3) and (**c**) the effect size of soil core transplanting in different periods. The effect size is the average difference in the ammonium and nitrate concentration at A2 or A3 with respect to the original site (A1). A2 = (A2 − A1)/A1 or A3 = (A3 − A1)/A1. A1: 3582-m altitude; A2: 3298-m altitude; A3: 3023-m altitude. EGS: early growing season; LGS: late growing season; OFP: onset of freezing period; FP: freezing period; TP: thawing period. * indicates a significant difference (*p* < 0.05) between different altitudes.

Table 2. Correlation analyses among the abundances of AOA, AOB and NH_4^+, NO_3^- at different altitudes.

Altitude		Abundance of AOB	Abundance of AOA	NH_4^+	NO_3^-
3582 m	Abundance of AOB	1	0.632 *	0.251	0.445
	Abundance of AOA		1	−0.323	−0.163
	NH_4^+			1	0.478
	NO_3^-				1
3298 m	Abundance of AOB	1	0.777 **	0.389	0.486
	Abundance of AOA		1	−0.093	0.076
	NH_4^+			1	0.741 *
	NO_3^-				1
3032 m	Abundance of AOB	1	−0.060	0.677 *	0.392
	Abundance of AOA		1	−0.493	−0.772 **
	NH_4^+			1	0.420
	NO_3^-				1

AOB is ammonia-oxidizing bacteria, AOA is ammonia-oxidizing archaea, NH_4^+ is ammonium, NO_3^- is nitrate. ** indicates significant difference at $p < 0.01$ (two-tailed). * indicates significant difference at $p < 0.05$ (two-tailed).

4. Discussion

Soil nitrogen cycling is one of the most important ecological processes in forest ecosystems [35]. Previous studies subdivided N cycling into decomposition processes, assimilative processes and dissimilative processes [36]. The uptake and utilization of ammonium or nitrate by plants and microorganisms for growth and replication were included in assimilative processes [36]. This soil core transplanting experiment was used to study the change and relationship between inorganic nitrogen (ammonium and nitrate) concentration and N-related (AOA and AOB) microorganisms. The observation in this study indicated that the soil transplant cooled A2 but warmed A3 winter soil temperature, respectively. This change may be related not only to the air temperature but also to the winter snow cover. The depth and duration of the snowpack is considered to be the important indirect effect of winter climate change [37,38]. Our previous study demonstrated that the snow depth decreased with decreasing altitude in this area [39]. The thickest snow cover in A1 may offer an ideal combination of moister and warmer soil conditions that can keep the soil warm [39,40]. However, the thinner and shorter duration of snow cover at A3 had a more sensitive response to solar radiation [41], which may produce a higher soil temperature.

Soil microorganisms are important drivers of soil quality and ecosystem function. Research on the spatial variations of AOB and AOA activity and their unique contributions to nitrification is needed [42]. Changes in environmental factors, such as elevation, N fertilization, temperature, and pH, may affect ammonia-oxidizing microorganisms in soil ecosystems. A previous study has pointed out that AOB were significantly higher than AOA during a soil warming and fertilization treatment [24], but the AOA were more abundant than AOB in a long-term fertilized soil [16]. This suggests that AOA may be more active than AOB in acidic soils, whereas this may be the opposite in alkaline soil [42]. However, the key factors are still difficult to assess [43]. A study at Mount Everest indicated that the AOA abundance increased along an altitudinal gradient decrease, whereas that of AOB did not shift significantly with altitude, suggesting that AOA may be more sensitive than AOB in response to elevated soil conditions [4]. In this study, we found that the samples at A2 had greater abundance of AOA and AOB than A1 at most sampling times, except for AOA in the early growing season and AOB in the freezing period in organic and mineral soils, respectively. Although previous studies have shown that a lower temperature may prevent microbial activity and even kill certain microbes [21,44], some tolerant or adaptive species may survive and replicate [45–47], to improve the microbial abundance of A2. At A3, the abundance of AOA and AOB was inconsistent in organic and mineral soils. In contrast to a previous study, temperature had a negative correlation with AOB but a positive correlation with AOA in a temperate beech forest soil [48]. However, AOA and AOB abundance was positively correlated with temperature [49]. Increased temperature at A3 resulted in a lower abundance of AOA and a greater abundance of AOB in the organic soil. This may be affected by the different responses to the environmental variation of AOA and AOB in alpine areas [4,50,51].

Inorganic N, as a common substrate, influenced AOA and AOB abundance. The concentration of ammonium in soil has been identified as an important factor driving the relative distributions of AOA and AOB [52]. In the investigated alpine forest, the abundance of AOA varied almost inversely with the measured ammonium concentration at all sampling altitudes, while that of AOB varied positively with ammonium concentration (Table 2), suggesting that ammonium may drive the separation between AOA and AOB [53]. In this study, the abundance of AOB was higher at A2 and A3 during most of the sampling period (Figure 4c). Higher ammonium concentration was observed only in the early growing season and later growing season, likely due to N mineralization over the winter [41]. This result suggests that inorganic N may determine the distribution of AOA and AOB in alpine forest soils by providing the substrate for microbial mineralization.

5. Conclusions

In summary, AOA and AOB abundance was recorded in winter in this alpine forest. Although the soil temperature was cooled at A2, higher microbial abundance was observed at A2. The increased temperature at A3 decreased the AOA abundance in the organic soil. The concentration of ammonium was positively correlation with AOB abundance and negatively with AOA abundance.

Author Contributions: W.Y. and F.W. conceived and designed the experiments; A.W., Y.L. and L.C. performed the experiments; L.Z., B.T., Y.Y. and Z.X. analyzed the data, W.Y. contributed reagents/materials/analysis tools; L.Z., and W.Y. wrote the paper.

Funding: This project was financially supported by the National Natural Science Foundation of China (Nos. 31570445, 31500509, 31570601).

Acknowledgments: We are very grateful to the insightful comments and useful suggestions from the reviewers.

Conflicts of Interest: The authors declare no conflict of interest. The founding sponsors had no role in the design of the study; in the collection, analyses, or interpretation of data; in the writing of the manuscript, and in the decision to publish the results.

References

1. Tan, B.; Wu, F.Z.; Yang, W.Q.; He, X.H. Snow removal alters soil microbial biomass and enzyme activity in a Tibetan alpine forest. *Appl. Soil Ecol.* **2014**, *76*, 34–41. [CrossRef]
2. Xu, Z.F.; Hu, R.; Xiong, P.; Wan, C.; Cao, G.; Liu, Q. Initial soil responses to experimental warming in two contrasting forest ecosystems, Eastern Tibetan Plateau, China: Nutrient availabilities, microbial properties and enzyme activities. *Appl. Soil Ecol.* **2010**, *46*, 291–299. [CrossRef]
3. Xu, Z.F.; Liu, Q.; Yin, H.J. Effects of temperature on soil net nitrogen mineralisation in two contrasting forests on the eastern Tibetan Plateau, China. *Soil Res.* **2014**, *52*, 562–567. [CrossRef]
4. Zhang, L.M.; Wang, M.; Prosser, J.I.; Zheng, Y.M.; He, J.Z. Altitude ammonia-oxidizing bacteria and archaea in soils of Mount Everest. *FEMS Microbiol. Ecol.* **2009**, *70*, 208–217. [CrossRef] [PubMed]
5. Edwards, A.C.; Scalenghe, R.; Freppaz, M. Changes in the seasonal snow cover of alpine regions and its effect on soil processes: A review. *Quat. Int.* **2007**, *162*, 172–181. [CrossRef]
6. Groffman, P.M.; Driscoll, C.T.; Fahey, T.J.; Hardy, J.P.; Fitzhugh, R.D. Colder soils in a warmer world: A snow manipulation study in a northern hardwood forest ecosystem. *Biogeochemistry* **2001**, *56*, 135–150. [CrossRef]
7. Williams, M.W.; Helmig, D.; Blanken, P. White on green: Under-snow microbial processes and trace gas fluxes through snow, Niwot Ridge, Colorado Front Range. *Biogeochemistry* **2009**, *95*, 1–12. [CrossRef]
8. Zhu, J.; Yang, W.; He, X. Temporal dynamics of abiotic and biotic factors on leaf litter of three plant species in relation to decomposition rate along a subalpine elevation gradient. *PLoS ONE* **2013**, *8*, e62073. [CrossRef] [PubMed]
9. Hungate, B.A.; Dukes, J.S.; Shaw, M.R.; Luo, Y.; Field, C.B. Atmospheric science. Nitrogen and climate change. *Science* **2003**, *302*, 1512–1513. [CrossRef] [PubMed]

10. Rustad, L.E.; Campbell, J.L.; Marion, G.M.; Norby, R.J.; Mitchell, M.J.; Hartley, A.E.; Cornelissen, J.H.C.; Gurevitch, J. A meta-analysis of the response of soil respiration, net nitrogen mineralization, and aboveground plant growth to experimental ecosystem warming. *Oecologia* **2001**, *126*, 543–562. [CrossRef] [PubMed]

11. Groffman, P.M.; Hardy, J.P.; Fashu-Kanu, S.; Driscoll, C.T.; Cleavitt, N.L.; Fahey, T.J.; Fisk, M.C. Snow depth, soil freezing and nitrogen cycling in a northern hardwood forest landscape. *Biogeochemistry* **2011**, *102*, 223–238. [CrossRef]

12. Campbell, J.L.; Mitchell, M.J.; Groffman, P.M.; Christenson, L.M.; Hardy, J.P. Winter in northeastern North America: A critical period for ecological processes. *Front. Ecol. Environ.* **2005**, *3*, 314–322. [CrossRef]

13. Freppaz, M.; Williams, B.L.; Edwards, A.C.; Scalenghe, R.; Zanini, E. Simulating soil freeze/thaw cycles typical of winter alpine conditions: Implications for N and P availability. *Appl. Soil Ecol.* **2007**, *35*, 247–255. [CrossRef]

14. Kreyling, J. Winter climate change: A critical factor for temperate vegetation performance. *Ecology* **2010**, *91*, 1939–1948. [CrossRef] [PubMed]

15. Prosser, J.I. Autotrophic nitrification in bacteria. *Adv. Microb. Physiol.* **1989**, *30*, 125–181. [CrossRef] [PubMed]

16. He, J.Z.; Shen, J.P.; Zhang, L.M.; Zhu, Y.G.; Zheng, Y.M.; Xu, M.G.; Di, H.J. Quantitative analyses of the abundance and composition of ammonia-oxidizing bacteria and ammonia-oxidizing archaea of a Chinese upland red soil under long-term fertilization practices. *Environ. Microbiol.* **2007**, *9*, 2364–2374. [CrossRef] [PubMed]

17. Könneke, M.; Bernhard, A.E.; de la Torre, J.R.; Walker, C.B.; Waterbury, J.B.; Stahl, D.A. Isolation of an autotrophic ammonia-oxidizing marine archaeon. *Nature* **2005**, *437*, 543–546. [CrossRef] [PubMed]

18. Leininger, S.; Urich, T.; Schloter, M.; Schwark, L.; Qi, J.; Nicol, W.; Prosser, J.I.; Schuster, S.C.; Schleper, C. Archaea predominate among ammonia-oxidizing prokaryotes in soils. *Nature* **2006**, *442*, 806–809. [CrossRef] [PubMed]

19. Reigstad, L.J.; Richter, A.; Daims, H.; Urich, T.; Schwark, L.; Schleper, C. Nitrification in terrestrial hot springs of Iceland and Kamchatka. *FEMS Microbiol. Ecol.* **2008**, *64*, 167–174. [CrossRef] [PubMed]

20. Auguet, J.C.; Nomokonova, N.; Camarero, L.; Casamayor, E.O. Seasonal changes of freshwater ammonia-oxidizing archaeal assemblages and nitrogen species in oligotrophic alpine lakes. *Appl. Environ. Microbiol.* **2011**, *77*, 1937–1945. [CrossRef] [PubMed]

21. Wang, A.; Wu, F.Z.; Yang, W.Q.; Wu, Z.C.; Wang, X.X.; Tan, B. Abundance and composition dynamics of soil ammonia-oxidizing archaea in an alpine fir forest on the eastern Tibetan Plateau of China. *Can. J. Microbiol.* **2012**, *58*, 572–580. [CrossRef] [PubMed]

22. Wang, A.; Wu, F.Z.; He, Z.H.; Xu, Z.F.; Liu, Y.; Tan, B.; Yang, W.Q. Characteristics of ammonia-oxidizing bacteria and ammonia-oxidizing archaea abundance in soil organic layer under the subslpine/alpine forest. *Acta Ecol. Sin.* **2012**, *32*, 4371–4378. [CrossRef]

23. Henry, H.A.L. Soil freeze–thaw cycle experiments: Trends, methodological weaknesses and suggested improvements. *Soil Biol. Biochem.* **2007**, *39*, 977–986. [CrossRef]

24. Long, X.; Chen, C.; Xu, Z.; Linder, S.; He, J. Abundance and community structure of ammonia oxidizing bacteria and archaea in a Sweden boreal forest soil under 19-year fertilization and 12-year warming. *J. Soils Sediments* **2012**, *12*, 1124–1133. [CrossRef]

25. Avrahami, S.; Conrad, R. Patterns of community change among ammonia oxidizers in meadow soils upon long-term inubation at different temperatures. *Appl. Environ. Microbiol.* **2003**, *69*, 6152–6164. [CrossRef] [PubMed]

26. Liu, X.; Chen, B. Climatic warming in the Tibetan Plateau during recent decades. *Int. J. Climatol.* **2000**, *20*, 1729–1742. [CrossRef]

27. Chen, H.; Zhu, Q.; Peng, C.; Wu, N.; Wang, Y.; Fang, X.; Gao, Y.; Zhu, D.; Yang, G.; Tian, J.; et al. The impacts of chimate change and human activities on biogeochemical cycles on the Qinghai-Tibetan Plateau. *Glob. Chang. Biol.* **2013**, *19*, 2940–2955. [CrossRef] [PubMed]

28. Yang, W.Q.; Feng, R.F.; Zhang, J.; Wang, K.Y. Carbon stock and biochemical properties in the organic layer and mineral soil under three subalpine forests in Western China. *Acta Ecol. Sin.* **2007**, *27*, 4157–4165. [CrossRef]

29. Feng, R.F.; Yang, W.Q.; Zhang, J. Review on biochemical property in forest soil organic layer and its responses to climate change. *Chin. J. Appl. Environ. Biol.* **2006**, *12*, 734–739. [CrossRef]

30. Gong, Z.T.; Zhang, G.L.; Chen, Z.C. *Pedogenesis and Soil Taxonomy*; Science Press: Beijing, China, 2007; ISBN 9787030196538.
31. Tan, B.; Wu, F.Z.; Yang, W.Q.; Liu, L.; Yu, S. Characteristics of soil animal community in the subalpine/alpine forests of western Sichuan during onset of freezing. *Acta Ecol. Sin.* **2010**, *30*, 93–99. [CrossRef]
32. Xu, W.; Li, W.; Jiang, P.; Wang, H.; Bai, E. Distinct temperature sensitivity of soil carbon decomposition in forest organic layer and mineral soil. *Sci. Rep.* **2014**, *4*, 6512. [CrossRef] [PubMed]
33. Francis, C.A.; Roberts, K.J.; Beman, J.M.; Santoro, A.E.; Oakley, B.B. Ubiquity and diversity of ammonia-oxidizing archaea in water columns and sediments of the ocean. *PNAS* **2005**, *102*, 14683–14688. [CrossRef] [PubMed]
34. Rotthauwe, J.H.; Witzel, K.P.; Liesack, W. The ammonia monooxygenase structural gene amoA as a functional marker: Molecular fine-scale analysis of natural ammonia-oxidizing populations. *Appl. Environ. Microbiol.* **1997**, *63*, 4704–4712. [CrossRef] [PubMed]
35. Binkley, D.; Hart, S.C. The Components of Nitrogen Availability Assessments in Forest Soils. In *Advances in Soil Science*, 1st ed.; Stewart, B.A., Ed.; Springer: New York, NY, USA, 1989; pp. 57–112. ISBN 978-1461387732.
36. Levy-Booth, D.J.; Prescott, C.E.; Grayston, S.J. Microbial functional genes involved in nitrogen fixation, nitrification and denitrification in forest ecosystems. *Soil Biol. Biochem.* **2014**, *75*, 11–25. [CrossRef]
37. Mellander, P.E.; Löfvenius, M.O.; Laudon, H. Climate change impact on snow and soil temperature in boreal Scots pine stands. *Clim. Chang.* **2007**, *85*, 179–193. [CrossRef]
38. Pederson, G.T.; Gray, S.T.; Woodhouse, C.A.; Betancourt, J.L.; Fagre, D.B.; Littell, J.S.; Emma, W.; Luckman, B.H.; Graumlich, L.J. The unusual nature of recent snowpack declines in the North American cordillera. *Science* **2011**, *333*, 332–335. [CrossRef] [PubMed]
39. Wu, Q. Effects of Forest Gap on Foliar Litter Decomposition in the Subalpine/Alpine Forest of Western Sichuan. Ph.D. Thesis, Sichuan Agricultural University, Chengdu, China, 2014. Available online: http://cdmd.cnki.com.cn/Article/CDMD-10626-1016050649.htm (accessed on 3 June 2014).
40. Freppaz, M.; Williams, M.W.; Seastedt, T.; Filippa, G. Response of soil organic and inorganic nutrients in alpine soil to a 16-year factorial snow and N-fertilization experiment, Colorado Front Range, USA. *Appl. Soil Ecol.* **2012**, *62*, 131–141. [CrossRef]
41. Yin, R.; Xu, Z.F.; Wu, F.Z.; Yang, W.Q.; Li, Z.P.; Xiong, L.; Xiao, S.; Wang, B. Effects of snow pack on wintertime soil nitrogen transformation in two subalpine forests of western Sichuan. *Acta Ecol. Sin.* **2014**, *34*, 2061–2067. [CrossRef]
42. Shen, J.P.; Zhang, L.M.; Di, H.J.; He, J.Z. A review of ammonia-oxidizing bacteria and archaea in Chinese soils. *Front. Microbiol.* **2012**, *3*, 296–302. [CrossRef] [PubMed]
43. Erguder, T.H.; Boon, N.; Wittebolle, L.; Marzorati, M.; Verstraete, W. Environmental factors shaping the ecological niches of ammonia-oxidizing archaea. *FEMS Microbiol. Rev.* **2009**, *33*, 855–869. [CrossRef] [PubMed]
44. Edwards, K.A.; McCulloch, J.; Peter Kershaw, G.; Jefferies, R.L. Soil microbial and nutrient dynamics in a wet Arctic sedge meadow in late winter and early spring. *Soil Biol. Biochem.* **2006**, *38*, 2843–2851. [CrossRef]
45. Walker, V.K.; Palmer, G.R.; Voordouw, G. Freeze-Thaw Tolerance and Clues to the Winter Survival of a Soil Community. *Appl. Environ. Microbiol.* **2006**, *72*, 1784–1792. [CrossRef] [PubMed]
46. Wilson, S.L.; Walker, V.K. Selection of low-temperature resistance in bacteria and potential applications. *Environ. Technol.* **2010**, *31*, 943–956. [CrossRef] [PubMed]
47. Drotz, S.H.; Sparrman, T.; Nilsson, M.B.; Schleucher, J.; Öquist, M.G. Both catabolic and anabolic heterotrophic microbial activity proceed in frozen soils. *PNAS* **2010**, *107*, 21046–21051. [CrossRef] [PubMed]
48. Rasche, F.; Knapp, D.; Kaiser, C.; Koranda, M.; Kitzler, B.; Zechmeister-Boltenstern, S.; Richter, A.; Sessitsch, A. Seasonality and resource availability control bacterial and archaeal communities in soils of a temperate beech forest. *ISME J.* **2011**, *5*, 389–402. [CrossRef] [PubMed]
49. Cao, H.; Li, M.; Hong, Y.; Gu, J.D. Diversity and abundance of ammonina-oxidizing archaea and bacteria in polluted mangrove sediment. *Syst. Appl. Microbiol.* **2011**, *34*, 513–523. [CrossRef] [PubMed]
50. Cavicchioli, R. Cold-adapted archaea. *Nat. Rev. Microbiol.* **2006**, *14*, 331–343. [CrossRef] [PubMed]
51. Tan, B.; Wu, F.; Yang, W.; Yu, S.; Liu, L.; Wang, A. The dynamics pattern of soil carbon and nutrients as soil thawing proceeded in the alpine/subalpine forest. *Acta Agric. Scand. B-SP* **2011**, *61*, 670–679. [CrossRef]

52. Regan, K.; Stempfhuber, B.; Schloter, M.; Rasche, F.; Prati, D.; Philippot, L.; Boeddinghaus, R.S.; Kandeler, E.; Marhan, S. Spatial and temporal dynamics of nitrogen fixing, nitrifying and denitrifying microbes in an unfertilized grassland soil. *Soil Biol. Biochem.* **2017**, *109*, 214–226. [CrossRef]

53. Prosser, J.I.; Nicol, G.W. Archaeal and bacterial ammonia-oxidisers in soil: The quest for niche specialization and differentiation. *Trends Microbiol.* **2012**, *20*, 523–531. [CrossRef] [PubMed]

![forests logo] *forests*

MDPI

Article

Characterization of Phosphorus in a Toposequence of Subtropical Perhumid Forest Soils Facing a Subalpine Lake

Yo-Jin Shiau [1], Chung-Wen Pai [2], Jeng-Wei Tsai [3], Wen-Cheng Liu [4], Rita S. W. Yam [5], Shih-Chieh Chang [6], Sen-Lin Tang [7] and Chih-Yu Chiu [7,*]

[1] Department of Safety, Health and Environmental Engineering, National Kaohsiung University of Science and Technology, Kaohsiung 81164, Taiwan; yshiau@ncsu.edu

[2] The Experimental Forest, College of Bio-Resource and Agriculture, National Taiwan University, Nantou 55743, Taiwan; cwpai724@yahoo.com.tw

[3] Department of Biological Science and Technology, China Medical University, Taichung 40402, Taiwan; tsaijw@mail.cmu.edu.tw

[4] Department of Civil and Disaster Prevention Engineering, National United University, Miaoli 36063, Taiwan; wcliu@nuu.edu.tw

[5] Department of Bioenvironmental Systems Engineering, National Taiwan University, Taipei 10617, Taiwan; ritayam@ntu.edu.tw

[6] Department of Natural Resources and Environmental Studies, National Dong Hwa University, Hualien 97401, Taiwan; scchang@mail.ndhu.edu.tw

[7] Biodiversity Research Center, Academia Sinica, Nangang, Taipei 11529, Taiwan; sltang@gate.sinica.edu.tw

* Correspondence: bochiu@sinica.edu.tw; Tel.: +886-2-2787-1180

Received: 1 May 2018; Accepted: 23 May 2018; Published: 25 May 2018

Abstract: The productivity of forests is often considered to be limited by the availability of phosphorus (P). Knowledge of the role of organic and inorganic P in humid subtropical forest soils is lacking. In this study, we used chemical fractionation and ^{31}P nuclear magnetic resonance (NMR) spectroscopy to characterize the form of P and its distribution in undisturbed perhumid Taiwan false cypress (*Chamaecyparis formosensis* Matsum.) forest soils. The toposequence of transects was investigated for the humic layer from summit to footslope and lakeshore. The clay layer combined with a placic-like horizon in the subsoil may affect the distribution of soil P because both total P and organic P (P_o) contents in all studied soils decreased with soil depth. In addition, P_o content was negatively correlated with soil crystalline Fe oxide content, whereas inorganic P (P_i) content was positively correlated with soil crystalline Fe oxide content and slightly increased with soil depth. Thus, P_i may be mostly adsorbed by soil crystalline Fe oxides in the soils. Among all extractable P fractions, the NaOH-P_o fraction appeared to be the major component, followed by NaHCO$_3$-P_o; the resin-P and HCl-P_i fractions were lowest. In addition, we found no typical trend for P_i and P_o contents in soils with topographical change among the three sites. From the ^{31}P-NMR spectra, the dominant P_o form in soils from all study sites was monoesters with similar spectra. The ^{31}P-NMR findings were basically consistent with those from chemical extraction. Soil formation processes may be the critical factor affecting the distribution of soil P. High precipitation and year-round high humidity may be important in the differentiation of the P species in this landscape.

Keywords: *Chamaecyparis* forest; humic substances; ^{31}P nuclear magnetic resonance spectroscopy (^{31}P NMR); P species; topography

1. Introduction

In terrestrial environments, mountainous forest is one of the canonical ecosystems that provide various ecosystem services such as maintaining biodiversity, water conservation [1] and carbon sequestration [2]. Meanwhile, it is also a vulnerable ecosystem that can be influenced by different soil nutrients [3]. Thus, understanding the soil nutrient distributions in a forest ecosystem is vital to maintain ecosystem functions and productivity [4]. In such ecosystems, phosphorus (P) can be a limiting element because unlike nitrogen (N), which is mainly deposited from the atmosphere [5], P is mostly acquired from weathering soil parent material and is continuously lost due to soil erosion [6–8]. The bioavailability of P in soils further relies on the chemical/physical conditions that fractionate the total P into different species [9,10]. Understanding P availability and its transformation among each fraction will help assess the P supply capacity of the soil over the long term and to adapt management practices [11].

Some isolated P fractionation pools have key functions in the P cycle and plant nutrition [11–15]. The organic and inorganic forms of P are usually separated and quantified by their plant availability. The mineralization of organic P is generally responsible for most of the P supply to plants [16,17], especially in mountain forest ecosystems [4,11,18]. Actually, the sequential fraction can provide a general indicator of how biological and geochemical forms of P change during soil weathering in mountain forest ecosystems. In general, organic P was found to be the dominant fraction of total P (TP) in these forest ecosystems, and available P content was determined by the mineralization processes of organic P [11,19]. Productivity, including plant growth and biomass production of trees in afforested mountain areas, is largely influenced by mineralization and microbial processes of organic P as well as soil organic matter.

However, identifying different fractions of P and evaluating their bioavailability are difficult because of spatial inconsistency and the complexity of geochemical properties. Several wet-chemical methods for determining P fractionations have been established and used in various ecosystems [20–23]. Also, ^{31}P-nuclear magnetic resonance (NMR) spectroscopy has been used for determining the composition of soil P [24–26]. De Feudis et al. [27] determined the P availability in subalpine forest soils in Italy and found organic P, bioavailable P contents and alkaline mono-phosphatase activity were all increased with altitude. Similarly, Doolette et al. [28] analyzed P composition in five alpine and subalpine forest soils and found that 54% to 66% of extractable P was contributed by organic P such as phosphomonoesters and inositol phosphonates, and the organic P composition was affected by temperature and soil moisture.

To our knowledge, studies of P fractionation in forests were mostly performed in temperate ecosystems [10,29,30], with relatively fewer studies from subtropical and tropical alpine forests. In this study, to evaluate the pedogenetic effects on the forms of P, we determined the composition of soil P with both chemical extraction and ^{31}P-NMR methods along a toposequence in a pristine subtropical subalpine forest. Because the changes in soil oxidation–reduction status affect the formation of iron (Fe) and consequently the P status in such humid forest soils, we hypothesized that the content of labile P associated with Fe oxides increases with changing topographic sequence from the summit to lakeshore because of leaching and soil erosion, whereas recalcitrant organic P, which is more complex-formed, will remain in the summit.

2. Materials and Methods

2.1. Study Sites

This study was conducted in the Yuanyang Lake forest ecosystem (24°35' N, 121°24' E) in northeastern Taiwan. The study sites covered an elevation of 1700 to 2000 m a.s.l., with an average annual temperature of 12.5 °C and a mean annual precipitation of more than 4000 mm. The ecosystem consists of a primary forest dominated by Taiwan false cypress (*Chamaecyparis obtusa* var. *formosana*) and an evergreen broadleaf shrub (*Rhododendron formosanum* Hemsl.). The bedrock of the study sites is composed of interbedded Tertiary shale and sandstone [31]. This locality can be described as a temperate, very wet and mountainous ecosystem. It has been established as a Nature Reserve and selected as one of the long-term ecological research sites in Taiwan. The forest soils are divided into three main groups, which are closely related to the topography. The soil at the summit, with a slope of about 15°, is classified as Typic Hapludult (Ultisols) [32], which is relatively well-drained and develops clear eluvial and illuvial boundaries. The footslope, with a slope of about 28°, is dominated by Typic Dystrochrept (Inceptisols), where poor drainage caused by the clay and silty clay mineral horizon beneath the organic layer limits the downward movement of soluble compounds and thus hampers the soil profile development. Lithic Medihemist (Histosols) stretches from the lakeshore to the toeslope, about 1.5 m above the lake and with a slope of about 10°, which is inundated by occasional storms. Details of the environment of this ecosystem are described elsewhere [33]. Briefly, the soils at the summit and footslope are derived from partially podzolized soils and pure peats, while the soils at the lakeshore are mostly derived from peat, and directly lie on the bedrock [33].

2.2. Soil Sampling

A pedon sample was collected from each site to the bedrock. However, the pedon sample at the lakeshore was limited to the O horizon because no mineral layers were developed on the bedrock. Each horizon in the pedon was collected separately to determine the basic soil physiochemical properties.

To further determine the P fractionation along the topography, soil samples were collected from three selected sites along a topographic sequence in the forest that covered the summit, footslope and lakeshore. At each sampling site, three composite samples, each containing five subsamples, were collected with a soil auger with 8 cm in diameter and 10 cm in depth (O_e and O_a horizons).

Visible coarse organic materials, such as roots and litter were manually removed before sieving. The remaining soil samples were air dried and sieved through a 2-mm sieve for chemical analysis.

2.3. General Soil Chemical Properties

Soil pH was measured at a soil:water ratio of 1:1. Total organic C (TOC) and total N (TN) contents in the soil were determined with an NCS Elemental Analyzer (Model NA1500 Fisons, Milan, Italy). Cation-exchange capacity was determined by the NH_4/Na exchange method [34]. Crystalline Fe (Fe_d) and Al (Al_d) oxide contents were calculated by difference between dithionite and oxalate contents with following the experiments of previous study [35]. Amorphous Fe (Fe_o) and Al (Al_o) oxide contents were measured by an ammonium oxalate extraction method [36].

2.4. Sequential Fractionation of P

Sequential fractionation was performed as described in [37]. The sequential fractionation procedure removes progressively less available P with each subsequent soil extraction [38]. The fractionation started with 0.5 g dried sieve soil. An anion exchange resin was used first to extract plant-available inorganic P (resin-P_i) [39]. Then, the other P_i content was determined directly in 0.5 M $NaHCO_3$, 0.1 M NaOH, 1 M HCl and concentrated HCl extractions. The extracted solutions were then digested with H_2SO_4 (97%) and H_2O_2 (30%) at 300 °C to determine the total dissolved P (P_d) content of each fraction. The organic P (P_o) content was calculated by subtracting P_i content from P_d content in each fraction ($P_o = P_d - P_i$). The remaining soil was digested with H_2SO_4 (97%) and H_2O_2 (30%) at 300 °C to determine the residual P content. All extracts and digestions obtained were measured colourimetrically by the malachite green procedure [40].

Summed P_i content was calculated as the sum of all analyzed P_i fractions including resin-P_i, $NaHCO_3$-P_i, NaOH-P_i, HCl-P_i and cHCl-P_i. Summed P_o content was calculated as the sum of all analyzed P_o fractions including $NaHCO_3$-P_o, NaOH-P_o and cHCl-P_o. Summed P content was calculated as the sum of P_i, P_o and residual P content. Total P content of the soil samples was determined by digestion with H_2SO_4 (97%) and H_2O_2 (30%) at 300 °C.

2.5. ^{31}P-NMR Measurements

Air-dried soil (5 g) was dispersed in 20 mL of 0.25 M NaOH-0.05 M EDTA for 2 h, and the suspension was centrifuged at $12,100\times g$ for 30 min. The extractant was then reacted with chelating resin for 6 h at room temperature to reduce the paramagnetic interference of iron and other metals in the NMR spectra. After being stirred, the resin was separated by filtration through Whatman 42 filter paper. The extract was freeze-dried for storage. A freeze-dried sample (0.1 g) of NaOH-EDTA extractant was dissolved in 0.5 mL of 0.5 M NaOH; then, 0.1 mL D_2O was added, and the solution was transferred to a 5-mm NMR tube for ^{31}P-NMR analysis [41]. The ^{31}P-NMR spectra were obtained at 242.86 MHz and 25 °C on a Bruker-600 NMR spectrometer with 60° pulse, 3.5-s delay and 0.33-s acquisition time. The ^{31}P-NMR spectra were proton-decoupled by using an inverse-gated pulse sequence to overcome the nuclear Overhauser enhancement and for quantification [40,41]. Depending on the P content in the alkaline extract, 500–2500 scans were used for an acceptable signal-to-noise ratio. Spectra were recorded with a line-broadening of 20 Hz. The chemical shift was measured relative to an external 85% H_3PO_4/D_2O standard. The assignment of signals was based on Newman and Tate [42], Dai et al. [43], Condron et al. [24], and Robinson et al. [41]. Contents of the various P components (phosphonate, inorganic orthophosphate, orthophosphate monoesters, orthophosphate diesters, pyrophosphate, polyphosphates) were determined according to relative resonance areas obtained by electronic integration. Inorganic orthophosphates and orthophosphate monoesters signals were separated by using a boundary determined from the valley between the two signals to the baseline [44].

2.6. Statistical Analyses

All extraction experiments were carried out in triplicate. Simple linear regression was used to compare the relation between soil P and soil Fe_d, Fe_o Al_d and Al_o concentrations. Differences in the P factions among the three sites were analyzed using one-way analysis of variance (One-Way ANOVA) and Tukey's honestly significant difference (HSD) test. JMP 11.0 (SAS Inc., Cary, NC, USA) was used for these statistical analyses. $p < 0.05$ was considered as statistically significant.

3. Results

The basic chemical properties of the studied soils are in Table 1. The soils were strongly acidic; pH values ranged from 3.3 to 4.5 in the three sampling sites. Both TOC and TN contents were high in the O horizon and decreased from the surface to the low horizons. Similarly, the cation-exchange capacity basically decreased from the surface to the low horizons. Total P content also decreased from the surface to the low horizons, but the difference was much less than for TOC and TN contents.

The Fe_o, Fe_d, Al_o and Al_d contents peaked in the Bt_2 horizon at the summit site and in the Bw_2 horizon at the footslope site. P_i content was associated with amorphous (Fe_o and Al_o) and crystalline (Fe_d and Al_d) Al and Fe oxide contents and migrated vertically through the horizons with illuviation. Moreover, P_i and Fe_d contents were positively correlated, and P_o and Fe_d as well as P_o and Fe_o contents were negatively correlated in the soil samples (Figure 1a). However, the relation between P_i and Fe_o contents was not statistically significant (Figure 1b). In addition, only P_i and Al_o contents were positively correlated; P_o and Al_d contents were negatively correlated but not P_i and Al_d nor P_o and Al_o contents (Figure 2a,b).

Contents of total P and summed P and P_i in the O horizons were greater at the footslope than the lakeshore, whereas the values in the summit site were in between those at the footslope and lakeshore. In addition, summed P_o and residual P contents were similar among the three sites (Table 2). HCl-P_i and cHCl-P_i contents were similar among the three sites, whereas cHCl-P_o content in the footslope soil was similar to that at the lakeshore but higher than that at the summit. NaOH-P_i content was higher at the footslope than the summit and lakeshore. In addition, NaOH-P_o content was higher at the summit than the footslope and lakeshore. NaHCO$_3$ extracted P_i content was higher at the summit than the footslope and lakeshore, whereas $NaHCO_3$-P_o content was the highest at the footslope and was similar at the summit and lakeshore. Resin-P_i content was higher at the lakeshore than the footslope, and resin-P_i content at the summit was in between that at the other two sites.

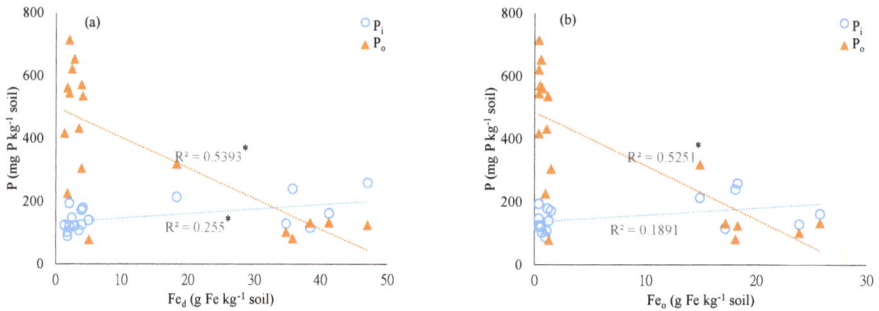

Figure 1. Correlations of contents of P_i and P_o with Fe_d (**a**) and Fe_o (**b**) in the pedon samples collected from the three sampling sites. * Statistically significant at $p < 0.05$.

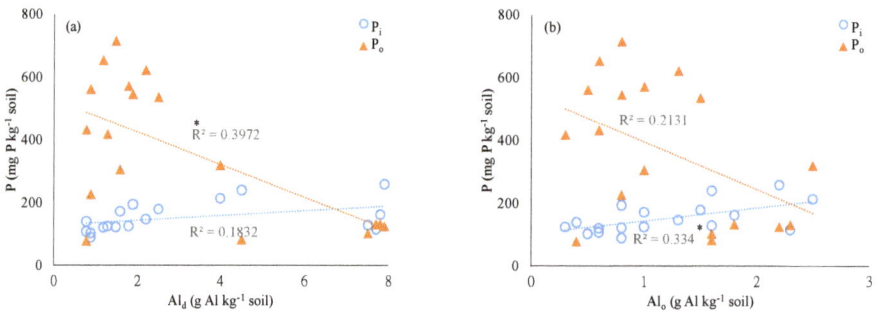

Figure 2. Correlations of contents of P_i and P_o with Al_d (**a**) and Al_o (**b**) in the pedon samples collected from the three sampling sites. * Statistically significant at $p < 0.05$.

Organic P was the dominant P fraction in the mountain forest soils of the three sites (Table 3). Moreover, $NaOH$-P_o represented the major P fraction and contributed to more than 40% of the summed P content at the three sampling sites. $NaHCO_3$-P_o was the second most abundant P fraction among the three sites and contributed more than 20% of the summed P content. Summed P_i content (resin-P_i + $NaHCO_3$-P_i + $NaOH$-P_i + HCl-P_i + $cHCl$-P_i) in surface soils of all study sites contained less than 18% of summed P, and $NaOH$-P_i was the major P_i fraction in total P_i.

Table 1. General chemical properties of studied soils.

Site	Horizon	Depth (cm)	pH	CEC (cmol(+) kg⁻¹)	Base Saturation (%)	TOC (g kg⁻¹)	Total N (g kg⁻¹)	Total P (mg kg⁻¹)	P_i (mg kg⁻¹)	P_o (mg kg⁻¹)	Fe_o (g kg⁻¹)	Fe_d (g kg⁻¹)	Al_o (g kg⁻¹)	Al_d (g kg⁻¹)
Summit	O_i	+10→+7	3.7	73.1	1.2	352.4	8.88	945.8	125.5	571.6	0.5	4.1	1.0	1.8
	O_e	+7→+2	3.5	124.9	3.3	492.5	17.49	1040.2	147.2	621.4	0.4	2.6	1.3	2.2
	O_a	+2→0	3.5	146.9	2.3	492.4	22.1	606.7	178.6	535.8	1.2	4.3	1.5	2.5
	A	0→1	3.5	60.1	1.8	207.3	11.9	667.8	171.7	305.2	1.5	4.1	1.0	1.6
	E	1→8	3.8	19.5	4.6	38.0	1.9	308.1	89.3	226.1	1.0	1.9	0.8	1.0
	B_{t1}	8→20	4.1	19.6	4.1	10.0	0.95	194.2	116.0	131.9	17.2	38.5	2.3	7.7
	B_{t2}	20→30	4.2	12.4	6.5	11.9	1.07	269.0	162.2	132.8	25.8	41.3	1.8	7.8
	BC	30→45	4.3	12.7	3.9	9.0	1.06	245.8	129.3	102.6	23.9	34.9	1.6	7.5
Footslope	O_i	+11→+8	3.8	98.9	10.6	541.7	12.8	905.0	122.3	714.5	0.4	2.2	0.8	1.5
	O_e	+8→+4	3.5	120.3	5.3	511.8	19.1	701.6	120.8	653.2	0.6	3.0	0.6	1.2
	O_a	+4→0	3.3	102.9	5.0	476.7	22.9	402.5	193.6	545.5	0.4	2.2	0.8	1.9
	A	0→5	3.5	65.3	2.6	237.6	12.9	305.9	213.4	318.9	14.9	18.3	2.5	4.0
	E	5→10	4	11.5	7.8	7.0	0.7	171.7	139.7	78.7	1.3	5.2	0.4	0.8
	Bw_1	10→23	4.1	16.6	2.4	10.0	1.1	283.8	239.7	82.0	18.1	35.9	1.6	4.5
	Bw_2	23→42	4.5	17.6	4.5	19.9	1.5	928.3	258.9	125.1	18.3	47.0	2.2	7.9
Lakeshore	O_i	+33→+22	3.5	36.3	5.7	363.1	12.7	934.1	103.0	561.8	0.7	1.9	0.5	0.9
	O_e	+22→+9	3.4	18.7	8	186.7	8.7	563.4	107.9	431.7	1.1	3.7	0.6	0.8
	O_a	+9→0	3.5	17.7	7.3	176.9	7.4	945.8	124.6	417.3	0.4	1.4	0.3	1.3

CEC: cation-exchange capacity; TOC: total organic C; Fe_o, Al_o: iron and aluminum extracted by the ammonium oxalate method; Fe_d, Al_d: iron and aluminum extracted by the citrate-bicarbonate–dithionite method.

Table 2. The fractionation of P (mg kg⁻¹) in humic soil samples (O horizon) in different topographic sites (sequential extraction).

Site	Inorganic P in Extracts					Summed Inorganic P (P_i)	Organic P in Extracts			Summed Organic P (P_o)	Residual-P	Summed P	Total P
	Resin-P_i	NaHCO₃-P_i	NaOH-P_i	HCl-P_i	cHCl-P_i		NaHCO₃-P_o	NaOH-P_o	cHCl-P_o				
Summit	4.8 [ab]	61.2 [a]	46.4 [b]	2.3	9.4	124.1 [ab]	124.4 [b]	415.6 [a]	20.6 [b]	560.6 [ab]	38.0	722.7 [ab]	757.0 [ab]
Footslope	2.4 [b]	35.3 [b]	83.3 [a]	4.4	4.7	130.1 [a]	272.7 [a]	335.3 [ab]	31.6 [a]	639.6 [a]	29.4	799.0 [a]	828.3 [a]
Lakeshore	7.3 [a]	25.4 [b]	36.4 [b]	5.9	9.9	84.8 [b]	177.5 [b]	306.5 [b]	27.6 [ab]	511.5 [b]	34.9	631.2 [b]	668.2 [b]

HCl-P_i: inorganic P extracted by 1.0 M HCl; cHCl-P_i: inorganic P extracted by concentrated HCl; cHCl-P_o: organic P extracted by concentrated HCl; Summed inorganic P: Resin-P_i + NaHCO₃-P_i + NaOH-P_i + HCl-P_i + cHCl-P_i; Summed organic P: NaHCO₃-P_o + NaOH-P_o + cHCl-P_o; Summed P: sum of P_i + P_o + residual P. Means followed by the same letters in the same column are not significantly different ($p > 0.05$) by Tukey's honestly significant difference test.

Table 3. Relative proportions of total P extracted for inorganic (P_i) and organic (P_o) forms in NaOH-EDTA extracts from humic soil samples (O horizon) determined by chemical extraction and ^{31}P-NMR spectroscopy.

Soil	P_i		P_o	
	Chemical [§]	NMR	Chemical [§]	NMR
		%		
Summit	17.2 ± 2.8	17.5 ± 1.2	77.5 ± 2.9	82.5 ± 3.7
Footslope	16.3 ± 2.1	17.3 ± 1.4	80.0 ± 2.4	82.7 ± 2.7
Lakeshore	16.9 ± 1.4	11.6 ± 1.7	81.1 ± 1.5	88.4 ± 5.7

[§] P_i: sum of resin-P_i + NaHCO$_3$-P_i + NaOH-P_i + HCl-P_i + cHCl-P_i. P_o: sum of NaHCO$_3$-P_o + NaOH-P_o + cHCl-P_o.

Spectra obtained from ^{31}P-NMR analysis of NaOH-EDTA extracts revealed inorganic orthophosphate, orthophosphate monoesters, orthophosphate diesters, pyrophosphates, and phosphonates in the soil extracts (Figure 3).

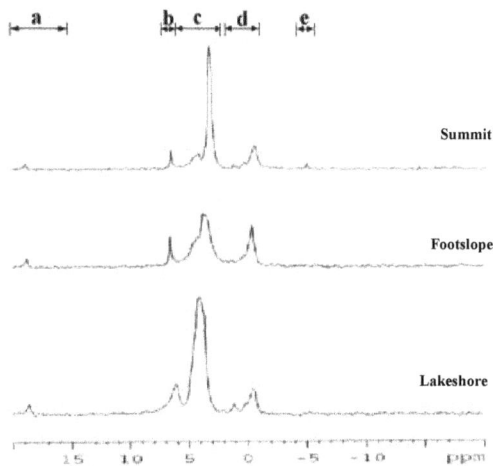

Figure 3. ^{31}P-NMR spectra for NaOH-EDTA extracts from soils at different sites. a: phosphonate, b: inorganic orthophosphate, c: orthophosphate monoesters, d: orthophosphate diesters, e: pyrophosphate.

Organic P compounds identified in the NaOH-EDTA extracts included orthophosphate monoesters, orthophosphate diesters, and phosphonates. Orthophosphate monoesters were the predominant species of extracted organic P in soil from all sites and contributed to more than 60% of the total P fractions (Figure 4). The proportion of orthophosphate diesters was much lower than that of orthophosphate monoesters and only contributed 15% to 20% of the total P pools. Content of phosphonates (18.7 ppm) ranged from only 2.2% to 4.0% of extracted P from the three sites, and the highest content was found at lakeshore, with waterlogged conditions.

Inorganic P compounds identified in the NaOH-EDTA extracts included orthophosphate and pyrophosphates. Inorganic orthophosphate signals at 6.1–6.3 ppm ranged from 11.6% to 17.3% of the spectral area for all study sites (Table 3). The highest inorganic P content was found at the footslope. In addition, small additional pyrophosphate resonance (−4.3 ppm) was observed only at the summit site.

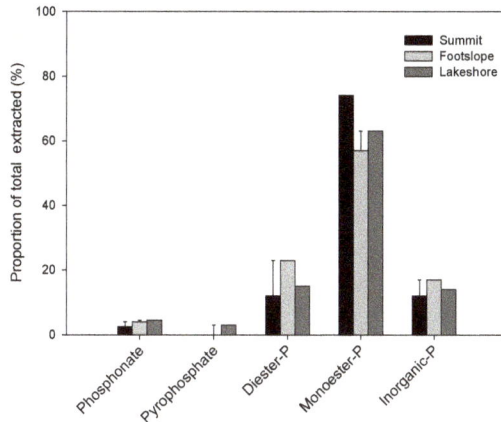

Figure 4. Proportion of extracted P in various classes from humic samples at different sites determined by ^{31}P-NMR spectroscopy. Error bars indicate standard deviation.

4. Discussion

4.1. Soil Physiochemical Properties and Chemical Extractable P

The soil in this study site contains high moisture because of the year-round high precipitation [45]. High soil moisture in mountainous forests retards decomposition of soil organic matter, and high precipitation increases the loss of cations, thereby resulting in decreased soil pH and Eh [38,46,47]. This explains our observations of low soil pH in the studied sites. Moreover, our previous study at the same sites revealed a clay and silt-clay layer under the organic layer [45]. This clay layer may retard the percolation [48] and therefore result in reduced soil TOC, TN and TP contents with increasing soil depth.

The high Fe_o, Fe_d, Al_o and Al_d contents in the B horizons implied that iron moved downwards and accumulated in the subsoil. Because high soil moisture and rich organic matter decreased soil redox potential in the surface layer, reduced Fe and Al ions moving from surface to the bottom layer were re-oxidized in the B horizons [49,50]. This formation of accumulated Fe, which may due to the redoximorphic process [51,52], created a placic-like horizon and resulted in slow permeability of P in such perhumid forest soil [49].

The positive correlations between P_i and Fe_d but not P_i and Fe_o contents implied that most inorganic P may be adsorbed by crystalline Fe oxides in soils, whereas the negative correlation between P_o and Fe_d as well as P_o and Fe_o contents implied that organic P was in a complex formation in the soils. In addition, the vertical increase in P_i content, with an opposite trend to P_o content, in each soil profile was significantly related to the content of amorphous and crystalline Fe oxides but not crystalline Al oxides. This observation implied that Fe oxides, rather than Al oxides, may chemically bind with P_i, and the accumulation of P_i in the subsoils could relate to the downward percolation and reoxidation of Fe. Sollins et al. [53] found that soil with high Fe hydrous oxides content tends to irreversibly fix polyvalent oxyanions such as phosphate because of chemosorption and occlusion. A similar trend was found in our previous study of subalpine forest soils [54], in which significant sorption of P_i to sesquioxides was via downward migration.

In addition, acidic soil conditions (pH < 4) typically facilitate Fe oxides reduction (i.e., Fe^{2+}), which may help the mobility of Fe in soil [55]. The acidic soils at this study site may further affect the mobility of P_i, thereby resulting in low P_i content at the three study sites.

The depth of O horizons increased from the summit to the lakeshore, which suggests a process of erosion–deposition. In addition, because the mineral clay layer and placic-like horizon reduced vertical percolation, most of the soil organic matter in the water flow is transported laterally [33,45]. This can help the downhill movement of soil nutrients in the O horizon [45,49] and may explain the increased soil TOC and TN contents from the summit to the footslope. In addition, because the soils at the summit and footslope were relatively well-drained while the soils at the lakeshore were poorly-drained [33], this may help the accumulation of organic matter and result in a thick O horizon at the lakeshore.

4.2. Chemical Extraction of Soil P

Because of the low overall P_i concentrations in the three study sites, the different chemical extractable P_i contents increased downhill but not significantly. In addition, the low labile P fractions, $NaHCO_3$-P_i and $NaOH$-P_i, at the lakeshore may also be due to the vigorous fluctuation of the water level of the lake after showers or storms, which could remove the suspended particles or detritus of litter with the flooding and reduce the accumulation of P in the soil near the lakeshore [30]. This was indirectly supported by the elevated TP concentration in the epilimnion of lake after medium rainfall events (256–620 mm) [56].

P_o appeared to be the predominant fraction in the perhumid forest and was mostly non-acid extractable. Because the soil is acidic at the study sites, most P_o fractions may not be labile and remained at higher values at the summit than at the lakeshore. The increase in total P_o content in the surface horizon has been attributed to the input and accumulation of organic matter [57] and factors affected by topography can influence the availability of soil P [58].

Extracted soil P compounds showed that the contents of highly labile ($NaHCO_3$-P_o), long-term P transformation ($NaOH$-P_o) and stable residual pool ($cHCl$-P_o) fractions changed between different sites, which showed that slope position affects the various P pools [59]. As shown in Table 2, the sum of the highly labile P_o fraction ($NaHCO_3$-P_o) and long-term P transformation ($NaOH$-P_o) contributed more than 75% of the total extractable P in soils of all study sites, so organic P was the major P source in these soils. The P_o accumulation in soil surfaces resulted from the biological cycling of P through the plant litter to the soil surface. In addition, we were aware that the density and height of understory *Rhododendron* were higher at the lakeshore than the footslope and summit, although we did not measure the abundance/density of the plants among the three sites. It could also be a potential factor causing the differences in P_o accumulation.

The $cHCl$-P_i extract has recalcitrant P forms associated with mainly Fe oxides and/or P derived from non-alkaline extractable debris, whereas $cHCl$-P_o may include both stable, little and/or bioavailable (non-alkaline extractable) P forms. However, the proportions of $cHCl$-extractable P_i and P_o were only about 4.5% to 7.5% of sequentially extracted total P in soils of all study sites. The residual P is associated with highly organic materials such as lignin and organometallic complexes [60], but we have no information on the composition of organic matter in this fraction.

4.3. Spectra of ^{31}P-NMR Analyses

Orthophosphate monoesters are the most common forms of organic P_o in soils [61–63]. Monoester P includes high proportions of inositol phosphate, sugar phosphate and choline phosphate primarily derived from plant, animal, and microbial residues [64].

Depending on the soil types, inositol phosphates are reported to be the predominant organic P forms in Podosols [28], whereas α- and β-glycerophosphate are the predominant organic P forms in Vertosols [65]. In addition, high inositol phosphate contents were reported from several studies with cold and wet climates [66–68].

Inositol phosphates are typically considered of limited bioavailability because of the complex structure with soil minerals, clays, and humic compounds [62,69]. Although ^{31}P-NMR analysis in our study had limited resolution to identify the inositol phosphates content, the low temperature and high precipitation of the study site may likely result in high inositol phosphates content in the soil.

Orthophosphate diesters at about 0 ppm [42] can be further classified into nucleic acids (−1–0 ppm) and phospholipids (0–2 ppm) [62]. Orthophosphate diesters, including nucleic and phospholipids, more frequently accumulate in cool and moist acidic forest soils with low microbial activities than in agricultural soils [70–72]. In acidic or wet soils, diester P proportion is between 10% and 36% of extracted P [72–74]. Our findings are consistent with previous studies because the diester P proportion was between 15% and 20% of extracted P in the study sites.

A higher proportion of diester P providing a labile source for available P [75] was found at the footslope site, with poor drainage, than at the other sites. The lower orthophosphate diesters than monoesters content at the three study sites may contribute to the complexity of the chemical compounds. Orthophosphate diesters are more rapidly mineralizable than monoesters because they are generally less adsorbed to soil colloids than monoesters and can be easily hydrolyzed [24,61,76].

The content of phosphonates in soil are due to bacteria such as *Bacillus cereus*, which has a phosphonatase enzyme that produces phosphonates, but the bacteria are less prevalent in acidic soil [44,74,77]. This observation may explain the low phosphonates concentrations observed at the study sites.

A small amount of pyrophosphate resonance (−4.3 ppm) was observed in spectra of soils at the three sites. Pyrophosphate is believed to be involved in biological P cycling in the soils and may be present in relatively well-drained soil that provides a proper environment for microbial activity and fungus [78]. In addition, pyrophosphate is contributed by fungal P compounds [79]. Because the surface soil of the three sites contained high soil organic matter and high soil moisture, it may provide a less favorable environment for microbial and fungal activities, resulting in low pyrophosphate concentrations.

Organic P represented between 82–88% and 77–81% of total P extracted by NaOH-EDTA and by chemical extraction from all sites studied, respectively. This result is similar to research byCade-Menun and Preston [80], who found 77% to 83% of P_0 in a low-pH forest perhaps because of the low decomposition of P_0 compounds in acidic forest soils that reduced the P_i concentrations [44].

The signal intensity of ^{31}P-NMR spectra caused by paramagnetic Al, Fe and Mn in soils may reduce the spectra quality [47,81,82]. Moreover, chemical hydrolysis of P_0 to P_i may occur during alkaline extractions [80,83]. However, in general, the results from NMR analysis were consistent with those of chemical fractionation in this study and other alpine and subalpine forest soils [28,49,54,68].

5. Conclusions

This study demonstrated that soil chemical extractable P_i and P_0 can be vertically affected by the formation of Fe oxides in soils. Because of a clay layer combined with a placic-like horizon in the subsoil in our test site, both total P and P_0 contents were decreased with increasing soil depth, with P_i content slightly increased in different soil horizons. The low permeable soil layer also favored downhill run-off, however, because P_i contents were relatively low compared with P_0 contents; the contents did not significantly differ among the three study sites. Because most of the P was in organic forms, a negligible amount of P_i may be released to the lake along the slope. Therefore, although topography and soil formation processes affect the distribution of soil P, high precipitation and year-round high humidity might be important for differentiation of the P species in this landscape. Moreover, the similarity of the ^{31}P-NMR spectra among the three sampling sites supports the alleviated differentiation of the P species in this landscape. Preserving plant abundance and plant litters should be able to maintain soil P cycles and prevent the P discharge from subtropical alpine forest ecosystems.

Author Contributions: C.-Y.C. conceived the methodology and experimental design; J.-W.T., W.-C.L., R.S.W.Y., S.-C.C. and C.-Y.C. performed the experiments and analyzed the data; Y.-J.S. and C.-W.P. wrote the original draft; C.-Y.C. reviewed and edited the manuscript.

Acknowledgments: The study was granted by Academia Sinica and Ministry of Sciences and Technology (MOST 106-2621-M-239-001), Taiwan.

Conflicts of Interest: The authors declare no conflict of interest.

References

1. Gutsch, M.; Lasch-Born, P.; Kollas, C.; Suckow, F.; Reyer, C.P.O. Balancing trade-offs between ecosystem services in Germany's forests under climate change. *Environ. Res. Lett.* **2018**, *13*, 045012. [CrossRef]
2. Pizzeghello, D.; Francioso, O.; Concheri, G.; Muscolo, A.; Nardi, S. Land use affects the soil C sequestration in alpine environment, NE Italy. *Forests* **2017**, *8*, 197. [CrossRef]
3. Wu, Y.H.; Zhou, J.; Yu, D.; Sun, S.Q.; Luo, J.; Bing, H.J.; Sun, H.Y. Phosphorus biogeochemical cycle research in mountainous ecosystems. *J. Mt. Sci.* **2013**, *10*, 43–53. [CrossRef]
4. Cassagne, N.; Remaury, M.; Gauquelin, T.; Fabre, A. Forms and profile distribution of soil phosphorus in alpine Inceptisols and Spodosols (Pyrenees, France). *Geoderma* **2000**, *95*, 161–172. [CrossRef]
5. Galloway, J.N.; Dentener, F.J.; Capone, D.G.; Boyer, E.W.; Howarth, R.W.; Seitzinger, S.P.; Asner, G.P.; Cleveland, C.C.; Green, P.A.; Holland, E.A.; et al. Nitrogen cycles: Past, present, and future. *Biogeochemistry* **2004**, *70*, 153–226. [CrossRef]
6. Amundson, R.; Jenny, H. On a state factor model of ecosystems. *BioScience* **1997**, *47*, 536–543. [CrossRef]
7. Vitousek, P.M.; Porder, S.; Houlton, B.Z.; Chadwick, O.A. Terrestrial phosphorus limitation: Mechanisms, implications, and nitrogen-phosphorus interactions. *Ecol. Appl.* **2010**, *20*, 5–15. [CrossRef] [PubMed]
8. Walker, T.W.; Syers, J.K. The fate of phosphorus during pedogenesis. *Geoderma* **1976**, *15*, 1–19. [CrossRef]
9. Chapin, F.I.; Mooney, H.; Chapin, M.; Matson, P. *Principles of Terrestrial Ecosystem Ecology*; Springer: New York, NY, USA, 2002.
10. Egli, M.; Filip, D.; Mavris, C.; Fischer, B.; Götze, J.; Raimondi, S.; Seibert, J. Rapid transformation of inorganic to organic and plant-available phosphorous in soils of a glacier forefield. *Geoderma* **2012**, *189–190*, 215–226. [CrossRef]
11. Achat, D.L.; Bakker, M.R.; Zeller, B.; Pellerin, S.; Bienaimé, S.; Morel, C. Long-term organic phosphorus mineralization in Spodosols under forests and its relation to carbon and nitrogen mineralization. *Soil Biol. Biochem.* **2010**, *42*, 1479–1490. [CrossRef]
12. Araújo, M.S.B.; Schaefer, C.E.R.; Sampaio, E.V.S.B. Soil phosphorus fractions from toposequences of semi-arid Latosols and Luvisols in northeastern Brazil. *Geoderma* **2004**, *119*, 309–321. [CrossRef]
13. Ciampitti, I.A.; García, F.O.; Picone, L.I.; Rubio, G. Phosphorus budget and soil extractable dynamics in field crop rotations in Mollisols. *Soil Sci. Soc. Am. J.* **2011**, *75*, 131. [CrossRef]
14. Zamuner, E.C.; Picone, L.I.; Echeverria, H.E. Organic and inorganic phosphorus in Mollisol soil under different tillage practices. *Soil Tillage Res.* **2008**, *99*, 131–138. [CrossRef]
15. Zhu, H.-J.; Sun, L.-F.; Zhang, Y.-F.; Zhang, X.-L.; Qiao, J.-J. Conversion of spent mushroom substrate to biofertilizer using a stress-tolerant phosphate-solubilizing *Pichia farinose* fl7. *Bioresour. Technol.* **2012**, *111*, 410–416. [CrossRef] [PubMed]
16. Cardoso, I.M.; Van der Meer, P.; Oenema, O.; Janssen, B.H.; Kuyper, T.W. Analysis of phosphorus by 31pnmr in Oxisols under agroforestry and conventional coffee systems in Brazil. *Geoderma* **2003**, *112*, 51–70. [CrossRef]
17. Vu, D.T.; Tang, C.; Armstrong, R.D. Transformations and availability of phosphorus in three contrasting soil types from native and farming systems: A study using fractionation and isotopic labeling techniques. *J. Soils Sediments* **2009**, *10*, 18–29. [CrossRef]
18. Achat, D.L.; Augusto, L.; Bakker, M.R.; Gallet-Budynek, A.; Morel, C. Microbial processes controlling P availability in forest Spodosols as affected by soil depth and soil properties. *Soil Biol. Biochem.* **2012**, *44*, 39–48. [CrossRef]
19. Zhang, Q.; Wang, Y.P.; Pitman, A.J.; Dai, Y.J. Limitations of nitrogen and phosphorous on the terrestrial carbon uptake in the 20th century. *Geophys. Res. Lett.* **2011**, *38*, L22701. [CrossRef]
20. Oberson, A.; Fardeau, J.C.; Besson, J.M.; Sticher, H. Soil phosphorus dynamics in cropping systems managed according to conventional and biological agricultural methods. *Biol. Fertil. Soils* **1993**, *16*, 111–117. [CrossRef]
21. Potter, R.L.; Jordan, C.F.; Guedes, R.M.; Batmanian, G.J.; Han, X.G. Assessment of a phosphorus fractionation method for soils: Problems for further investigation. *Agric. Ecosyst. Environ.* **1991**, *34*, 453–463. [CrossRef]
22. Ruttenberg, K.C. Development of a sequential extraction method for different forms of phosphorus in marine sediments. *Limnol. Oceanogr.* **1992**, *37*, 1460–1482. [CrossRef]
23. Tiessen, H.; Moir, J.O. Characterisation of available P by sequential extraction. In *Soil Sampling and Methods of Analysis*; Lewis Publisher: Boca Raton, FL, USA, 1993; pp. 75–86.

24. Condron, L.M.; Frossard, E.; Tiessen, H.; Newmans, R.H.; Stewart, J.W.B. Chemical nature of organic phosphorus in cultivated and uncultivated soils under different environmental conditions. *J. Soil Sci.* **1990**, *41*, 41–50. [CrossRef]

25. Kizewski, F.; Liu, Y.-T.; Morris, A.; Hesterberg, D. Spectroscopic approaches for phosphorus speciation in soils and other environmental systems. *J. Environ. Qual.* **2011**, *40*, 751. [CrossRef] [PubMed]

26. Preston, C.M. Applications of nmr to soil organic matter analysis: History and prospects. *Soil Sci.* **1996**, *161*, 144–166. [CrossRef]

27. De Feudis, M.; Cardelli, V.; Massaccesi, L.; Bol, R.; Willbold, S.; Cocco, S.; Corti, G.; Agnelli, A. Effect of beech (*Fagus sylvatica* L.) rhizosphere on phosphorous availability in soils at different altitudes (central Italy). *Geoderma* **2016**, *276*, 53–63. [CrossRef]

28. Doolette, A.L.; Smernik, R.J.; McLaren, T.I. The composition of organic phosphorus in soils of the Snowy Mountains region of south-eastern Australia. *Soil Res.* **2017**, *55*, 10–18. [CrossRef]

29. Huang, W.; Liu, J.; Wang, Y.P.; Zhou, G.; Han, T.; Li, Y. Increasing phosphorus limitation along three successional forests in southern China. *Plant Soil* **2012**, *364*, 181–191. [CrossRef]

30. Xiao, R.; Bai, J.; Gao, H.; Huang, L.; Deng, W. Spatial distribution of phosphorus in marsh soils of a typical land/inland water ecotone along a hydrological gradient. *Catena* **2012**, *98*, 96–103. [CrossRef]

31. Ho, C.S. *An Introduction to the Geology of Taiwan: Explanatory Text of the Geologic Map of Taiwan*, 2nd ed.; Central Geological Survey: Taipei, Taiwan, 1988.

32. Soil Survey Staff. *Keys to Soil Taxonomy*, 12 ed.; Agricultural Handbook No 436; United States Department of Agriculture: Washington, DC, USA, 2014.

33. Chiu, C.-Y.; Lai, S.-Y.; Lin, Y.-M.; Chiang, H.-C. Distribution of the radionuclide ^{137}Cs in the soils of a wet mountainous forest in Taiwan. *Appl. Radiat. Isot.* **1999**, *50*, 1097–1103. [CrossRef]

34. Rhoades, J.D. Soluble salts. In *Methods of Soil Analysis Part 2, Chemical and Microbiological Properties*; Page, A.L., Miller, R.H., Kenney, D.R., Eds.; Agronomy Monograph: Kincaid, DC, USA, 1982; Volume 9.

35. Mehra, O.P.; Jackson, M.L. Iron oxide removal from soils and clays by a dithionite-citrate system buffered with sodium bicarbonate. In *Clays and Clay Minerals*; Elsevier: New York, NY, USA, 2013; pp. 317–327.

36. McKeague, J.A.; Day, J.H. Dithionite- and oxalate-extractable Fe and Al as aids in differentiating various classes of soils. *Can. J. Soil Sci.* **1966**, *46*, 13–22. [CrossRef]

37. Hedley, M.J.; Stewart, J.W.B.; Chauhan, B.S. Changes in inorganic and organic soil phosphorus fractions induced by cultivation practices and by laboratory incubations. *Soil Sci. S. Am. J.* **1982**, *46*, 970. [CrossRef]

38. Richter, D.D.; Allen, H.L.; Li, J.; Markewitz, D.; Raikes, J. Bioavailability of slowly cycling soil phosphorus: Major restructuring of soil P fractions over four decades in an aggrading forest. *Oecologia* **2006**, *150*, 259–271. [CrossRef] [PubMed]

39. Rheinheimer, D.S.; Anghinoni, I.; Flores, A.F. Organic and inorganic phosphorus as characterized by phosphorus-31 nuclear magnetic resonance in subtropical soils under management systems. *Commun. Soil Sci. Plant Anal.* **2002**, *33*, 1853–1871. [CrossRef]

40. Lajtha, K.; Driscoll, C.; Jarrell, W.; Elliott, E. Soil phosphorous: Characterization and total element analysis. In *Standard Soil Methods for Long-Term Ecological Research*; Roberston, G., Coleman, D., Bledsoe, C., Sollins, P., Eds.; Oxford University Press: Oxford, UK, 1999.

41. Robinson, J.S.; Johnston, C.T.; Reddy, K.R. Combined chemical and ^{31}P-NMR spectroscopic analysis of phosphorus in wetland organic soils. *Soil Sci.* **1998**, *163*, 705–713. [CrossRef]

42. Newman, R.H.; Tate, K.R. Soil phosphorus characterisation by ^{31}P nuclear magnetic resonance. *Commun. Soil Sci. Plant Anal.* **1980**, *11*, 835–842. [CrossRef]

43. Preston, C.M. Review of solution NMR of humic substances. In *NMR of Humic Substances and Coal*; Wershaw, R.L., Mikita, M.A., Eds.; Lewis Publisher: Chelsea, MI, USA, 1987; pp. 3–32.

44. Dai, K.O.H.; David, M.B.; Vance, G.F.; Krzyszowska, A.J. Characterization of phosphorus in a spruce-fir Spodosol by phosphorus-31 nuclear magnetic resonance spectroscopy. *Soil Sci. Soc. Am. J.* **1996**, *60*, 1943. [CrossRef]

45. Chen, J.-S.; Chiu, C.-Y. Effect of topography on the composition of soil organic substances in a perhumid sub-tropical montane forest ecosystem in taiwan. *Geoderma* **2000**, *96*, 19–30. [CrossRef]

46. Ohno, T.; Fernandez, I.J.; Hiradate, S.; Sherman, J.F. Effects of soil acidification and forest type on water soluble soil organic matter properties. *Geoderma* **2007**, *140*, 176–187. [CrossRef]

47. Prietzel, J.; Dümig, A.; Wu, Y.; Zhou, J.; Klysubun, W. Synchrotron-based P k-edge XANES spectroscopy reveals rapid changes of phosphorus speciation in the topsoil of two glacier foreland chronosequences. *Geochim. Cosmochim. Acta* **2013**, *108*, 154–171. [CrossRef]

48. Candler, R.; Zech, W.; Alt, H.G. A comparison of water soluble organic substances in acid soils under beech and spruce in NE-Bavaria. *Z. Pflanz. Bodenk.* **1989**, *152*, 61–65. [CrossRef]

49. Jien, S.H.; Baillie, I.; Hu, C.-C.; Chen, T.-H.; Iizuka, Y.; Chiu, C.-Y. Forms and distribution of phosphorus in a placic podzolic toposequence in a subtropical subalpine forest, Taiwan. *Catena* **2016**, *140*, 145–154. [CrossRef]

50. Jien, S.H.; Hseu, Z.Y.; Iizuka, Y.; Chen, T.H.; Chiu, C.Y. Geochemical characterization of placic horizons in subtropical montane forest soils, northeastern Taiwan. *Eur. J. Soil Sci.* **2010**, *61*, 319–332. [CrossRef]

51. Hseu, Z.-Y.; Chen, Z.-S.; Wu, Z.-D. Characterization of placic horizons in two subalpine forest Inceptisols. *Soil Sci. Soc. Am. J.* **1999**, *63*, 941. [CrossRef]

52. Wu, S.P.; Chen, Z.S. Characteristics and genesis of Inceptisols with placic horizons in the subalpine forest soils of Taiwan. *Geoderma* **2005**, *125*, 331–341. [CrossRef]

53. Sollins, P.; Robertson, G.P.; Uehara, G. Nutrient mobility in variable- and permanent-charge soils. *Biogeochemistry* **1988**, *6*, 181–199. [CrossRef]

54. Chiu, C.-Y.; Pai, C.-W.; Yang, K.-L. Characterization of phosphorus in sub-alpine forest and adjacent grassland soils by chemical extraction and phosphorus-31 nuclear magnetic resonance spectroscopy. *Pedobiologia* **2005**, *49*, 655–663. [CrossRef]

55. Spark, D.L. *Environmental Soil Chemistry*; Academic Press: Cambridge, MA, USA, 1995.

56. Tsai, J.-W.; Kratz, T.K.; Hanson, P.C.; Kimura, N.; Liu, W.-C.; Lin, F.-P.; Chou, H.-M.; Wu, J.-T.; Chiu, C.-Y. Metabolic changes and the resistance and resilience of a subtropical heterotrophic lake to typhoon disturbance. *Can. J. Fish. Aquat. Sci.* **2011**, *68*, 768–780. [CrossRef]

57. Smeck, N.E. Phosphorus. *Soil Sci.* **1973**, *115*, 199–206. [CrossRef]

58. Smeck, N.E. Phosphorus dynamics in soils and landscapes. *Geoderma* **1985**, *36*, 185–199. [CrossRef]

59. Agbenin, J.O.; Tiessen, H. Phosphorus forms in particle-size fractions of a toposequence from northeast Brazil. *Soil Sci. Soc. Am. J.* **1995**, *59*, 1687. [CrossRef]

60. Schlichting, A.; Leinweber, P.; Meissner, R.; Altermann, M. Sequentially extracted phosphorus fractions in peat-derived soils. *J. Plant Nutr. Soil Sci.* **2002**, *165*, 290–298. [CrossRef]

61. Fox, T.R.; Miller, B.W.; Rubilar, R.; Stape, J.L.; Albaugh, T.J. Phosphorus nutrition of forest plantations: The role of inorganic and organic phosphorus. In *Soil Biology*; Springer: Berlin/Heidelberg, Germany, 2010; pp. 317–338.

62. Turner, B.L.; Mahieu, N.; Condron, L.M. Quantification of myo-inositol hexakisphosphate in alkaline soil extracts by solution ^{31}P NMR spectroscopy and spectral deconvolution. *Soil Sci.* **2003**, *168*, 469–478. [CrossRef]

63. Turner, B.L.; Mahieu, N.; Condron, L.M.; Chen, C.R. Quantification and bioavailability of scyllo-inositol hexakisphosphate in pasture soils. *Soil Biol. Biochem.* **2005**, *37*, 2155–2158. [CrossRef]

64. Magid, J.; Tiessen, H.; Condron, L.M. Dynamics of organic phosphorus in soils under natural and agricultural ecosystems. In *Humic Substances in Terrestrial Ecosystems*; Elsevier: New York, NY, USA, 1996; pp. 429–466.

65. McLaren, T.I.; Smernik, R.J.; Guppy, C.N.; Bell, M.J.; Tighe, M.K. The organic P composition of Vertisols as determined by ^{31}P NMR spectroscopy. *Soil Sci. Soc. Am. J.* **2014**, *78*, 1893. [CrossRef]

66. Ahlgren, J.; Djodjic, F.; Börjesson, G.; Mattsson, L. Identification and quantification of organic phosphorus forms in soils from fertility experiments. *Soil Use Manag.* **2013**, *29*, 24–35. [CrossRef]

67. Turner, B.L.; Cheesman, A.W.; Godage, H.Y.; Riley, A.M.; Potter, B.V.L. Determination of neo- and D-chiro-inositol hexakisphosphate in soils by solution ^{31}P NMR spectroscopy. *Environ. Sci. Technol.* **2012**, *46*, 4994–5002. [CrossRef] [PubMed]

68. Vincent, A.G.; Vestergren, J.; Gröbner, G.; Persson, P.; Schleucher, J.; Giesler, R. Soil organic phosphorus transformations in a boreal forest chronosequence. *Plant Soil* **2013**, *367*, 149–162. [CrossRef]

69. Turner, B.L.; Newman, S.; Cheesman, A.W.; Reddy, K.R. Sample pretreatment and phosphorus speciation in wetland soils. *Soil Sci. Soc. Am. J.* **2007**, *71*, 1538. [CrossRef]

70. Cade-Menun, B.J. Characterizing phosphorus in environmental and agricultural samples by ^{31}P nuclear magnetic resonance spectroscopy. *Talanta* **2005**, *66*, 359–371. [CrossRef] [PubMed]

71. Cade-Menun, B.J.; Berch, S.M.; Preston, C.M.; Lavkulich, L.M. Phosphorus forms and related soil chemistry of podzolic soils on northern Vancouver Island. I. A comparison of two forest types. *Can. J. For. Res.* **2000**, *30*, 1714–1725. [CrossRef]

72. Makarov, M.I.; Guggenberger, G.; Zech, W.; Alt, H.G. Organic phosphorus species in humic acids of mountain soils along a toposequence in the northern Caucasus. *Z. Pflanz. Bodenk.* **1996**, *159*, 467–470. [CrossRef]

73. Forster, J.C.; Zech, W. Phosphorus status of a soil catena under liberian evergreen rain forest: Results of ^{31}P NMR spectroscopy and phosphorus adsorption experiments. *Z. Pflanz. Bodenk.* **1993**, *156*, 61–66. [CrossRef]

74. Zech, W.; Alt, H.G.; Haumaier, L.; Blasek, R. Characterization of phosphorus fractions in mountain soils of the Bavarian Alps by ^{31}P NMR spectroscopy. *Z. Pflanz. Bodenk.* **1987**, *150*, 119–123. [CrossRef]

75. Miltner, A.; Haumaier, L.; Zech, W. Transformations of phosphorus during incubation of beech leaf litter in the presence of oxides. *Eur. J. Soil Sci.* **1998**, *49*, 471–475. [CrossRef]

76. Condorn, L.; Frossard, E.; Newman, R.H. Use of ^{31}P NMR in the study of soils and the environment. In *Nuclear Magnetic Resonance Spectroscopy in Environmental Chemistry*; Oxford University Press: New York, NY, USA, 1997; pp. 247–271.

77. Tate, K.R.; Newman, R.H. Phosphorus fractions of a climosequence of soils in New Zealand tussock grassland. *Soil Biol. Biochem.* **1982**, *14*, 191–196. [CrossRef]

78. Rousk, J.; Brookes, P.C.; Baath, E. Contrasting soil ph effects on fungal and bacterial growth suggest functional redundancy in carbon mineralization. *Appl. Environ. Microbiol.* **2009**, *75*, 1589–1596. [CrossRef] [PubMed]

79. Makarov, M.I.; Haumaier, L.; Zech, W.; Marfenina, O.E.; Lysak, L.V. Can ^{31}P NMR spectroscopy be used to indicate the origins of soil organic phosphates? *Soil Biol. Biochem.* **2005**, *37*, 15–25. [CrossRef]

80. Cade-Menun, B.J.; Preston, C.M. A comparison of soil extraction procedures for ^{31}P NMR spectroscopy. *Soil Sci.* **1996**, *161*, 770–785. [CrossRef]

81. Bol, R.; Amelung, W.; Haumaier, L. Phosphorus-31–nuclear magnetic–resonance spectroscopy to trace organic dung phosphorus in a temperate grassland soil. *J. Plant Nutr. Soil Sci.* **2006**, *169*, 69–75. [CrossRef]

82. Hedges, J.I.; Oades, J.M. Comparative organic geochemistries of soils and marine sediments. *Org. Geochem.* **1997**, *27*, 319–361. [CrossRef]

83. Leinweber, P.; Haumaier, L.; Zech, W. Sequential extractions and ^{31}P-NMR spectroscopy of phosphorus forms in animal manures, whole soils and particle-size separates from a densely populated livestock area in northwest Germany. *Biol. Fertil. Soils* **1997**, *25*, 89–94. [CrossRef]

forests

Article

Regional Scale Determinants of Nutrient Content of Soil in a Cold-Temperate Forest

Shusheng Yuan [1,2], Tongtong Tang [3], Minchao Wang [2,4], Hao Chen [2], Aihua Zhang [2] and Jinghua Yu [2,*]

1   State Engineering Laboratory of Bio-Resource Eco-Utilization (Heilongjiang), Northeast Forestry University, Harbin 150040, China; dove_treasure@163.com
2   Institute of Applied Ecology, Chinese Academy of Sciences, Shenyang 110016, China; wmc0084@163.com (M.W.); m18810041511@163.com (H.C.); zah0130@163.com (A.Z.)
3   Institute of Soil Science, Chinese Academy of Sciences, Nanjing 210008, China; doveyuan1010@163.com
4   School of Life Science, Shenyang Normal University, Shenyang 110016, China
*   Correspondence: yujh@iae.ac.cn; Tel.: +86-24-8397-0345

Received: 1 February 2018; Accepted: 22 March 2018; Published: 30 March 2018

Abstract: The effect of climatic factors on soil nutrients is significant. Identifying whether soil nutrients respond to local climate and how the forest types modulate this responsiveness is critical for forest management. Therefore, six soil nutrients from five main forest types found for a range of sites within the Daxing'an Mountains, China, were investigated. Climatic factors were obtained from the WorldClim dataset. Pearson correlations and stepwise regressions were employed to elucidate and model the response of the six soil nutrients to the four different climatic factors in this study. On the whole, climate was correlated with all the nutrients. Further, from stepwise regressions, climatic factors could affect soil nutrients in distinct forests. Our findings suggest that climatic factors are instrumental in affecting soil nutrients in different forest types. Identifying the relationships between soil nutrients, climatic factors and forest types, as suggested in this research, can provide theoretical foundations to further comprehend nutrient cycling in the forest ecosystem.

Keywords: Daxing'an Mountains; climatic factors; soil nutrients; forest types; principal component analyses

1. Introduction

Climate changes have significant effects on ecosystems. In the present paper, with the primary focus on the links between ecosystems and climate change, gradients of natural climate are noteworthy in studying the interactions between climate and variation in forest ecosystem processes. Terrestrial ecosystems play a dominant and irreplaceable role, due to the functions of releasing and absorbing greenhouse gases in such climate-feedbacks, while storing a great deal of carbon in vegetation and soil, thus serving as the global carbon sink [1]. Some studies have shown that there are strong linkages between climate change and soil. The study of Brittany et al. showed that the gradient of climates (precipitation and temperature) has obvious regulating effects on the physical and chemical properties of soil, such as pH, Mg^{2+}, N, P and K content [2]. The effects of climatic factors on SOC (soil organic carbon) density were obvious and stronger than those of grassland and farmland [3]. Furthermore, regression analysis showed that temperature has a negative correlation with SOC content, and precipitation has a positive correlation with SOC content, but using multiple regression analysis, temperature and precipitation explained 43% of total variance in the SOC variables [4,5]. Soil organic matter related to SOC, total nitrogen (TN), total phosphorus (TP), total potassium (TK), available P (AP) and available K (AK) has been extensively used to evaluate soil quality [6–9]. Moreover, forest types can impact the cycling and amounts of the nutrients, and nutrients have been confirmed to be influenced by the upper layer [10]. However, the impacts of tree species upon soil nutrients varied depending upon the type of bedrock, climate and forest management [11]. Therefore, understanding the relationships between soil nutrients and climate change in different forest types will provide more reliable information to prudently manage forest resources and promote sustainable forestry development under climate change in the future.

The Daxing'an Mountains forest area is in the mid-latitude and high-latitude area that is extremely sensitive to global warming [12]. The Daxing'an Mountains forest area is the main forest in China. It plays an important role in carbon sequestration management and ecological environment construction. Nevertheless, under the influence of climate change, the edge of the forest has retreated 140 km over the past century in this region [13]. Therefore, the soil nutrients of different vegetation types in this region have attracted widespread attention. Jiang et al. [14] studied the soil nutrients of different forests. However, there is less research focused on the soil nutrient characteristics in different forest types in the Daxing'an Mountains forest area. Although the distribution of SOC, N, P, and K in the Liaodong Mountains area [15] and the correlations between SOC, inorganic carbon and soil nutrients in the northeast of China [16] have been studied, studies reporting research related to the comprehensive evaluation of soil nutrients from different forest types in the Daxing'an Mountains forest area are scarce.

In this study, the soil nutrients of a total of 230 sample plots collected from five main forest types were measured from the Daxing'an Mountains, and four bioclimatic variables (mean annual temperature (MAT), temperature seasonality (TS), mean annual precipitation (MAP) and precipitation seasonality (PS)) were obtained from the WorldClim dataset. We hypothesized that climatic factors could affect soil nutrients in different forest types. Thus, identifying whether soil nutrients respond to local climate and how the forest types modulate this responsiveness is critical for forest management.

2. Methods

2.1. Site Description

The forest in the Daxing'an Mountains is one of the most important areas in China: the lush natural forest is distributed widely. It is an important production base for forest trees in China, and also an important ecological barrier in northeastern China. The study area comprises about 86,000 km^2 and belongs to the cool coniferous forests. The investigated forest plots are shown in detail in Figure 1 [17]. Five main forest types were chosen, including pure *Larix gmelinii* (Rupr.) Kuzen forest (PL) (87 samples, altitude: 235–1023 m), pure *Betula platyphylla* Suk. forest (PB) (64 samples, altitude: 160–1003 m), pure *Quercus mongolica Fisch.* ex Ledeb. forest (PQ) (36 samples, altitude: 240–771 m), *Larix gmelinii* (Rupr.) Kuzen and *Betula platyphylla* Suk. mixed forest (MLB) (25 samples, altitude: 247–1038 m) and pure *Pinus sylvestris* L. var. *mongolica* Litv. forest (PP) (18 samples, altitude: 296–905 m). The study was conducted in the eastern forest zones of the Daxing'an Mountains area (45°59′–53°19′ N, 119°47′–130°53′ E), Heilongjiang Province and Inner Mongolia Autonomous Region. This region has a continental monsoon climate, and receives a mean annual precipitation (MAP) of 764 mm. The temperature varies between −41 °C in January–February and 35 °C in July–August, with a mean annual temperature (MAP) of −2.8 °C.

2.2. Field Soil Sampling and Preliminary Analysis

The forest-covered area of the Daxing'an Mountains was systematically divided into 30 km × 30 km grids using ArcGIS 10.0 (Esri, Redlands, CA, USA) as the meshing tool. The exact latitude and longitude for each grid were recorded with a GPS system (Google, Mountain, CA, USA) [18,19]. Soil sample depth was 0–20 cm [3,4], taken from 3–7 plots (30 m × 30 m each) in each 30 km × 30 km grid (total grids = 52), and 3–7 plots were chosen based on the investigation areas. As much as possible, we chose plots from the central region of the grid; the distance of each plot to the edge of the grid must be more than 15% of the length on the side of the grid. A total of 230 sample plots were included in this study, and the geometric center coordinates for each sample plot were input into Excel, saved in CSV (Comma Separated Value) format, and the ArcGIS 10.0 software was used to extract the climatic data for each sample plot [19].

SOC was determined by external heating with the potassium dichromate oxidation method; TN was determined by the Semi-micro Kjeldahl method; TP and TK were determined by the method of the NaOH melt—Mo-Sb Colorimetry; AP was determined by the method of the HCl-NaOH extracts; AK was determined using the flame photometry method [20,21].

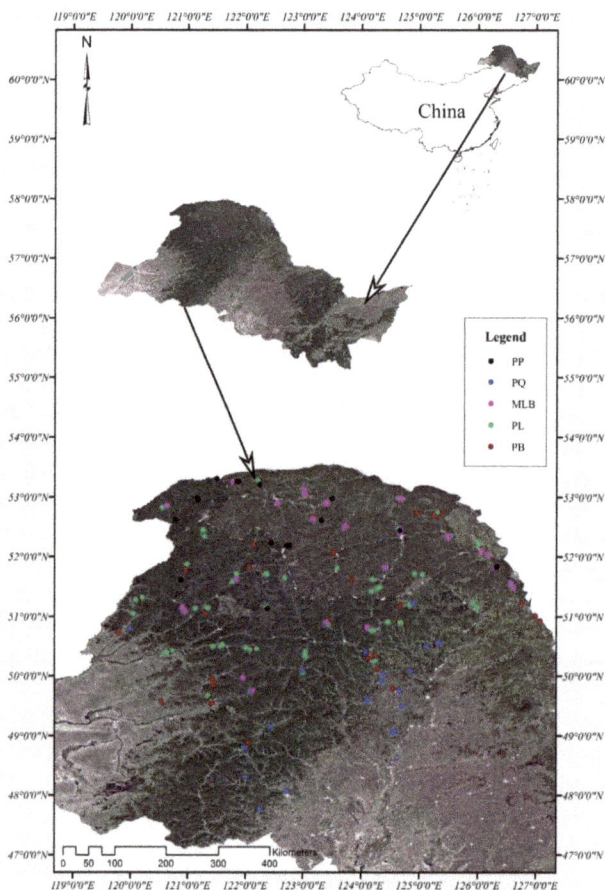

Figure 1. Map of the study area and investigated plots of five forest types in the Daxing'an Mountains. PP = pure *Pinus sylvestris* L. var. *mongolica* Litv. forest, PQ = pure *Quercus mongolica Fisch.* ex Ledeb. forest, MLB = *Larix gmelinii* (Rupr.) Kuzen and *Betula platyphylla* Suk. mixed forest, PL = pure *Larix gmelinii* (Rupr.) Kuzen forest, PB = pure *Betula platyphylla* Suk. forest.

2.3. Climatic Data

Climatic data were obtained from the WorldClim database (http://www.worldclim.org/), the accuracy class of which is a spatial resolution of approximately 1 km². The WorldClim data are collected from weather stations across the globe, which include altitude, temperature, and rainfall (period 1950–2000) [22]. In the present study, four bioclimatic variables (MAT, TS, MAP and PS) were considered to assess the current climatic conditions. The IPCC 4th assessment data provided information for the future climate projections [23].

2.4. Statistical Analyses

The multivariate statistical analysis method has been employed to determine the minimum dataset under the hypothesis that soil nutrients significantly impact forest type. Principal component analysis (PCA) has previously been applied in different research fields to identify nutrients in semiarid soils [24,25] and soil pollutant sources [26] as well as to assess the effect of tillage on soil quality and yield [27–30]. Dimension reduction analysis by using the PCA method to reduce the dimensional data and eliminate the redundant data [31,32]. In our research, we built a hypothesis about which principal components (PCs) possess the highest eigenvalues, variables, and absolute eigenvectors and may best express the minimum dataset.

Pearson correlation coefficients were employed to evaluate the correlations between climatic factors and soil nutrients. Analyses of regression are helpful for inspecting differences among group comparisons; therefore, they are suitable for assessing the variation of soil nutrients under diverse climatic factors. To test whether the climatic factors (MAT, TS, MAP, and PS) affected the soil nutrients (SOC, TN, TP, TK, AP, and AK), a simple linear regression was used for each biological element of the 230 sites with the four climatic factors. To study forest types, specifically the response to climate changes, the climate-change response trends were compared among the forest types. The slopes of the regression lines were used to indicate the different responses of forest types to the climate changes. Linear models compared with non-linear models (Spearman Rank Correlation) gave the best regression results. In addition, stepwise regression between climatic factors and soil nutrients in five main forest types was also analyzed (F-to-enter $p \leq 0.05$, F-to remove $p \geq 0.10$).

The statistical analyses were conducted using SPSS 17.0 software, while the graphs were made using OriginPro 9.0 software (OriginLab Corporation, Northampton, MA, USA).

3. Results

3.1. Variation of Soil Nutrients and Climatic Factors in Different Forest Types

In Figure 2, the contents of SOC, TN, TK, TP, AK and AP in five main forest types averaged at 28.23 g·kg^{-1}, 4.03 g·kg^{-1}, 26.23 g·kg^{-1}, 1.70 g·kg^{-1}, 158.68 mg·kg^{-1} and 21.46 mg·kg^{-1}, respectively. In particular, SOC, TN, TK, TP, AK and AP contents in the PQ were lower than those in the other four forest types.

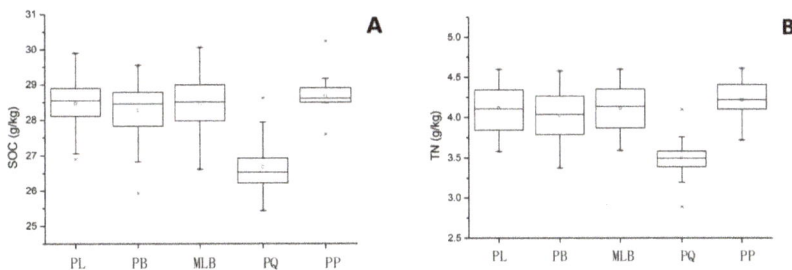

Figure 2. *Cont.*

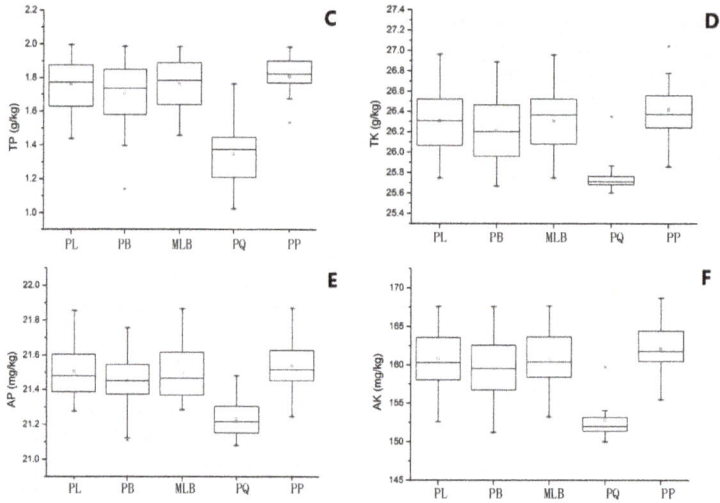

Figure 2. Variation of soil properties in different forest types. SOC = soil organic carbon, TN = total nitrogen, TP = total phosphorus, TK = total potassium, AP = available phosphorus, and AK = available potassium. PP = pure *Pinus sylvestris* L. var. *mongolica* Litv. forest, PQ = pure *Quercus mongolica Fisch. ex Ledeb.* forest, MLB = *Larix gmelinii* (Rupr.) Kuzen and *Betula platyphylla* Suk. mixed forest, PL = pure *Larix gmelinii* (Rupr.) Kuzen forest, PB = pure *Betula platyphylla* Suk. forest. **A**, **B**, **C**, **D**, **E** and **F** were respectively represent for SOC, TN, TP, TK, AP and AK content of five forest types. "□" = average value, "×" = outliter.

The averages of the MAT and MAP were −1.74 °C and 534.93 mm, respectively; the ranges of the TS and PS were 14,772–16,885 and 95–116, respectively (Figure 3). MAT and PS were higher in the PQ than in the other four forest types, while TS showed the opposite trend (Figure 3).

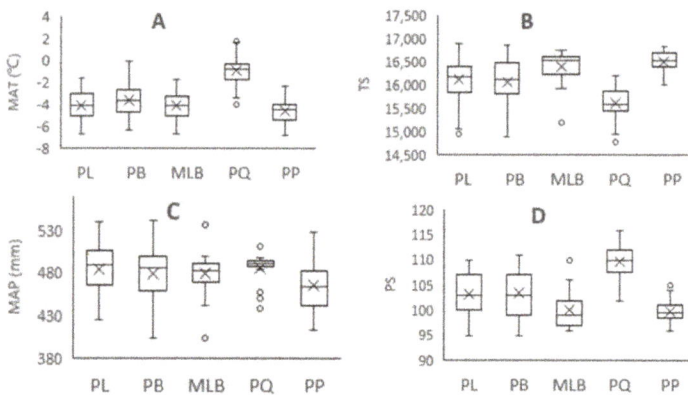

Figure 3. Variation of climatic factors in different forest types. MAT = mean annual temperature, TS = temperature seasonality, MAP = mean annual precipitation, and PS = precipitation seasonality. PL = pure *Larix gmelinii* (Rupr.) Kuzen forest, PB = pure *Betula platyphylla* Suk. Forest, MLB = *Larix gmelinii* (Rupr.) Kuzen and *Betula platyphylla* Suk. mixed forest, PQ = pure *Quercus mongolica Fisch. ex Ledeb.* forest, PP = pure *Pinus sylvestris* L. var. *mongolica* Litv. Forest. **A**: MAT of five forest types, **B**: TS of five forest types, **C**: MAP of five forest types, **D**: PS of five forest types. "×" = average value, "○" = outliter.

3.2. Principal Component Analysis of the Different Forest Types and the Soil Nutrients

The results of the PCA showed the variables that characterized the soil nutrients of the different forest types (Figure 4). PCs 1–6 explained 100.0% of the variation, and can be broken down as follows: 95.55%, 2.22%, 1.13%, 0.73%, 0.24% and 0.12%, respectively, as shown in Table 1. Principal component 1 can reflect most of the variation; it includes TN (0.296), TK (0.816), TP ($-$0.263), AP (0.749), AK (0.447) and SOC ($-$1.302). As shown in Figure 4, five main forest types showed PQ separated from other types.

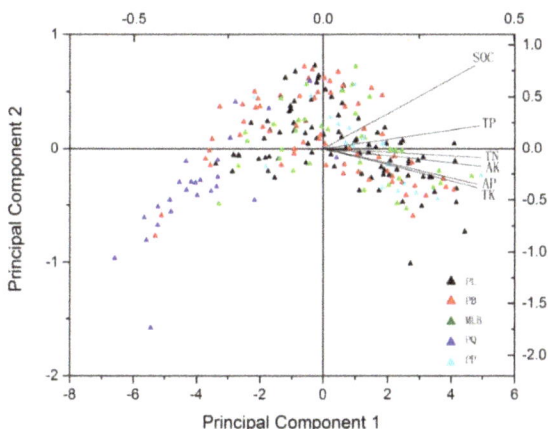

Figure 4. Principal component analysis (PCA) of the different forest types and the soil nutrients. PL = pure *Larix gmelinii* (Rupr.) Kuzen forest, PB = pure *Betula platyphylla* Suk. Forest, MLB = *Larix gmelinii* (Rupr.) Kuzen and *Betula platyphylla* Suk. mixed forest, PQ = pure *Quercus mongolica Fisch.* ex Ledeb. forest, PP = pure *Pinus sylvestris* L. var. *mongolica* Litv. Forest. SOC = soil organic carbon, TN = total nitrogen, TP = total phosphorus, TK = total potassium, AP = available phosphorus, and AK = available potassium.

Table 1. Total variance explained.

Component	Initial Eigenvalues			Extraction Sums of Squared Loadings			Rotation Sums of Squared Loadings		
	Total	% of Variance	Cumulative %	Total	% of Variance	Cumulative %	Total	% of Variance	Cumulative %
1	5.733	95.551	95.551	5.733	95.551	95.551	3.363	56.058	56.058
2	0.133	2.217	97.768	0.133	2.217	97.768	2.503	41.710	97.768
3	0.068	1.134	98.903						
4	0.044	0.733	99.635						
5	0.015	0.244	99.880						
6	0.007	0.120	100.000						

Extraction Method: Principal Component Analysis.

3.3. Correlations between Climatic Factors and Soil Nutrients

The relationships between soil nutrients and climatic factors are shown in Table 2. MAT was negatively correlated with SOC, TN, TK, AK and AP, while it was positively correlated with TP ($r = 0.187$) ($p < 0.01$). TS was positively correlated with SOC, TN, TK, AK and AP ($r = 0.307$–0.417), while it was negatively correlated with TP ($r = -0.405$) ($p < 0.01$). In contrast to TS, the relationships between PS and soil nutrients showed the opposite trend ($p < 0.01$). MAP was positively correlated with SOC, TN, AK and AP, while it was negatively correlated with TK and TP ($p < 0.01$).

Table 2. Correlation coefficient matrix for soil nutrients and climatic factors.

Nutrient	MAT	TS	MAP	PS
SOC	−0.223 **	0.417 **	0.311 **	−0.4700 **
TN	−0.101	0.341 **	0.411 **	−0.414 **
TK	−0.052	0.312 **	−0.411 **	−0.378 **
TP	0.187 **	−0.405 **	−0.383 **	0.471 **
AK	−0.106	0.365 **	0.384 **	−0.425 **
AP	−0.050	0.307 **	0.385 **	−0.367 **

$N = 230$. SOC = soil organic carbon, TN = total nitrogen, TP = total phosphorus, TK = total potassium, AP = available phosphorus, and AK = available potassium. MAT = mean annual temperature, TS = temperature seasonality, MAP = mean annual precipitation, and PS = precipitation seasonality. (** $p < 0.01$).

Moreover, relationships between soil nutrients and climatic factors in the five main forest types were also observed. In Table 3, p (*, **) value represents whether climatic factors are correlated with soil nutrient contents, so as to determine whether there is statistical significance. The r value indicated the correlation between climate factors and soil nutrient contents. MAT was significantly and positively correlated with nutrients in PL, PB, and MLB, except for SOC in PB. TS had no effect on almost all nutrients but positively correlated with SOC in PB. MAP was similar to MAT in PL, PB, and MLB; in addition, MAP was also significantly correlated with TN, TK, TP, AK, AP in PP and TN, TP in PQ. Although the correlation between PS and nutrients was unimpressive compared with MAT and MAP in five forest types, it was negatively correlated with SOC in PL, TN, TP and TK in MLB, as well as SOC and AK in PP.

Table 3. Pearson correlations (r) between soil nutrients and climatic factors in the five main forest types.

Types	Nutrient	MAT	TS	MAP	PS	Types	Nutrient	MAT	TS	MAP	PS
PL	SOC	0.287 **	0.130	0.543 **	−0.235 *	PB	SOC	0.188	0.261 *	0.542 **	−0.223
	TN	0.380 **	0.010	0.603 **	−0.140		TN	0.364 **	0.176	0.600 **	−0.148
	TK	0.376 **	0.030	0.579 **	−0.142		TK	0.387 **	0.191	0.586 **	−0.165
	TP	0.349 **	0.040	0.614 **	−0.173		TP	0.284 *	0.235	0.609 **	−0.222
	AK	0.379 **	0.058	0.586 **	−0.172		AK	0.345 **	0.227	0.550 **	−0.173
	AP	0.485 **	−0.06	0.542 **	−0.061		AP	0.365 **	0.204	0.519 **	−0.154
MLB	SOC	0.362 *	0.151	0.358 *	−0.274	PQ	SOC	0.199	0.301	0.322	−0.271
	TN	0.46 **	0.151	0.455 **	−0.337 *		TN	0.295	0.202	0.426 *	−0.215
	TK	0.546 **	0.157	0.407 *	−0.316		TK	0.121	0.047	0.201	0.059
	TP	0.549 **	0.197	0.434 **	−0.384 *		TP	0.141	0.383	0.454 *	−0.332
	AK	0.537 **	0.188	0.418 *	−0.349 *		AK	0.219	0.171	0.219	−0.128
	AP	0.556 **	0.060	0.437 **	−0.246		AP	0.168	0.329	0.337	−0.227
PP	SOC	0.309	0.126	0.360	−0.470 *						
	TN	0.206	0.069	0.629 **	−0.453						
	TK	0.270	0.044	0.614 **	−0.336						
	TP	0.176	0.088	0.621 **	−0.436						
	AK	0.273	0.056	0.619 **	−0.485 *						
	AP	0.205	0.028	0.611 **	−0.464						

SOC = soil organic carbon, TN = total nitrogen, TP = total phosphorus, TK = total potassium, AP = available phosphorus, and AK = available potassium. MAT = mean annual temperature, TS = temperature seasonality, MAP = mean annual precipitation, and PS = precipitation seasonality. PL = pure *Larix gmelinii* (Rupr.) Kuzen forest, PB = pure *Betula platyphylla* Suk. Forest, MLB = *Larix gmelinii* (Rupr.) Kuzen and *Betula platyphylla* Suk. mixed forest, PQ = pure *Quercus mongolica* Fisch. ex Ledeb. forest, PP = pure *Pinus sylvestris* L. var. *mongolica* Litv. Forest. (* $p < 0.05$; ** $p < 0.01$).

3.4. Stepwise Regressions between Climatic Factors and Soil Nutrients

Step regression between soil nutrients and climatic factors in the five main forest types is shown in Table 4. In the case of PB, the four climatic factors could affect all the six soil nutrients, and MAP was the first parameter entered into the model. In the case of MLB and PP, MAP and MAT were the key factors for influencing the five soil nutrients (TN, TP, TK, AP and AK). For PL, MAT, MAP and TS mainly affected SOC, TN, TP, TK and AK. However, for PQ, MAP was the key factor for TN and TP, and no parameters were entered into the model of SOC, TK, AP and AK. In all, we found that there were different influencing factors in various forest types.

Table 4. Step regressions between soil nutrients and climatic factors in the five main forest types.

Forest Types	Soil Nutrients	R	R^2	Adjusted R^2	Standard Error of the Estimate
PL	SOC	0.715	0.512	0.488	0.45880
	TN	0.741	0.549	0.539	0.19099
	TP	0.734	0.538	0.527	0.10257
	TK	0.717	0.513	0.502	0.21375
	AP	0.759	0.576	0.566	0.08648
	AK	0.726	0.527	0.516	2.61645
PB	SOC	0.762	0.580	0.552	0.50973
	TN	0.852	0.727	0.708	0.16591
	TP	0.829	0.687	0.665	0.10871
	TK	0.853	0.728	0.709	0.17232
	AP	0.812	0.660	0.637	0.08950
	AK	0.844	0.713	0.694	2.29320
MLB	SOC	0.606	0.367	0.307	0.60923
	TN	0.803	0.644	0.611	0.17465
	TP	0.809	0.654	0.621	0.09086
	TK	0.767	0.588	0.550	0.20503
	AP	0.740	0.548	0.520	0.09897
	AK	0.784	0.614	0.578	2.40780
PQ	SOC	–	–	–	–
	TN	0.782	0.612	0.557	0.15610
	TP	0.454	0.206	0.172	0.16556
	TK	–	–	–	–
	AP	–	–	–	–
	AK	–	–	–	–
PP	SOC	0.470	0.221	0.172	0.50111
	TN	0.787	0.619	0.569	0.16243
	TP	0.759	0.576	0.520	0.08790
	TK	0.817	0.667	0.520	0.16327
	AP	0.768	0.590	0.535	0.10037
	AK	0.822	0.676	0.633	2.13592

Note: SOC = soil organic carbon, TN = total nitrogen, TP = total phosphorus, TK = total potassium, AP = available phosphorus, and AK = available potassium. PL = pure *Larix gmelinii* (Rupr.) Kuzen forest, PB = pure *Betula platyphylla* Suk. Forest, MLB = *Larix gmelinii* (Rupr.) Kuzen and *Betula platyphylla* Suk. mixed forest, PQ = pure *Quercus mongolica* Fisch. ex Ledeb. forest, PP = pure *Pinus sylvestris* L. var. *mongolica* Litv. Forest, "–" = there is no value.

4. Discussion

4.1. Forest Types Influence Soil Nutrient Contents

The relationships between soil nutrients and forest types have been presented previously [3]. In addition, northeastern China has been considered as one of the regions with the most abundant soil nutrition [16,33]. To confirm the influencing trends of the variation of soil nutrient contents to forest types, PCA was carried out on the collected data. From the distribution of the loading plot in the PCA space, it was found that the forest type influenced the soil nutrients. For instance, SOC, TP, TN, AK, AP and TK were higher in PL, PB, MLB and PP in this study, while these soil nutrients showed the opposite trend. So, we suspect that there were certain correlations between soil nutrients and forest type. In addition, the differences caused by vegetation effects in the responses of the nutrients may be due to slight distinctions in parent material in different forest types [34,35].

From the above results, it was found that the distributions of soil nutrients from different forests were different. These values were influenced by the forest site conditions, advantageous tree species and different amounts of forest litter as well as the composition and decomposition levels, so the differences in the forest soil nutrients are very obvious. For instance, the distribution of SOC was ranged in order PL > MLB > PB > PP > PQ with the SOC average content being 28.68, 28.46, 28.43, 28.28 and 26.68 g·kg^{-1} respectively. From the viewpoint of succession, PL, MLB and PP were in the top stage of succession—the complexity of the tree species composition increased the possibilities of the accumulation of organic matter [36]—but PB and PQ were the secondary forests which were disturbed more frequently in recent years. So, the SOC content of PL, MLB and PP should be larger than that of PB or PQ [37]. However, the SOC content in PP was minimal, even lower than in PB. This phenomenon was unexpected, and it is possible that it is related to the terrain: the slope is large, litter does not accumulate as much, and in addition to the soil acidity, the litter layer was difficult to decompose; therefore, the conditions are not conducive to the formation of organic matter, which means that the SOC content is low [3]. This also means that SOC stock will continue to increase if the interference is ended and the forest is developed toward the climax community; otherwise, the forest can turn into PQ and the stock of SOC will decrease, especially in the rich PB forest region.

Statistically, the distributions of TN and SOC were identical. A large number of data analysis results show that the TN was positively correlated with SOC. The order was PL > MLB > PB > PP > PQ with the contents being 4.2, 4.12, 4.11, 4.02 and 3.50 g·kg^{-1} respectively. The distribution of TN identified in this study was in accordance with that presented by Zu et al. [16] and Jiang et al. [14]. The order of the TP content was PP > PL = MLB > PB > PQ and the contents were 1.81, 1.76, 1.76, 1.70 and 1.35 g·kg^{-1} respectively. The correlations of the AP and SOC were identical but opposite to that of TP. The same phenomenon appeared with AK and TK, and this could be explained by the composition of TP and TK, which is very complex, with the existence of inorganic and organic states, and AP and AK being only part of them. This trend may have been due to the influence of various factors such as the climate. In addition, although AP content decreased with the MAP and MAT increasing, TP content increased [32,33]. This confirms that the vegetation type is a key factor that affects the soil nutrients of the Daxing'an Mountains ecosystems.

4.2. Soil Nutrient Responses to Climatic Factors

The study of Harradine, F. et al. indicated that climate (especially precipitation and temperature) has significant effects on pedogenesis and macronutrient cycling in soil [38]. It is a challenge to isolate each of the individual soil forming factors such as climate, vegetation, parent material and so on, due to the frequent co-variance of many factors [39]. For instance, changes in species vary with the climate and location. In the present paper, with the primary focus on the links between ecosystems and climate change, gradients of natural climate are noteworthy in studying the interactions between climate and variation in forest ecosystem processes.

On the whole, it was found that SOC decreased with increasing MAT, and SOC increased with increasing MAP (Table 1). This trend may be due to the hydrothermal conditions of Daxing'an Mountains area. The MAT in Daxing'an Mountains is −3.69 °C and the MAP is 481.2 mm. The region is rich in forest resources, and rainfall is abundant which is conducive to the growth of plants, while the low temperature is beneficial to the accumulation of biomass. Yimer, F. et al. suggested that some other factors, such as erosion, leaching of cations and variations in biomass production may influence soil property [39]. Our results were consistent with previous research which showed that increasing temperature leads to the growth of microorganisms [40], thus increasing the decomposition rate of SOC [41,42]. Precipitation change will affect the content of plant-available water and the length of the growing season; a reduction in precipitation can limit plant growth [30], and the soil microbial number will surge after rain [43], thus reducing the SOC content in soil. In this study, a similar conclusion can be drawn.

More than half of the soil nutrients were significantly linked with variations in the climatic factors, but PQ had weak correlations between climatic factors and soil nutrients, showing that the soil nutrient distribution characteristics were affected by forest types. TN was affected by MAP, MAT and TS, while TP was susceptible to MAP in PQ (Table 3). However, MAT and MAP were the key factors for most soil nutrients in PB, PL, MLB and PP (Table 3), indicating that MAP and MAT played key roles in the accumulation of biomass matter in this region. In addition, the weak correlations between soil nutrients and the climatic factors (PS, TS) indicated that the changes in temperature and precipitation affected the time scale of soil nutrients.

The wide distribution of the forested land resulted in the higher storage of soil nutrient elements in the Daxing'an Mountains area than in other areas of China, even though the contents vary between the different forest types. Although determining the mechanism through which climate acted on the forest types proved difficult, the spatial distribution of the soil nutrients was related to vegetation in the Daxing'an Mountains.

5. Conclusions

The results of this study revealed obvious differences in the variation of soil nutrients. The content of each nutrient in PP was minimal, in obvious contrast to other forest types. Correlations between the soil nutrients and climatic factors were found in this paper. Climatic factors could affect soil nutrients in different forest types. We confirmed that climatic factors (MAT and MAP) are instrumental in affecting soil nutrients (SOC, TN, TP, TK, AP and AK) in five main forest types in the Daxing'an Mountains. Identifying the relationships between soil nutrients, climatic factors and forest types, as suggested in this research, can provide theoretical foundations to further comprehend nutrient cycling in the forest ecosystem.

Acknowledgments: This study was supported by the National Natural Science Foundation of China (41330530).

Author Contributions: S.Y. and T.T. collected the samples and analyzed the data; M.W. and A.Z. performed the experiments; H.C. consulted the literatures; S.Y. wrote the paper; J.Y. designed the experiments.

Conflicts of Interest: The founding sponsors had no role in the design of the study; in the collection, analyses, or interpretation of data; in the writing of the manuscript, and in the decision to publish the results.

References

1.   Schimel, D.S.; Braswell, B.H.; Holland, E.A.; Mckeown, R.; Ojima, D.S.; Painter, T.H.; Parton, W.J.; Townsend, A.R. Climatic, edaphic, and biotic controls over storage and turnover of carbon in soils. *Glob. Biogeochem. Cycles* **1994**, *8*, 279–294. [CrossRef]
2.   Brittany, G.J.; Paul, J.V.; John, A.A. Effects of climate and vegetation on soil nutrients and chemistry in Great Basin studied along a latitudinal-elevational climate gradient. *Plant Soil* **2014**, *382*, 151–163.
3.   Duan, X.W.; Rong, L.; Hu, J.M.; Zhang, G.L. Soil organic carbon stocks in the Yunnan Plateau, southwest China: Spatial variations and environmental controls. *J. Soil Sediments* **2014**, *14*, 1643–1658. [CrossRef]
4.   Campos, C.A.; Auilar, G.S.; Landgrave, R. Soil organic carbon stocks in Veracruz State (Mexico) estimated using the 1:250,000 soil database of INEGI: Biophysical contributions. *J. Soil Sediments* **2014**, *14*, 860–871. [CrossRef]
5.   Hobley, E.; Wilson, B.; Wilkie, A.; Gray, J.; Koen, T. Drivers of soil organic carbon storage and vertical distribution in Eastern Australia. *Plant Soil* **2015**, *390*, 111–127. [CrossRef]
6.   Solomon, D.; Fritzsche, F.; Lehmann, J.; Tekalign, M.; Zech, W. Soil organic matter dynamics in the subhumid agroecosystems of the Ethiopian Highlands. *J. Organomet. Chem.* **2002**, *66*, 35–42.
7.   Chandra, L.R.; Gupta, S.; Pande, V.; Singh, N. Impact of forest vegetation on soil characteristics: A correlation between soil biological and physico-chemical properties. *3 Biotech.* **2016**, *6*, 188. [CrossRef] [PubMed]
8.   Prietzel, J.; Christophel, D.; Traub, C.; Kolb, E.; Schubert, A. Regional and site-related patterns of soil nitrogen, phosphorus, and potassium stocks and Norway spruce nutrition in mountain forests of the Bavarian Alps. *Plant Soil* **2015**, *386*, 151–169. [CrossRef]
9.   Jiang, F.; Wu, X.; Xiang, W.H.; Fang, X.; Zeng, Y.L.; Ouyang, S.; Lei, P.F.; Deng, X.W.; Peng, C.H. Spatial variations in soil organic carbon, nitrogen and phosphorus concentrations related to stand characteristics in subtropical areas. *Plant Soil* **2017**, *413*, 289–301. [CrossRef]
10.   Gosz, J.R. Nitrogen cycling in coniferous ecosystems. *Ecol. Bull. (Sweden)* **1981**, *33*, 405–426.
11.   Augusto, L.; Ranger, J.; Dan, B.; Rothe, A. Impact of several common tree species of European temperate forests on soil fertility. *Ann. For. Sci.* **2002**, *59*, 233–253. [CrossRef]
12.   Wu, J.; Liu, Q. Charcoal-recorded climate changes from Moon Lake in Late Glacial. *J. China Univ. Geosci.* **2012**, *37*, 947–954.

13. Wang, X.Y.; Zhao, C.; Yu, J. Impacts of Climate Change on Forest Ecosystems in Northeast China. *Adv. Clim. Chang. Res.* **2013**, *4*, 230–241.

14. Jiang, C.Q.; Xu, Q.; Jiang, P.K. Integrated evaluation of soil chemical and biochemical fertility under different vegetations. *For. Res.* **2002**, *15*, 700–705.

15. Qu, K.Y.; Dai, L.M.; Feng, H.M.; Zhang, H.S. Soil fertility characteristics of main forest types in Easten Mountain areas of Liaoning. *Chin. J. Soil Sci.* **2009**, *40*, 558–562.

16. Zu, Y.G.; Li, R.; Wang, W.J.; Su, D.X.; Wang, Y.; Qiu, L. Soil organic and inorganic carbon contents in relation to soil physicochemical properties in northeastern China. *Acta Ecol. Sin.* **2011**, *31*, 5207–5216.

17. Zhou, Y.L. *Geography of the Vegetation in Northeast China*; Science Press: Beijing, China, 1997.

18. Yu, J.H.; Wang, C.J.; Wan, J.Z.; Han, S.J.; Wang, Q.G.; Nie, S.M. A model-based method to evaluate the ability of nature reserves to protect endangered tree species in the context of climate change. *For. Ecol. Manag.* **2014**, *327*, 48–54. [CrossRef]

19. Wan, J.; Wang, C.; Han, S.; Yu, J. Planning the priority protected areas of endangered orchid species in northeastern China. *Biodivers. Conserv.* **2014**, *23*, 1395–1409. [CrossRef]

20. Avery, B.W.; Bascomb, C.L. *Soil Survey Laboratory Methods*; Rothamsted Experimental Station, Lawes Agricultural Trust: Harperden, UK, 1974.

21. Bao, S.D. *Soil and Agricultural Chemistry Analysis*; China Agriculture Press: Beijing, China, 2000.

22. Kumar, P. Assessment of impact of climate change on Rhododendrons in Sikkim Himalayas using Maxent modelling: Limitations and challenges. *Biodivers. Conserv.* **2012**, *21*, 1251–1266. [CrossRef]

23. Pachauri, R.K.; Reisinger, A. *Climate Change 2007: Synthesis Report*; Contribution of Working Groups I, II and III to the Fourth Assessment Report of the Intergovernmental Panel on Climate Change; IPCC: Paris, France, 2007; p. 104.

24. Blecker, S.W.; Stillings, L.L.; Amacher, M.C.; Ippolito, J.A.; Decrappeo, N.M. Development of vegetation based soil quality indices for mineralized terrane in arid and semi-arid regions. *Ecol. Indic.* **2012**, *20*, 65–74. [CrossRef]

25. Bautista-Cruz, A.; León-González, F.D.; Carrillo-González, R.; Robles, C. Identification of soil quality indicators for maguey mezcalero (*Agave angustifolia* Haw.) plantations in Southern Mexico. *Afr. J. Agric. Res.* **2011**, *6*, 4795–4799.

26. Ouyang, W.; Huang, H.; Hao, F.; Shan, Y.; Guo, B. Evaluating spatial interaction of soil property with non-point source pollution at watershed scale: The phosphorus indicator in Northeast China. *Sci. Total Environ.* **2012**, *432*, 412–421. [CrossRef] [PubMed]

27. Sharma, K.L.; Grace, J.K.; Mishra, P.K.; Venkateswarlu, B.; Nagdeve, M.B.; Gabhane, V.V.; Sankar, G.M.; Korwar, G.R.; Chary, G.R.; Rao, C.S. Effect of soil and nutrient-management treatments on soil quality indices under cotton-based production system in Rainfed Semi-arid Tropical Vertisol. *Commun. Soil Sci. Plant* **2011**, *42*, 1298–1315. [CrossRef]

28. Batey, T. Soil compaction and soil management—A review. *Soil Use Manag.* **2010**, *25*, 335–345. [CrossRef]

29. Jagadamma, S.; Lal, R.; Hoeft, R.G.; Nafziger, E.D.; Adee, E.A. Nitrogen fertilization and cropping system impacts on soil properties and their relationship to crop yield in the central Corn Belt, USA. *Soil Tillage Res.* **2008**, *98*, 120–129. [CrossRef]

30. Chang, J.; Clay, D.E.; Dalsted, K.; Clay, S.; O'neill, M. Corn (L.) yield prediction using multispectral and multidate reflectance. *Agron. J.* **2003**, *95*, 1447–1453. [CrossRef]

31. Hothorn, T.; Everitt, B.S. *A Handbook of Statistical Analyses Using R*, 3rd ed.; CRC Press: Boca Raton, FL, USA, 2014; Volume 37, p. 434.

32. Andrews, S.S.; Carroll, C.R. Designing a soil quality assessment tool for sustainable agroecosystem management. *Ecol. Appl.* **2011**, *11*, 1573–1585. [CrossRef]

33. Wang, F.; Huang, M.; Sun, X.H.; Gong, Y.Z.; Wang, J.B. Evaluation of soil nutrients for different forest types in Xing'an Mountains forest area. *Bull. Soil Water Conserv.* **2013**, *33*, 182–187.

34. Christensen, J.H.; Carter, T.R.; Rummukainen, M.; Amanatidis, G. Evaluating the performance and utility of regional climate models: The PRUDENCE project. *Clim. Chang.* **2007**, *81*, 1–6. [CrossRef]

35. Gavazov, K.S. Dynamics of alpine plant litter decomposition in a changing climate. *Plant Soil* **2010**, *337*, 19–32. [CrossRef]

36. Shanin, V.; Komarov, A.; Khoraskina, Y.; Bykhovets, S.; Linkosalo, T.; Mäkipää, R. Carbon turnover in mixed stands: Modelling possible shifts under climate change. *Ecol. Model.* **2013**, *251*, 232–245. [CrossRef]

37. Deng, L.; Wang, K.B.; Chen, M.L.; Shangguan, Z.P.; Sweeney, S. Soil organic carbon storage capacity positively related to forest succession on the Loess Plateau, China. *CATENA* **2013**, *110*, 1–7. [CrossRef]

38. Harradine, F.; Jenny, H. Influence of parent material and climate on texture and nitrogen and carbon contents of virgin California soils—Texture and nitrogen contents of soils. *Soil Sci.* **1985**, *85*, 235–243. [CrossRef]

39. Yimer, F.; Ledin, S.; Abdelkadir, A. Soil property variations in relation to topographic aspect and vegetation community in the south-eastern highlands of Ethiopia. *For. Ecol. Manag.* **2006**, *232*, 90–99. [CrossRef]

40. Körner, C. Ecological impacts of atmospheric CO_2 enrichment on terrestrial ecosystems. *Philos. Trans.* **2003**, *361*, 2023–2041. [CrossRef] [PubMed]

41. Davidson, E.A.; Trumbore, S.E.; Amundson, R. Soil warming and organic carbon content. *Nature* **2000**, *408*, 789–790. [CrossRef] [PubMed]

42. Lenton, T.M.; Huntingford, C. Global terrestrial carbon storage and uncertainties in its temperature sensitivity examined with a simple model. *Glob. Chang. Biol.* **2003**, *9*, 1333–1352. [CrossRef]

43. Orchard, V.A.; Cook, F.J. Relationship between soil respiration and soil moisture. *Soil Biol. Biochem.* **1983**, *15*, 447–453. [CrossRef]

forests

MDPI

Article

Soil Degradation and the Decline of Available Nitrogen and Phosphorus in Soils of the Main Forest Types in the Qinling Mountains of China

Xiaofeng Zheng [1,†], Jie Yuan [1,†], Tong Zhang [1], Fan Hao [1], Shibu Jose [2] and Shuoxin Zhang [1,3,*]

1   College of Forestry, Northwest A&F University, Yangling 712100, China; xfzheng1991@gmail.com (X.Z.);
    yuanjie@nwsuaf.edu.cn (J.Y.); zt9401@gmail.com (T.Z.); haofan@nwafu.edu.cn (F.H.)
2   School of Natural Resources, University of Missouri, Columbia, MO 65211, USA; joses@missouri.edu
3   Qinling National Forest Ecosystem Research Station, Huoditang, Ningshan 711600, China
*   Correspondence: sxzhang@nwsuaf.edu.cn; Tel./Fax: +86-29-8708-2993
†   These authors contributed equally to this work.

Received: 6 November 2017; Accepted: 20 November 2017; Published: 21 November 2017

Abstract: Soil degradation has been reported worldwide. To better understand this degradation, we selected *Pinus armandii* and *Quercus aliena* var. *acuteserrata* forests, and a mixed forest of *Q. aliena* var. *acuteserrata* and *P. armandii* in the Qinling Mountains in China for our permanent plots and conducted three investigations over a 20-year period. We determined the amounts of available nitrogen (N) and phosphorus (P) in the soil to track the trajectory of soil quality and compared these with stand characteristics, topographic and climatic attributes to analyze the strength of each factor in influencing the available N and P in the soil. We found that the soil experienced a severe drop in quality, and that degradation is continuing. Temperature is the most critical factor controlling the soil available N, and species composition is the main factor regulating the soil available P. Given the huge gap in content and the increasing rate of nutrients loss, this reduction in soil quality will likely negatively affect ecosystem sustainability.

Keywords: soil degradation; soil available nitrogen; soil available phosphorus; temperature; stand density

1. Introduction

Soil degradation is a global threat to land ecosystems [1]. The Qinling Mountains in China are not immune to this global trend [2,3], with an average soil loss of about 42 Mg/ha/y for crop lands and as high as 300 Mg/ha/y for individual fields [4]. The pressure from changing climate and biota population, together with other abiotic, biotic, and political factors, have already resulted in severe soil degradation, which appears to be worsening [5–7]. In a forest ecosystem, soil degradation causes trees, shrubs, and herbs to be more vulnerable. Soil degradation may alter the stability of the forest ecosystem due to the progression in vegetation cover, and changes in community diversity and density [8]. Increasing knowledge of soil degradation in forest ecosystems is critical for maintaining these ecosystems.

Soil degradation is defined as the decline in soil quality. Soil nitrogen (N) and phosphorus (P), especially the amounts available for plant uptake, are essential indicators of soil nutrient status [9]. N and P are present in many forms, including inorganic, organic, and particulate forms. Atmospheric N is the main reservoir for N and phosphate is the main reservoir for P, and the majority of these elements are unavailable to most plants. N and P are available to plants as ammonium, nitrate, and phosphate ions, but N and P also exist in complex organic forms and the residue requires several years in soil before N and P are available for plant use [10]. The abundance of soil available N and P can be the determining limitation to plant growth and productivity.

Analyzing the abundance of soil available N and P in forest ecosystems is used to determine plant N and P limitation. N limitation has been reported in many ecosystems, such as grasslands, freshwater systems, and in forest systems [11–13]. P limitation and N–P coupled limitation have been repeatedly reported [14,15]. A wide variety of ecosystems have demonstrated the importance of N and P limitation to plants. In forest ecosystems without human interference, changing climate and community population are the most important factors, which affect the soil available N and P. Generally, an increase in temperature will accelerate the decay of soil organic matter, accumulating more soil available nutrients [16]. However, in some regions, with global warming, the available nutrients for plant uptake continue to decrease [17]. The aim of this work was to determine how the soil available N and P are influenced by different factors.

The major objective of this work was to investigate the changes in soil available N and P over time. The Qinling Mountains are a natural boundary between North and South China and support a wide variety of plant and animal life, some of which is found nowhere else on earth. The selected forest plots in the Qinling Mountains can be an ideal choice for study. The goals of this study were to (i) investigate the abundance of soil available N and P in the Qinling Mountains forest system; (ii) to determine the status of the former reported soil degradation in the Qinling Mountains forest system; and (iii) to explore the factors influencing the abundance of soil available N and P.

2. Materials and Methods

2.1. Site Description

This was a long-term, local-scale study, and the experimental platform was based on the Qinling National Forest Ecosystem Research Station and on the Huoditang Experimental Forest of Northwest Agriculture and Forestry University, Shaanxi Province, China, with a total area of 2037 ha. The study area is located at 33°18′–33°28′ N and 108°21′–108°39′ E. The altitude is 800–2500 m, with a mean annual precipitation of 900–1200 mm, and a mean annual temperature of 8–10 °C. The frost-free period is 170 days, and the climate is categorized as warm temperate. The topography mainly consists of granite and gneiss, with a 35° mean slope. The soils of the region are acidic and the soil units are Cambisols, Umbrisols and Podzols. Most forests are secondary natural forests with a mean age of 60 years comprising several age classes. The study forests underwent intensive selective logging in the 1960s and 1970s, but since then, no significant anthropogenic disturbances have occurred, except for lacquer tree tapping and occasional illegal tree felling. Since a natural forest protection project was initiated in 1998, anthropogenic disturbances of any kind have been absent in the region.

Pinus armandii and *Quercus aliena* var. *acuteserrata* are the most widely distributed and dominant tree species in the study area. A total of 18 forest plots were chosen with characteristics shown in Table 1, cited from Yuan [18]. Six were *P. armandii* and *Q. aliena* var. *acutiserrata* mixed species-dominated forest sample plots, and six were pure single species-dominated forest sample plots of each of these two species. In this study, the investigated forests were natural secondary forests, all 40 years old in 1996. The *P. armandii* forest was dominated by *P. armandii*, with a forest canopy density of 65%. The mean stand height and diameter at breast height (DBH) were 14 m and 21 cm, respectively. The *Q. aliena* var. *acuteserrata* forest was dominated by *Q. aliena* var. *acuteserrata*, with a forest canopy density of 75%. The mean stand height and DBH were 12 m and 18 cm, respectively. In the shrub layer, height varied from 45 cm to 650 cm, and the percent cover was 25%. The mixed forest was dominated by *Q. aliena* var. *acuteserrata* (40% of the trees) and *P. armandii* (50% of the trees), with a forest canopy density of 70%. The mean stand height was 14 m, and mean DBH was 20 cm.

Table 1. Plot characteristics in the main types of forests at the Huoditang Experimental Forest Farm in the Qinling Mountains, China.

Sample Plot	Latitude and Longitude	Altitude (m)	Aspect	Slope	Stand Density (Trees/ha)
Q. aliena var. *acuteserrata* 1#	33°20'55" N 108°23'54" E	1597	318°	32°	1183
Q. aliena var. *acuteserrata* 2#	33°19'20" N 108°25'48" E	1641	14°	28°	1482
Q. aliena var. *acuteserrata* 3#	33°20'49" N 108°25'58" E	1620	277°	26°	1232
Q. aliena var. *acuteserrata* 4#	33°19'10" N 108°28'48" E	1640	260°	23°	1024
Q. aliena var. *acuteserrata* 5#	33°20'08" N 108°28'12" E	1671	240°	30°	1086
Q. aliena var. *acuteserrata* 6#	33°20'42" N 108°29'21" E	1534	218°	18°	1584
P. armandii 1#	33°19'26" N 108°27'10" E	1410	288°	34°	1628
P. armandii 2#	33°19'30" N 108°27'54" E	1460	198°	32°	1486
P. armandii 3#	33°22'54" N 108°28'02" E	1483	245°	27°	1712
P. armandii 4#	33°23'10" N 108°28'10" E	1532	194°	22°	1426
P. armandii 5#	33°21'10" N 108°32'39" E	1540	243°	29°	1267
P. armandii 6#	33°22'01" N 108°32'19" E	1983	245°	15°	1834
Mixed forest 1#	33°22'24" N 108°27'10" E	1580	257°	22°	1354
Mixed forest 2#	33°19'01" N 108°30'24" E	1481	204°	24°	1288
Mixed forest 3#	33°20'10" N 108°30'42" E	1424	255°	32°	1078
Mixed forest 4#	33°21'05" N 108°24'48" E	1582	73°	26°	1041
Mixed forest 5#	33°21'20" N 108°26'50" E	1798	182°	31°	1121
Mixed forest 6#	33°23'18" N 108°25'40" E	1627	202°	21°	1311

2.2. Sampling Work

In the summer of 1996, we selected *P. armandii*, *Q. aliena* var. *acuteserrata* forests, and a mixed forest of *Q. aliena* var. *acuteserrata* and *P. armandii* for our permanent plots, and randomly established six sample plots with an area of 60 m × 60 m in each of the forest types. To reduce disturbance, these sample plots were protected by an enclosure. Each sample plot was located at least 50 m from the forest edge and was separated from other plots by a buffer strip of at least 100 m. We assumed that the annual average precipitation was the same in these sample plots. The annual average temperature and precipitation were sourced from the unpublished Qinling long-term ecological monitoring database and were measured using a HMP45C weather station (Vaisala, Helsinki, Finland) located 1612 m above sea level in this region at 33°20'16" N and 108°26'45" E. The 2017 mean annual climate data were calculated by combining those of 2016 and the existing record of 2017. We also assumed a standard lapse rate of 0.7 °C in this forest area [18], and estimated the annual average temperatures in these sample plots.

We dug soil pits and then sampled the soil with a soil corer (Φ50.46 × 50 mm). Five profiles were sampled at fixed depths (0–10, 10–20, 20–40, and 40–60 cm) in each plot, and every sample was weighed immediately on site to calculate the soil water content. Each subsample contained the same amount of soil taken from these four fixed depths. We pooled the five subsamples to form one composite sample per plot. The samples were crushed, after separating the stones and root fragments that had been mistakenly collected. A 2-mm sieve was used for further sieving. Soil available N and P was analyzed following the methods described by Liu [19]. Plot characteristics and soil nutrient measurements were initially conducted in July of 1996. Two other soil nutrient measurements were conducted in July 2012, and July 2017, providing a total of three measurements during this 20-year study.

2.3. Data Analysis

R Studio version 3.2.3 and Origin 9.0 were used to perform the data analysis and to construct the figures. The significant differences over time were tested using one-way ANOVA with SAS 8.0 software (SAS Inc., Cary, NC, USA). All soil nutrient variables, referring to the abundance of soil available N and P at each site along the time scale, were displayed by box plot with standard errors. Stand density (SD); stand types (ST, referring to tree species), mean annual temperature (MAT); mean annual total precipitation (MAP); and topography status (TG), including aspect, slope, and elevation were considered as factors influencing the abundance of soil available N and P. When analyzing the effect of each of these factors on the abundance of soil available N and P, partial methods of redundancy analysis (RDA), relying on the R function "Vegan", were used to remove the effect of conditioning or background variables before completing RDA. We used an RDA model (X ~ Y + Z) to extract components of variance and calculate the attributes of factor Y, cleansed of the effect of Z. By dividing the total inertia value by the constrained inertia value of Y, we obtained the explanatory influence of Y expressed as a percentage.

3. Results

3.1. Change in Soil Available N and P over Time

In the Qinling Mountains forest ecosystem, the abundance of soil available N has significantly decreased over time, especially in the *Q.* var. *acuteserrata* related forest (Figure 1). The soil available N in the *Q.* var. *acuteserrata* and mixed forest, between *Q.* var. *acuteserrata* and *P. armandii*, ranged from 34 mg/kg to 58 mg/kg in 1996, from 16 mg/kg to 26 mg/kg in 2012, and finally from 11 mg/kg to 17 mg/kg in 2017. The soil available N in the *P. armandii* forest ranged 18 mg/kg to 30 mg/kg, 15 mg/kg to 24 mg/kg, and 13 mg/kg to 20 mg/kg at the same sampling points in time, respectively. The soil available N loss in the *Q.* var. *acuteserrata* related forest was high between 1996 and 2012. An obvious decline in the concentration of soil available N can be seen in Figure 1. In the *P. armandii* forest, the decline in soil available N concentration is also clear.

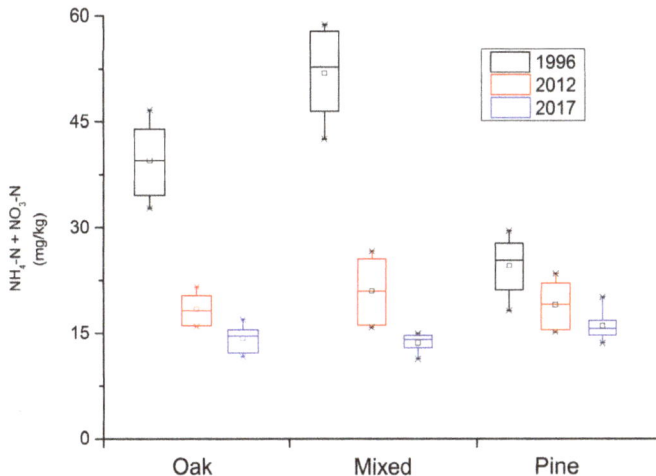

Figure 1. The loss in soil available nitrogen (N) along a 20-year time scale in the Qinling Mountains forest ecosystem. "AN" represents the soil available N at a given time, "Pinus" represents the *P. armandii* forest, "Mingled" represents the mixed *Q.* var. *acuteserrata and P. armandii* forest, and "Quercus" represents the *Q.* var. *acuteserrata* forest.

When examining the amount of P available in the soil in the Qinling Mountains forest ecosystem (Figure 2), the *Q.* var. *acuteserrata* and the mixed *Q.* var. *acuteserrata* and *P. armandii* forest ranged from 5.8 mg/kg to 8.0 mg/kg in 1996, from 5.0 mg/kg to 7.2 mg/kg in 2012, and finally from 0.8 mg/kg to 3 mg/kg in 2017. The soil available P in the *P. armandii* forest ranged 8.5 mg/kg to 9.4 mg/kg, 6.5 mg/kg to 7.1 mg/kg, and 1.5 mg/kg to 2.8 mg/kg along this same time scale, respectively. Overall, a trend in soil nutrient degradation in the soil available P was also demonstrated. Numerically, the abundance loss is not that significant, but the decreasing scale is large. Notably, in the 5-year period from 2012 to 2017, more available P in the soil was lost than in the prior 15-year period, from 1996 to 2012.

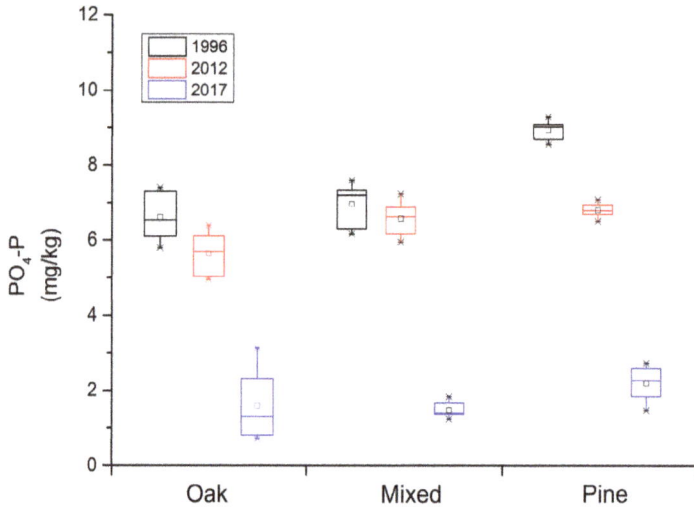

Figure 2. The loss of soil available phosphorus (P) along a 20-year time scale in the Qinling Mountains forest ecosystem. "AP" represents the soil available P, "Pinus" represents the *P. armandii* forest, "Mingled: represents the mixed *Q. var. acuteserrata and P. armandii* forest, and "Quercus" represents the *Q. var. acuteserrata* forest.

The loss of soil available N and P over time is obvious. We can see large differences between the different sampling dates (Table 2). The one-way ANOVA results showed that time had a significant effect on the decline of soil available N and P in all the plots except the decline of soil available P over the first period from 1996 to 2012 in the mixed forest (capital letters in Table 2). In the next period, from 2012 to 2017, the abundance of soil available P in the mixed forest also experienced a significant decline and a huge loss of soil available P.

Table 2. The one-way ANOVA results of soil available N and P loss over time.

Nutrient Elements	Forest Types	Year		
		1996	2012	2017
	Q. aliena var. *acuteserrata*	39.50 (5.68) A	18.36 (1.54) B	14.21 (1.02) C
N	Mixed forest	51.75 (6.82) A	20.87 (5.34) B	13.54 (1.34) C
	P. armandii	24.67 (4.44) A	19.14 (1.78) B	16.15 (1.23) C
	Q. aliena var. *acuteserrata*	6.62 (0.64) A	5.65 (0.58) B	1.59 (0.95) C
P	Mixed forest	6.95 (0.59) A	6.57 (0.49) A	1.46 (0.22) B
	P. armandii	8.94 (0.27) A	6.81 (0.20) B	2.20 (0.49) C

Standard errors are shown in parentheses, and different capital letters within the same row indicate significant differences among means ($p < 0.05$).

3.2. Factors Influencing the Abundance of Soil Available N and P

Good climate data records and topography information were available for all forest plots used in this study. We analyzed the strength of the TG, including plot elevation, slope, and aspect; climate factors, including MAP and MAT, together with the ST and SD, influencing the abundance of soil available N and P.

MAT has an influential role in soil available N loss (Figure 3). MAT explained 2% of the loss in soil available N in 1996, 5% in 2012, and 15% of the loss in 2017. The SD had a slightly increasing role in explaining the amount of soil available N, from 14% to 17%, over the study period. The TG, MAP, and ST did not show a uniform trend in affecting soil available N. The influence of all the unconstrained factors decreased gradually from 30% to 27%, and finally to 16% in 2017. We conclude that MAT was the main driver of the decline in soil available N concentrations, whereas the SD contributed minimally.

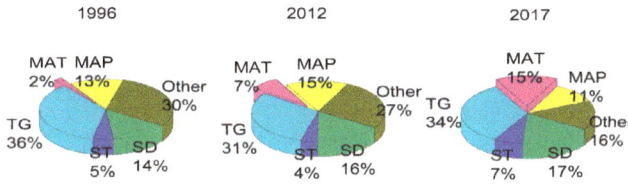

Figure 3. Influence of factors on soil available N throughout the sampling period. MAP represents mean annual precipitation, MAT represents mean annual temperature, TG represents topography indicators including aspect, slope and elevation. ST represents stand type, SD represents the stand density, "Other" represents all the unconstrained factors.

In explaining the concentrations of soil available P, the SD was the dominant factor (Figure 4). The effect of SD on soil available P was 3% in 1996, 5% in 2012 and 14% in 2017. All other factors, including the TG, MAP, ST and Other, did not show a uniform trend in affecting the soil available P content. The influential strength of all the unconstrained factors increased from 18% to 24% and then decreased to 20%. Although not strong, the influence of SD became more important with time, so we conclude that the changing SD is the main driver of soil available P loss.

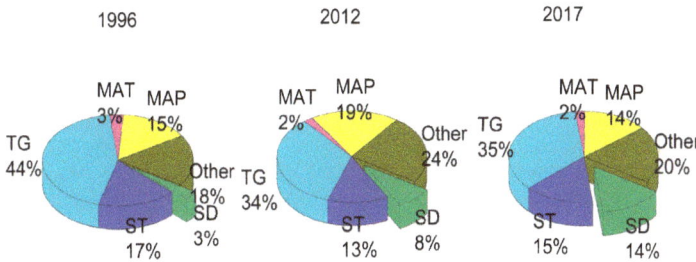

Figure 4. Influence of factors on soil available P throughout the 20-year time span, where "MAP" denotes mean annual precipitation, "MAT" represents mean annual temperature, and "TG" represents topography indicators including aspect, slope, and elevation. "ST" represents stand type, "SD" represents the stand density, and "Other" represents all the unconstrained factors.

4. Discussion

This study has two main advantages in understanding the change in concentrations of soil available N and P in forests of the Qinling Mountains. This work quantitatively demonstrated the change in abundance of soil available N and P during the 20-year period and demonstrated that the trend in soil degradation with respect to available N and P is continuing in the forests of the Qinling Mountains. We also calculated the strength of various factors influencing the abundance of soil available N and P. A longer study period would be more persuasive in describing the trajectory of the change in soil available N and P contents. The interactions of unknown factors, classified as "Other", was ignored and might affect the veracity of the results when calculating the specific strength of the factors.

In our study, the mixed forests had more soil available N than the oak-dominated forest and had much higher amounts than in the pine-dominated forests. In terms of soil available P, the pine-dominated forests had the highest concentrations, while the mixed forests and oak-dominated forests had slightly lower amounts. Climate will have a profound effect on the decline of soil available N and P [20]. Temperature may enhance the soil microbial process and soil biochemical processes, such as denitrification, which are positively related to temperature [21]. Tree species can have a strong influence on soil properties [22]. Different vegetation types can have varying influence on the rate of soil N and P input, and these can affect the soil nutrient accumulation and loss [23,24].

Soil N and P use efficiency can vary among plant species, and the abundance of soil available N and P in this study was influenced by exploitation of soil nutrients by different species. This is consistent with our findings that the concentrations of soil available N and P were species-dependent. Discussions about the pros and cons of tree species richness for soil nutrients and other ecosystem aspects date back to the early 19th century [25,26], when higher levels of soil N were found in forests with more tree species [27]. The decomposition rate of soil N has been found to be directly affected by tree species [28], which may result in the difference in the abundance of soil available N in different areas. Our findings for the soil available N content are consistent with these findings. The soil available N content was positively correlated with the species richness.

Over the studied time period, the forest soil experienced a degradation in available nutrient concentrations. The decreasing amount of soil available N and P will limit the forest's productivity, threatening the stability of the ecosystem. Global climate change has complicated effects on global N and P cycles [29]. Soil N can be lost through emissions of ammonia, nitrous oxide, and nitric oxide. Increases in soil enzyme activity, related to soil microbial metabolism, can leads to soil N loss. N-binding agents (especially tannins) have been reported as enzyme inhibitors, so we speculated that the influence of tannins from leaf litter of some species could inhibit mineralization of organic N [30]. Other factors like nitrification and subsequent leaching can be also responsible for soil available N loss [31]. Soil P loss is predominately caused by surface runoff. Overland flow and leaching contribute to the loss of soil P [32]. As a consequence of soil available P consumption (such as plant uptake), declines in sources of organic P and loss of soil available P are inevitable in these forests.

The losses of soil available N and P, which govern the N and P cycling patterns in forest systems, have a wide range of controlling factors, including climate factors, soil homeostasis, stand traits, and vegetation type [33,34]. The abundance of soil available N and P is probably dependent on all of these mentioned factors. In 1996, MAT was found to explain only 2% of the variation in soil available N. The influence increased over the 20-year period. When examining the trend in global warming, declines in soil available N and P may exhibit similar trends. Aspect and slope also regulate overland runoff, light availability and radiation. Temperature and precipitation are also related to elevation, which is why topographic factors have an important influence on the concentrations of available N and P. The decreasing influential effect of topographic factors over the study period implies they may continue to decline in the future. The advantages of higher N and P input and higher organic matter decomposition rate, caused by increasing temperature, may be offset by increasing uptake by the forest biomass. As a result, the increasing influence of stand density on available P concentration over time is understandable.

5. Conclusions

In the forests of the Qinling Mountains, the concentrations of soil available N and P are species-related and are positively related to species richness. The reduction in soil quality is concerning, especially given the large decrease in available N and P contents and the continuing decline over the study period. Mean annual temperature will be a critical factor regulating soil available N, and stand density will be the critical factor affecting soil available P in the Qinling Mountains forest system.

Acknowledgments: We are grateful to the Qinling National Forest Ecosystem Research Station for providing some data and the experimental equipment. We wish to thank our academic editor and two anonymous reviewers for improving this manuscript. Moreover, our appreciation goes to the guest editor, Robert G. Qualls, for exhaustive instructions. This research was funded by the project "Technical management system for increasing the capacity of carbon sink and water regulation of mountain forests in the Qinling Mountains" (201004036) of the State Forestry Administration of China.

Author Contributions: J.Y., X.Z. and S.Z. conceived and designed the experiments; X.Z., J.Y. and T.Z. performed the experiments; J.Y., X.Z. and S.Z. analyzed the data; X.Z., J.Y., T.Z. and F.H. contributed reagents/materials/analysis tools; J.Y., X.Z., S.Z. and S.J. wrote the paper.

Conflicts of Interest: The authors declare no conflict of interest.

References

1. Oldeman, L.R.; Hakkeling, R.U.; Sombroek, W.G. *World Map of the Status of Human-Induced Soil Degradation: An Explanatory Note*; International Soil Reference and Information Centre: Wageningen, The Netherlands, 1991.

2. Wang, T.; Yan, C.Z.; Song, X.; Li, S. Landsat images reveal trends in the Aeolian desertification in a source area for sand and dust storms in China's Alashan plateau (1975–2007). *Land Degrad. Dev.* **2013**, *24*, 422–429. [CrossRef]

3. Zhang, S.; Lei, R.; Liu, G.; Dang, K.; Shang, L.; Zhang, Y. Nutrient cycle in main types of forests at Huoditang Forest Region in the Qinling Mountains. *J. Northwest For. Coll.* **1996**, *11*, 115–120. (In Chinese with English Abstract)

4. Hurni, H. Land degradation, famine, and land resource scenarios in Ethiopia. In *World Soil Erosion and Conservation*; Pimentel, D., Ed.; Cambridge University Press: Cambridge, UK, 1993; pp. 27–62.

5. Shapiro, J. Environmental Degradation in China under Mao and Today: A Comparative Reflection. *Glob. Environ.* **2016**, *9*, 440–457. [CrossRef]

6. Chen, J.; Chen, J.Z.; Tan, M.Z.; Gong, Z.T. Soil degradation: A global problem endangering sustainable development. *J. Geogr. Sci.* **2002**, *12*, 243–252.

7. Smil, V. *The Bad Earth. Environmental Degradation in China*; ME Sharpe, Inc.: Armonk, NY, USA, 1984; Volume 9, pp. 332–333.

8. Le Houérou, H.N. Climate change, drought and desertification. *J. Arid Environ.* **1996**, *34*, 133–185. [CrossRef]

9. Schoenholtz, S.H.; Van Miegroet, H.; Burger, J.A. A review of chemical and physical properties as indicators of forest soil quality: Challenges and opportunities. *For. Ecol. Manag.* **2000**, *138*, 335–356. [CrossRef]

10. Lemanceau, P.; Maron, P.A.; Mazurier, S.; Mougel, C.; Pivato, B.; Plassart, P.; Ranjard, L.; Revellin, C.; Tardy, V.; Wipf, D. Understanding and managing soil biodiversity: A major challenge in agroecology. *Agron. Sustain. Dev.* **2015**, *35*, 67–81. [CrossRef]

11. Elser, J.J.; Andersen, T.; Baron, J.S.; Bergström, A.K.; Jansson, M.; Kyle, M.; Hessen, D.O. Shifts in lake N: P stoichiometry and nutrient limitation driven by atmospheric nitrogen deposition. *Science* **2009**, *326*, 835–837. [CrossRef] [PubMed]

12. Reich, P.B.; Hobbie, S.E.; Lee, T.; Ellsworth, D.S.; West, J.B.; Tilman, D.; Trost, J. Nitrogen limitation constrains sustainability of ecosystem response to CO_2. *Nature* **2006**, *440*, 922–925. [CrossRef] [PubMed]

13. Hu, S.; Chapin, F.S.; Firestone, M.K.; Field, C.B.; Chiariello, N.R. Nitrogen limitation of microbial decomposition in a grassland under elevated CO_2. *Nature* **2001**, *409*, 188–191. [CrossRef] [PubMed]

14. Vitousek, P.M.; Porder, S.; Houlton, B.Z.; Chadwick, O.A. Terrestrial phosphorus limitation: Mechanisms, implications, and nitrogen–phosphorus interactions. *Ecol. Appl.* **2010**, *20*, 5–15. [CrossRef] [PubMed]

15. Koerselman, W.; Meuleman, A.F. The vegetation N: P ratio: A new tool to detect the nature of nutrient limitation. *J. Appl. Ecol.* **1996**, *33*, 1441–1450. [CrossRef]

16. Conant, R.T.; Ryan, M.G.; Ågren, G.I.; Birge, H.E.; Davidson, E.A.; Eliasson, P.E.; Hyvönen, R. Temperature and soil organic matter decomposition rates–synthesis of current knowledge and a way forward. *Glob. Chang. Biol.* **2011**, *17*, 3392–3404. [CrossRef]

17. Durán, J.; Morse, J.L.; Groffman, P.M.; Campbell, J.L.; Christenson, L.M.; Driscoll, C.T.; Mitchell, M.J. Climate change decreases nitrogen pools and mineralization rates in northern hardwood forests. *Ecosphere* **2016**, *7*, e01251. [CrossRef]

18. Yuan, J.; Jose, S.; Zheng, X.F.; Cheng, F.; Hou, L.; Li, J.X.; Zhang, S.X. Dynamics of Coarse Woody Debris Characteristics in the Qinling Mountain Forests in China. *Forests* **2017**, *8*, 403. [CrossRef]

19. Liu, G.S.; Jiang, N.H.; Zhang, L.D.; Liu, Z.L. *Soil Physical and Chemical Analysis and Description of Soil Profiles*; China Standard Methods Press: Beijing, China, 1996; pp. 24–266. (In Chinese with English Abstract)

20. Jeppesen, E.; Kronvang, B.; Olesen, J.E.; Audet, J.; Søndergaard, M.; Hoffmann, C.C.; Andersen, H.E.; Lauridsen, T.L.; Liboriussen, L.; Larsen, S.E.; et al. Climate change effects on nitrogen loading from cultivated catchments in Europe: Implications for nitrogen retention, ecological state of lakes and adaptation. *Hydrobiologia* **2011**, *663*, 1–21. [CrossRef]

21. Veraart, A.J.; De Klein, J.J.; Scheffer, M. Warming can boost denitrification disproportionately due to altered oxygen dynamics. *PLoS ONE* **2011**, *6*, e18508. [CrossRef] [PubMed]

22. Zheng, X.; Wei, X.; Zhang, S. Tree species diversity and identity effects on soil properties in the Huoditang area of the Qinling Mountains, China. *Ecosphere* **2017**, *8*, e01732. [CrossRef]

23. Binkley, D.; Sollins, P.; Bell, R.; Sachs, D.; Myrold, D. Biogeochemistry of adjacent conifer and alder-conifer stands. *Ecology* **1992**, *73*, 2022–2033. [CrossRef]

24. Lee, M.R.; Flory, S.L.; Phillips, R.P. Positive feedbacks to growth of an invasive grass through alteration of nitrogen cycling. *Oecologia* **2012**, *170*, 457–465. [CrossRef] [PubMed]

25. Duffy, J.E.; Richardson, J.P.; Canuel, E.A. Grazer Diversity Effects on Ecosystem Functioning in Seagrass beds. *Ecol. Lett.* **2003**, *6*, 637–645. [CrossRef]

26. Hector, A.; Bagchi, R. Biodiversity and ecosystem multifunctionality. *Nature* **2007**, *448*, 188–190. [CrossRef] [PubMed]

27. Gamfeldt, L.; Snäll, T.; Bagchi, R.; Jonsson, M.; Gustafsson, L.; Kjellander, P.; Mikusiński, G. Higher levels of multiple ecosystem services are found in forests with more tree species. *Nat. Commun.* **2013**, *4*, 1340. [CrossRef] [PubMed]

28. Hansson, K.; Fröberg, M.; Helmisaari, H.S.; Kleja, D.B.; Olsson, B.A.; Olsson, M.; Persson, T. Carbon and nitrogen pools and fluxes above and below ground in spruce, pine and birch stands in southern Sweden. *For. Ecol. Manag.* **2013**, *309*, 28–35. [CrossRef]

29. Bouwman, L.; Goldewijk, K.K.; Van Der Hoek, K.W.; Beusen, A.H.; Van Vuuren, D.P.; Willems, J.; Rufino, M.C.; Stehfest, E. Exploring global changes in nitrogen and phosphorus cycles in agriculture induced by livestock production over the 1900–2050 period. *Proc. Natl. Acad. Sci. USA* **2013**, *110*, 20882–20887. [CrossRef] [PubMed]

30. Adamczyk, B.; Karonen, M.; Adamczyk, S.; Engström, M.T.; Laakso, T.; Saranpää, P. Tannins can slow-down but also speed-up soil enzymatic activity in boreal forest. *Soil Biol. Biochem.* **2017**, *107*, 60–67. [CrossRef]

31. Wang, J.; Zhu, B.; Zhang, J.; Mueller, C.; Cai, Z. Mechanisms of soil N dynamics following long-term application of organic fertilizers to subtropical rain-fed purple soil in China. *Soil Biol. Biochem.* **2015**, *91*, 222–231. [CrossRef]

32. Bai, Z.; Li, H.; Yang, X.; Zhou, B.; Shi, X.; Wang, B.; Oenema, O. The critical soil P levels for crop yield, soil fertility and environmental safety in different soil types. *Plant Soil* **2013**, *372*, 27–37. [CrossRef]

33. Niu, S.; Classen, A.T.; Dukes, J.S.; Kardol, P.; Liu, L.; Luo, Y.; Rustad, L.; Sun, J.; Tang, J.; Templer, P.H.; et al. Global patterns and substrate-based mechanisms of the terrestrial nitrogen cycle. *Ecol. Lett.* **2016**, *19*, 697–709. [CrossRef] [PubMed]

34. Smyth, C.E.; Titus, B.; Trofymow, J.A.; Moore, T.R.; Preston, C.M.; Prescott, C.E.; CIDET Working Group. Patterns of carbon, nitrogen and phosphorus dynamics in decomposing wood blocks in Canadian forests. *Plant Soil* **2016**, *409*, 459–477. [CrossRef]

forests

MDPI

Article

Soil Chemical Attributes, Biometric Characteristics, and Concentrations of N and P in Leaves and Litter Affected by Fertilization and the Number of Sprouts per the *Eucalyptus* L'Hér. Strain in the Brazilian Cerrado

Natasha M. I. Godoi [1], Sabrina N. dos S. Araújo [2], Salatiér Buzetti [2], Rodolfo de N. Gazola [1], Thiago de S. Celestrino [1], Alexandre C. da Silva [3], Thiago A. R. Nogueira [2] and Marcelo C. M. Teixeira Filho [2,*]

[1] Postgraduate Program in Agronomy School of Engineering (FEIS), São Paulo State University (UNESP), Ilha Solteira 15385-000, São Paulo State, Brazil; natashagodoi@bol.com.br (N.M.I.G.); rngazola@gmail.com (R.d.N.G.); thiagocelestrino@yahoo.com.br (T.d.S.C.)

[2] Department of Plant Protection, Rural Engineering and Soils, Ilha Solteira School of Engineering (FEIS), São Paulo State University (UNESP), Ilha Solteira 15385-000, São Paulo State, Brazil; sabrina@agr.feis.unesp.br (S.N.d.S.A.); sbuzetti@agr.feis.unesp.br (S.B.); tarnogueira@agr.feis.unesp.br (T.A.R.N.)

[3] Department of Forestry, Cargill International SA, Três Lagoas 79610-090, Mato Grosso do Sul State, Brazil; Alexandre_C_Silva@cargill.com

* Correspondence: mcmteixeirafilho@agr.feis.unesp.br; Tel.: +55-17-3743-1940

Received: 31 March 2018; Accepted: 20 May 2018; Published: 24 May 2018

Abstract: Given the lack of recommendations for the fertilization of *Eucalyptus* clones in the second production cycle, the effects of fertilizer rates and the number of sprouts per strain in terms of the soil chemical attributes, biometric characteristics, and the concentrations of N and P in the leaves and in the litter of *Eucalyptus* L'Hér. in the Brazilian Cerrado were evaluated. The experimental design was a randomized block with four replicates, arranged in a 2 × 4 factorial scheme: one or two sprouts per strain; four fertilizer rates (0, 50, 100, or 200% of 200 kg ha^{-1} of the formula 06-30-06 + 1.5% Cu + 1% Zn) applied immediately after sprout definition. The option of one sprout per strain yielded higher contents of organic matter (K, S, B, and Mn) in the 0.20–0.40-m layer, the leaf chlorophyll index, the diameter at breast height, and the height of the *Eucalyptus* 44 months after the definition of sprouts. However, N and P leaf concentrations and the wood volume did not differ as a function of the sprout numbers. The fertilizer dosage did not influence the wood volume, even in sandy soil with low fertility. Approximately 86% of the wood volume was obtained from the supply of soil and root nutrient reserves and 14% of this productivity is due to fertilization minerals. The adequate fertilization in the first cycle of the *Eucalyptus* supplies almost the entire nutritional demand of the forest in the second production cycle.

Keywords: *Eucalyptus* sp.; wood volume; second production cycle; annual increment average; soil fertility; nutrient cycling

1. Introduction

Brazil is a country known for its forestry. It is estimated that the area occupied by forests planted in 2016 was 7.74 million hectares, about 0.9% of the national territory and a growth of 0.5% compared to the area in 2015 [1]. *Eucalyptus* L'Hér. planted in 2016 amounted to 5.7 million hectares in the states of Minas Gerais (24%), São Paulo (17%), and Mato Grosso do Sul (15%). There was an increase of 400 thousand hectares of *Eucalyptus* in the period between 2015 and 2016 [1].

Eucalyptus presents a high mobilization of nutrients due to its rapid growth. The harvesting of wood in Brazil is generally carried out when trees are seven years old and in cycles ranging from 7 to 21 years (one to three production cycles during these periods). The removal of biomass results in a large decreases in nutrients, consequently reducing their availability for future plantations. The *Eucalyptus* tree trunk is the largest biomass accumulator among all the parts of the plant, accounting for about 65–80% of the total accumulated biomass [2]. It is, therefore, responsible for the largest removal of nutrients. The total biomass extracted and its compartmentalization determine the degree of the removal of nutrients [2]. Faria et al. [3] found an average decrease of 52% in productivity from the first to the second production cycle of the *Eucalyptus*, which they attributed to the nutrient removal, especially of K, in the previous cycle. Rocha et al. [4] reported that even if forest residues from the first to the second *Eucalyptus* production cycle are maintained, there may be a 6% reduction in the productivity of the wood, and it may take up to 16 years for a site to recover completely when the plant residues are removed.

This situation is made worse by the fact that most plantations in the Brazilian Cerrado region are concentrated in soils with low natural fertility [5,6], where the nutritional deficit is mainly accentuated for N, P, K, Ca, Mg, S, B, and Zn [7]. These generally sandy soils present high levels of aluminum and low water availability, which may compromise the *Eucalyptus'* productivity over time. Therefore, the maintenance of the productivity of the forest requires the replacement of nutrients removed by the harvest and lost by other processes such as leaching. According to Costa et al. [8], fertilization leads to a 30–50% increase in the productivity of *Eucalyptus* forest sites.

Studies have shown that the application of fertilizers to *Eucalyptus* plantations can represent gains ranging from 5% to 90% in the volume of wood (VW). This response to fertilizers, especially N fertilizers, is even more significant in future rotations with more productive genetic materials and the removal of large amounts of nutrients from stands grown in sandy soils [9].

The lack of nutrient fertilization is one of the causes of the reduced productivity of *Eucalyptus* plantations in areas with low fertility soils, such as those in the Brazilian Cerrado. In most of these areas with low soil fertility, the responses to fertilization with N, P, K, S, and B are positive [10]. The application of fertilizer alters the tree's growth, the cycling and nutrient stocks in biomass of the trees, the understory, and the soil [11].

The sprouting of strains is an interesting and common technique in *Eucalyptus* plantations; it can be about 40% more economical than planting seedlings. The management of forests via clear cutting and regeneration through the sprouting of the strains has the advantage of having a high initial growth rate of sprouts compared with plants of the first production cycle, particularly since these first plants have a root system that contains organic and inorganic reserves that can be readily used [12,13]. These roots also facilitate the uptake of water and nutrients. Additionally, because of the high root to shoot ratio, the assimilates are preferably allocated with regard to the shoot formation, thereby increasing the differences in the growth rate compared to plants of the first rotation. This rapid initial growth of shoots may result in maximum wood yields earlier than those of the first rotation. Despite this apparent advantage, it has been recorded that many *Eucalyptus* forests in subsequent cycles have exhibited a decrease in productivity that is not always associated with the reduction in the number of trunks of the original settlement [12].

Given that nutrient cycling and productivity are favored by a greater availability of nutrients, the second production cycle of *Eucalyptus* has a root system that is almost completely formed. However, though the use of root reserves can be the key to the success of the second *Eucalyptus* production cycle, their impact on mature roots and sprouts is uncertain.

Due to the lack of recommendations for the fertilization of *Eucalyptus* clones in the second production cycle in the Brazilian Cerrado, it is necessary to define the best dose of fertilizers for one or two sprouts per *Eucalyptus* strain because the nutritional requirements can vary. The nutrient demand in the management of two sprouts per *Eucalyptus* strain, as well as the response to mineral fertilization, tends to be higher. The objective of this research was to evaluate the effect of mineral fertilizer rates and the number of sprouts per strain on the chemical attributes of the three layers of soil, the biometric characteristics, and the concentrations of the N and P in the leaves and in the litter of *Eucalyptus* during the second production cycle in the Brazilian Cerrado. Additionally, another objective of this research was to better understand what contributes the most in this system to the wood volume productivity and the supply of soil and roots with nutrient reserves or fertilization minerals.

2. Material and Methods

2.1. Location and Climate

The experiment set up in a commercial *Eucalyptus* field, managed by Cargill Agrícola S/A and located in the municipality of Três Lagoas in the state of Mato Grosso do Sul at a latitude of 20°45′ South, a longitude of 51°40′ West (Figure 1), and an altitude of approximately 320 m. It is in this Brazilian Cerrado region that the *Eucalyptus* cultivation proceeds the fastest. The largest paper and pulp industries in the world are located here. Prior to the first cycle production of the *Eucalyptus* in the sample area, there was a degraded brachiaria (*Urochloa brizantha* (Hochst. ex A.Rich.) R.Webster) pasture. The soil of the experimental area is an Arenosols, according to the World Reference Base for Soils (WRB) or an orthosic Quartzarenic Neosol (Entisols) according to the Embrapa classification system [14].

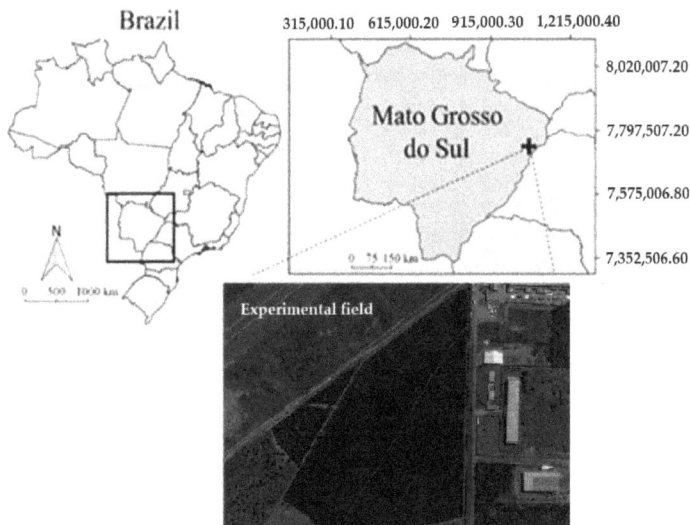

Figure 1. The study area at the Cargill Agrícola S/A (The municipality of Três Lagoas, state of Mato Grosso do Sul, Brazil; 20°45′ S and 51°40′ W, altitude 320 m).

The results of the soil chemical analysis of the experimental area at a depth of 0.00–0.20 m were determined prior to the installation of the experiment, according to the methodology proposed by Raij et al. [15]. We noted the following results: P_{resin} = 4 mg dm^{-3}, pH_{CaCl2} = 4.0; organic matter (OM) = 11 g dm^{-3}; K, Ca, Mg, H + Al = 0.4, 1.0, 1.0, and 25.0 $mmol_c$ dm^{-3}, respectively. The contents of S-SO_4, B, Cu, Fe, Mn, and Zn (diethylentriamene pentaacetate—DTPA) were 3.0, 0.23, 0.5, 24.0, 2.6, and 0.6 mg dm^{-3}, respectively, with a base saturation (V%) of 7% and an aluminum saturation (m%) of 83%. We observed that the soil was acidic with a low content of OM. The soil was deficient in the contents of macro and micronutrient (that is, below that which is considered as adequate for the good development of *Eucalyptus*). Fertility levels like those of this soil are commonly found in this region.

The climate in the region is Aw, according to the Köppen's classification system [16], characterized as humid tropical with a rainy season in the summer and a dry season in the winter.

2.2. Treatments and Experimental Design

The experimental design was a randomized block with eight treatments and four replicates arranged in a 2 × 4 factorial scheme. The first number in the scheme being the number (one or two) of sprouts per strain in the second cycle. It has four fertilizer rates (0, 50, 100, or 200% of 200 kg ha^{-1} of the formula 06-30-06 + 1% Ca + 3% S + 1% Mg + 1.5% Cu + 1% Zn, which is commonly used in the region) applied soon after the definition of the sprout, in April 2013. The scheme of the eight treatments are listed below:

- one sprout per *Eucalyptus* strain without fertilization;
- one sprout per *Eucalyptus* strain with 50% fertilization;
- one sprout per *Eucalyptus* strain with 100% fertilization;
- one sprout per *Eucalyptus* strain with 200% fertilization;
- two sprouts per *Eucalyptus* strain without fertilization;
- two sprouts per *Eucalyptus* strain with 50% fertilization;
- two sprouts per *Eucalyptus* strain with 100% fertilization;
- two sprouts per *Eucalyptus* strain with 200% fertilization.

Each plot was composed of 49 plants distributed in seven rows with seven plants in each row, with a line of *Eucalyptus* plants as a border at the ends. In all, the total plot area with a border was 367.5 m^2.

2.3. Management History

The *Eucalyptus urophylla* S.T.Blake clone (*Eucalyptus urophylla* × *Eucalyptus grandis* W. Hill ex Maiden) is the most commonly planted *Eucalyptus* species in the region and in Brazil. It is commonly planted with a spacing of 3.0 × 2.5 m. In the first cycle of *Eucalyptus* production in the entire experimental area, the following operations were performed: (a) sampling and soil analysis in the 0–0.20-m and 0.20–0.40-m depth layers; (b) the control of ants; (c) chemical weeding over the total area; (d) liming with 1.5 t ha^{-1} of limestone (Relative total neutralizing power—RTNP = 88%); (e) the application of 250 kg ha^{-1} of gypsum (with 14% S and 17% Ca) after liming; (f) felling: furrows were opened with a depth of 0.50 m in the planting line (on 13 December 2006); (g) the planting of 200 kg ha^{-1} of formula 06-30-06 fertilizer, enriched with 1.0% Ca + 3.0% S + 1.0% Mg + 1.5% Cu + 1.0% Zn, in the planting groove at a depth of 0.15 m; (h) the first topdressing fertilization using 120 kg ha^{-1} of formula 18-00-18 + 6% S + 0.5% B (300 kg of potassium chloride + 282 kg of ammonium sulfate + 368 kg of ammonium nitrate + 50 kg of Borogran with 10% B) 60 days after planting, applied manually in the form of a crown or semicircle 0.30 m away from the seedling neck on the soil and without incorporation into the soil; (i) the second topdressing fertilization using 270 kg ha^{-1} of formula 18-00-18 + 6% S + 0.5% B (300 kg of potassium chloride + 282 kg of ammonium sulfate + 368 kg of ammonium nitrate + 50 kg of Borogran) 10 months after planting, applied mechanically in a continuous fillet on the ground about 0.60 m away from the stem of the plant; (j) the third topdressing fertilization using 350 kg ha^{-1} of formula 15-00-20 + 10% S + 0.5% B (333 kg of potassium chloride + 417 kg of ammonium sulfate + 368 kg of ammonium nitrate + 50 kg of Borogran) 14 months after planting, applied mechanically in a continuous fillet on the ground about 0.60 m away from the stem of the plant; and (k) the *Eucalyptus* harvest was carried out 6 years after planting. The average productivity was 392 m^3 ha^{-1} of wood.

2.4. Experiment Management

Based on the soil analysis (Section 2.1) and the requirement of the *Eucalyptus* crop, two months before the *Eucalyptus* harvest of the first production cycle or rotation, the application of liming and gypsum were performed again in the experimental area. We applied 2 t ha^{-1} of limestone (RTNP = 88%) in order to increase the saturation of the bases to 60% in the total area and without the incorporation of this correction. After liming, 700 kg ha^{-1} of gypsum (with 14% S and 17% Ca) was applied next to the planting line in a range of 0.7–1.0 m on the soil surface and without incorporation into the soil.

After the harvesting of *Eucalyptus* during the rainy season, the strains that were covered with vegetal residue were cleaned within a 15 cm radius of the border of the strain in order to avoid impairing the emission of the sprouts. The thinning was conducted in April 2013 when the sprouts were, on average, 2.5–3.0 m long (the diameter at breast height was between 6.0 and 9.0 cm).

With respect to the selection of sprouts, regardless of the choice of one or two sprouts (Figure 2), we selected more vigorous sprouts located at the top of the strain (the upper side of the strain). In the plots in which the plants exhibited two sprouts, the sprouts were chosen in opposite positions, if possible, so that the opposition of one sprout to the other was in the direction of the planting line. The purpose here was to avoid or reduce the breakage or tipping of the sprouts when in contact with the machines and equipment that transit between the lines of the plantations during the fertilization operations and the maintenance of the forest.

Figure 2. The two sprouts (**A**) and one sprout (**B**) per *Eucalyptus* strain. Três Lagoas, state of Mato Grosso do Sul, Brazil, 2017.

The mineral fertilizer treatments described above were performed manually and in a semi-circle in April 2013. Weed control and plant pest control were performed when necessary.

2.5. Evaluations

2.5.1. Soil Chemical Analysis

Forty-four months after the definition of sprouts and after mineral fertilization, which corresponds to four years of *Eucalyptus* harvest cultivated in the first cycle, soil samples were obtained at depths of 0–0.20, 0.20–0.40, and 0.40–1.00 m in the *Eucalyptus* planting line. The planting line was fertilized in a semi-circle at eight points per plot to form a composite sample. We evaluated the alteration of the chemical attributes of the soil according to the protocol by Raij et al. [15]. The OM content in the soil was estimated by the Walkley–Black method. The available amounts of P, Ca^{2+}, K$^+$, and Mg^{2+}

in the soil were estimated by an ion-exchange resin procedure with B in hot water and Cu, Fe, Mn, and Zn in DTPA. The concentration of P, Al, and B in the soil extracts was quantified by a colorimetric method. The Ca, Mg, Cu, Fe, Mn, and Zn concentrations were determined by an atomic absorption spectrophotometer (AAS) (VARIAN SpectrAA 220FS) and the K using a flame-photometer (METEOR NAK-II). The available amount of S-SO$_4$ was estimated using a solution of calcium phosphate (Ca (H$_2$PO$_4$) 0.01 mol L^{-1}) and the quantification was determined by turbidimetry. The exchangeable aluminum was extracted with a 1 M KCl solution and determined by titration with 0.025 M of NaOH. The total acidity (H + Al) was extracted with a buffer solution of calcium acetate with a pH of 7.0 and determined by titration with ammonium hydroxide (0.025 M). From these values, the sum of bases (SB) {SB = Ca^{2+} + Mg^{2+} + K$^+$}, the total cation exchange capacity (CEC) at a pH of 7.0 (CEC = SB + (H + Al)), the saturation of exchangeable cations (V%) {V% = SB × 100)/CEC}, and the aluminum saturation (m%) {m% = (Al^{3+} × 100)/(SB + Al^{3+})} were obtained.

2.5.2. Concentrations of N and P in Leaves and Litter

At the same time (that is, 44 months after the definition of the sprouts and the mineral fertilization), samples (150 g each) of the *Eucalyptus* leaves and litter were collected from six representative trees or from near these plants. From the leaflet samples deposited on the soil and the biomass of the leaves of the *Eucalyptus* trees, the N and P concentrations were determined using the methodology described by Malavolta et al. [17].

2.5.3. The Height of the Plants

The height of the plants (H) is an essential piece of information for determining the VW of the trees. At an age of 44 months, the height of the three representative plants, sectioned at the level of the soil per plot, was measured using a scale. In order to better measure the height, 10 plants per plot were measured using a Forestor Vertex apparatus, which is composed of a hypsometer and a transponder [18].

2.5.4. Diameter at Breast Height

We used a graduate student and a forest compass to measure the diameter at a breast height (DBH) of 1.30 m. This evaluation was also performed 44 months after the definition of the sprouts and the mineral fertilization in the three representative plants sectioned at the soil level per plot. In addition, in order to better measure the DBH, 10 plants per plot were measured.

2.5.5. Total Wood Volume with Bark

Based on the plant height and the DBH evaluations noted above, the total VW with the bark (m^3 ha^{-1}) was calculated using the following formulae:

$$VW = \sum \frac{VW_i}{A_i} 1000 \qquad (1)$$

$$VW_i = \frac{\pi (DBH_i)^2 \cdot ff \cdot H}{4} \qquad (2)$$

where VW$_i$ = the volume of the wood with the bark from each tree *i*; A = the area of the useful plot (367.5 m^2); VW = the total volume with bark (m^3 ha^{-1}); DBH$_i$ = the DBH from each tree (m); ff = the form factor. In this case, a value of 0.5 was assigned and regionally defined for the clone used. Additionally, H$_i$ = the total height of each tree (m).

2.5.6. Average Annual Increment

The average annual increment (AAI) in m^3 ha^{-1} $year^{-1}$ was calculated using the following formula:

$$AAI = \frac{VW}{t} \tag{3}$$

where AAI = the average annual increment (m^3 ha^{-1} $year^{-1}$); VW = the total volume with bark (m^3 ha^{-1}), and t = the time (year).

2.5.7. Leaf Chlorophyll Index

We determined the leaf chlorophyll index (LCI) indirectly at the same time as the other variables by examining the last fresh leaves of the middle third of the plant in 10 leaves per plot, using a portable digital chlorophyllometer (Falker Agricultural Automation, Porto Alegre, Brazil).

2.6. Statistical Analysis

We analyzed the results using a variance analysis (F test) and Tukey's test at a 5% probability level to compare the number of sprouts. A polynomial regression was applied to verify the effect of the fertilizer rates. All the statistical analyses including the Pearson correlation ($p < 0.05$) were performed using the SAS system (SAS Institute Inc., Cary, NC, USA).

3. Results

3.1. Soil Chemical Analysis

The chemical attributes of the soil at a depth of 0–0.20 m at 44 months after choosing the *Eucalyptus* sprouts are listed in Table 1. The pH values of the soil under the *Eucalyptus* plants indicated the acid reaction and did not vary as a function of the sprout numbers per strain or the fertilization rates. Such variation was noted for the other parameters studied, with the exception of the K and Mn concentrations. We verified that there were higher contents of K and Mn when we selected one sprout per strain. However, we observed that an increase in the fertilization rate decreased the Mg content, the sum of bases (SB), the CEC, and the V% and that it increased the iron content up to 106.6 kg ha^{-1} at a depth of 0–0.20 m. However, there was an increasing linear response for the P and Cu contents of the soil, as well as an increase in m%.

Table 1. The soil chemical attributes at a depth 0–0.20 m, 44 months after choosing the *Eucalyptus* sprouts as a function of the number of sprouts per strain and the fertilization rates.

	P Resin	O.M.	pH CaCl2	K	Ca	Mg	H + Al	SB	CEC
	(mg dm^{-3})	(g dm^{-3})				(mmol$_c$ dm^{-3})			
Sprouts per strain									
1	6.83a	12.92a	4.12a	0.63a	4.92a	4.00a	24.92a	9.55a	34.47a
2	7.67a	12.42a	4.09a	0.42b	3.83a	3.75a	24.83a	8.01a	32.84a
S.M.D. (5%)	1.89	1.34	0.20	0.14	1.79	1.52	2.93	2.83	3.19
Fertilizing (%) [+]									
0	4.00 [1]	13.67	4.15	0.58	4.67	4.83 [2]	25.33	10.08 [3]	35.42 [4]
50	5.00	12.33	4.20	0.52	5.33	5.17	23.83	11.02	34.85
100	5.33	12.50	4.07	0.43	4.00	3.17	25.33	7.60	32.93
200	14.67	12.17	4.00	0.58	3.50	2.33	25.00	6.42	31.42
F Test									
Sprout (S)	0.89 ns	0.64 ns	0.07 ns	10.10 **	1.69 ns	0.12 ns	0.00 ns	1.37 ns	1.19 ns
Fertilizing (F)	31.84 **	1.19 ns	0.86 ns	1.18 ns	0.92 ns	3.61 *	0.27 ns	2.62 *	1.51 ns
S x F	1.37 ns	0.12 ns	0.76 ns	0.92 ns	0.30 ns	0.31 ns	0.64 ns	0.41 ns	0.51 ns
C.V. (%)	29.80	12.04	5.71	30.34	46.66	44.92	13.46	36.80	10.84
Average overall	7.25	12.67	4.10	0.53	4.38	3.88	24.88	8.78	33.65
Treatment	V	m	S-SO$_4$	B	Cu	Fe	Mn	Zn	

Table 1. *Cont.*

	P Resin	O.M.	pH CaCl2	K	Ca	Mg	H + Al	SB	CEC
	(%)	(%)	(mg dm^{-3})			(mg dm^{-3})			
Sprouts per strain									
1	27.71a	42.33a	8.08a	0.54	1.06a	25.67a	16.73a	0.53a	
2	24.26a	42.42a	10.17a	0.47	1.10a	24.92a	9.38b	0.61a	
S.M.D. (5%)	7.18	19.29	6.55	0.17	0.23	4.64	5.21	0.32	
Fertilizing (%) [+]									
0	28.64 [5]	25.67 [6]	9.17	0.46	0.90 [7]	29.83 [8]	14.10	0.52	
50	31.72	36.83	7.67	0.49	0.95	23.33	11.22	0.57	
100	23.02	54.50	8.17	0.50	1.00	20.67	12.50	0.35	
200	20.57	52.50	11.50	0.57	1.47	27.33	14.42	0.85	
F Test									
Sprout (S)	1.06 ns	0.00 ns	0.46 ns	0.56 ns	0.15 ns	0.12 ns	9.14 **	0.25 ns	
Fertilizing (F)	2.32 *	2.31 *	0.31 ns	0.37 ns	5.98 **	3.56 *	0.37 ns	1.93 ns	
S x F	0.54 ns	0.26 ns	0.32 ns	1.37 *	0.83 ns	0.52 ns	0.38 ns	0.42 ns	
C.V. (%)	31.57	51.98	52.00	39.28	24.28	20.97	45.61	54.21	
Average overall	25.99	42.38	9.12	0.50	1.08	25.29	13.06	0.57	

B: determined in hot water; Cu, Fe Mn, and Zn: determined in DTPA. The chemical analysis was performed at the UNESP/FEIS Soil Fertility Laboratory. [+] The percentage refer to 200 kg ha^{-1} of formula 06-30-06 + 1.0% Ca + 3.0% S + 1.0% Mg + 1.5% Cu + 1.0% Zn. Means followed by the same letters in the column are not significantly different by the Tukey's test at $p < 0.05$. * Significant at $p < 0.05$. ** Significant at $p < 0.01$. ns = not significant. [1] $Y = 2.5333 + 0.0539x$ ($R^2 = 0.86$ **); [2] $Y = 5.1333 - 0.0144x$ ($R^2 = 0.83$ **); [3] $Y = 10.6933 - 0.0219x$ ($R^2 = 0.76$ *); [4] $Y = 35.4933 - 0.0210x$ ($R^2 = 0.96$ *); [5] $Y = 30.3600 - 0.0500x$ ($R^2 = 0.70$ *); [6] $Y = 30.5333 + 0.1353x$ ($R^2 = 0.72$ *); [7] $Y = 0.8267 + 0.0029x$ ($R^2 = 0.89$ **); [8] $Y = 29.8470 - 0.1705x + 0.0008x^2$ ($R^2 = 1.00$ ** e PM = 106.6%).

The interaction of the fertilizer dosage for a given number of sprouts per strain for the content of B in the soil at a depth of 0–0.20 m, 44 months after the setting of the sprouts (Table 2) modulated the increasing linear function only when one sprout was chosen per strain.

Table 2. The results of the interaction between the number of sprouts per strain and the fertilization rates for the B content in soil at a depth of 0.00–0.20 m, 44 months after the setting of the *Eucalyptus* shoots.

	B (mg dm^{-3})	
	Sprouts of Strain	
Fertilizing (%) [+]	1 [1]	2
0	0.42a	0.50a
50	0.44a	0.54a
100	0.56a	0.44a
200	0.73a	0.42a

B: determined in hot water; [+] the percentage refers to 200 kg ha^{-1} of formula 06-30-06 + 1.0% Ca + 3.0% S + 1.0% Mg + 1.5% Cu + 1.0% Zn. Means followed by the same letters in the column are not significantly different by the Tukey's test at $p < 0.05$. [1] $Y = 0.3920 + 0.0016x$ ($R^2 = 0.97$ *).

At a soil depth of 0.20–0.40 m, 44 months after choosing the *Eucalyptus* sprouts (Table 3), the contents of OM, K, S-SO$_4$, B, and Mn were higher when there was only one sprout per strain. These findings are consistent with those for the contents of K and Mn in the 0–0.20 m layer. On the other hand, for the management of two sprouts per strain, no higher contents of these nutrients were verified compared to the chemical attributes of the soil at this depth.

The contents of P, S-SO$_4$, B, Cu, Mn, and Zn, as well as H + Al and CEC, increased linearly at a depth of 0.20–0.40 m in the soil with increasing fertilization rates (200 kg ha^{-1} of formula 06-30-06 + 1.0% Ca + 3.0% S + 1.0% Mg + 1.5% Cu + 1.0% Zn) (Table 3). For the pH, the decreasing linear function was adjusted according to the increment of the fertilizer rates noted above.

Table 3. The soil chemical attributes at a depth of 0.20–0.40 m, 44 months after the choice of *Eucalyptus* sprouts as a function of the number of sprouts per strain and fertilization rates.

	P $_{Resin}$	O.M.	pH $_{CaCl2}$	K	Ca	Mg	H + Al	SB	CEC
	(mg dm^{-3})	(g dm^{-3})				(mmol$_c$ dm^{-3})			
Sprouts per strain									
1	3.50a	10.33a	3.90a	0.39a	1.25a	1.50a	24.58a	3.14a	27.72a
2	3.42a	9.67b	3.92a	0.27b	1.83a	1.25a	23.33a	3.35a	26.68a
S.M.D. (5%)	1.62	0.64	0.08	0.10	0.91	0.53	2.40	1.22	2.54
Fertilizing (%) [+]									
0	2.50 [1]	10.17	3.97 [2]	0.33	1.50	1.83	22.67 [3]	3.67	26.33 [4]
50	2.50	9.67	3.93	0.33	1.00	1.67	22.50	2.50	25.00
100	3.33	10.33	3.90	0.33	1.67	1.33	24.50	3.33	27.83
200	5.50	9.83	3.83	0.32	2.00	1.17	26.17	3.48	29.65
F Test									
Sprout (S)	0.01 ^{ns}	4.92 *	0.23 ^{ns}	7.68 *	1.91 ^{ns}	1.03 ^{ns}	1.24 ^{ns}	0.13 ^{ns}	0.78 ^{ns}
Fertilizing (F)	3.53 *	1.03 ^{ns}	3.63 *	0.03 ^{ns}	0.97 ^{ns}	1.64 ^{ns}	3.68 *	0.82 ^{ns}	3.86 *
S x F	0.24 ^{ns}	0.20 ^{ns}	1.43 ^{ns}	0.22 ^{ns}	0.45 ^{ns}	0.12 ^{ns}	0.93 ^{ns}	0.38 ^{ns}	0.83 ^{ns}
C.V. (%)	53.40	7.36	2.20	33.56	57.14	43.82	11.46	42.96	10.65
Average overall	3.46	10.00	3.91	0.33	1.54	1.38	23.96	3.25	27.20
Treatment	V	m	S-SO$_4$	B	Cu	Fe	Mn	Zn	
	(%)	(%)	(mg dm^{-3})			(mg dm^{-3})			
Sprouts per strain									
1	11.49a	79.00a	12.17a	0.57a	0.95a	13.83a	5.12a	0.28a	
2	12.42a	78.50a	9.00b	0.43b	0.87a	13.67a	2.88b	0.28a	
S.M.D. (5%)	4.08	5.85	2.88	0.12	0.11	2.67	1.25	0.10	
Fertilizing (%) [+]									
0	13.73	74.33	7.67 [5]	0.45 [6]	0.87 [7]	13.83	2.55 [8]	0.18 [9]	
50	10.20	80.83	9.17	0.42	0.80	10.67	3.55	0.27	
100	12.04	81.67	13.67	0.52	0.92	14.83	5.15	0.30	
200	11.87	78.17	11.83	0.62	1.05	15.67	4.77	0.37	
F Test									
Sprout (S)	0.24 ^{ns}	0.03 ^{ns}	5.56 *	6.30 *	2.77 ^{ns}	0.02 ^{ns}	14.74 *	0.03 ^{ns}	
Fertilizing (F)	0.57 ^{ns}	1.46 ^{ns}	3.99 *	3.54 *	4.46 *	3.10 ^{ns}	4.12 *	4.43 *	
S x F	0.54 ^{ns}	0.62 ^{ns}	0.97 ^{ns}	2.74 ^{ns}	1.96 ^{ns}	0.41 ^{ns}	0.68 ^{ns}	0.18 ^{ns}	
C.V. (%)	38.96	8.48	31.09	27.69	13.51	22.15	35.72	42.90	
Average overall	11.96	78.75	10.58	0.50	0.91	13.75	4.00	0.28	

B: determined in hot water; Cu, Fe, Mn, and Zn: determined in DTPA. The chemical analysis was performed at the UNESP/FEIS Soil Fertility Laboratory. [+] The percentage refers to 200 kg ha^{-1} of formula 06-30-06 + 1.0% Ca + 3.0% S + 1.0% Mg + 1.5% Cu + 1.0% Zn. Means followed by the same letters in the column are not significantly different by the Tukey's test at $p < 0.05$. * Significant at $p < 0.05$. ** Significant at $p < 0.01$. ^{ns} = not significant. [1] Y = 2.0667 + 0.0159x (R^2 = 0.92 **); [2] Y = 3.9667 − 0.0007x (R^2 = 1.00 *); [3] Y = 22.2667 + 0.0193x (R^2 = 0.91 *); [4] Y = 25.4367 + 0.0202x (R^2 = 0.74 *); [5] Y = 8.6333 + 0.0223x (R^2 = 0.50 *); [6] Y = 0.4157 + 0.0010x (R^2 = 0.87 *); [7] Y = 0.8133 + 0.0011x (R^2 = 0.77 **); [8] Y = 3.0267 + 0.0112x (R^2 = 0.65 *); [9] Y = 0.2033 + 0.0009x (R^2 = 0.94 *).

For the chemical attributes of the soil at a depth of 0.40–1.00 m, 44 months after choosing the *Eucalyptus* sprouts listed in Table 4, we verified a higher S-SO$_4$ content in the one sprout per strain condition. This same finding was noted for the Mn content.

Much like the data recorded at a depth of 0.20–0.40 m, the contents of P, Cu, and Mn increased linearly. For the content of P, the linear equation was adjusted (Table 4). At 200% of the recommended fertilization (200 kg ha^{-1} of formula 06-30-06 + 1.0% Ca + 3.0% S + 1.0% Mg + 1.5% Cu + 1.0% Zn), we observed a higher content of this P. This result indicates that the increase in the dose contributed to the greater leaching of P in the 0.40–1.00 m layer.

The OM content decreased along the soil profile (Table 1, Table 3, and Table 4) and the lowest content was noted in the 0.40–1.00 m layer (Table 4).

Table 4. The soil chemical attributes at a depth of 0.40–1.00 m, 44 months after choosing the *Eucalyptus* sprouts as a function of the number of sprouts per strain and the fertilization rates.

	P Resin #	M. O.	pH CaCl2	K #	Ca#	Mg	H + Al	SB #	CEC
	(mg dm^{-3})	(g dm^{-3})				(mmol$_c$ dm^{-3})			
Sprouts per strain									
1	4.08a	9.42a	3.92a	0.33a	1.25a	1.08	24.75a	2.67a	27.42
2	7.83a	9.33a	3.87a	0.26a	1.83a	1.00	22.25a	3.09a	25.34
S.M.D. (5%)	7.07	0.66	0.06	0.10	0.96	0.18	3.56	1.01	3.63
Fertilizing (%)[+]									
0	1.33 [(1)]	9.33	3.88	0.27	1.50	1.00	23.50	2.77	26.27
50	2.67	9.17	3.93	0.35	1.00	1.00	21.33	2.35	23.68
100	6.67	9.33	3.88	0.27	1.83	1.00	23.50	3.10	26.60
200	13.16	9.67	3.87	0.30	1.83	1.17	25.67	3.30	28.97
F Test									
Sprout (S)	1.29 [ns]	0.07 [ns]	2.90 [ns]	2.37 [ns]	1.68 [ns]	1.00 [ns]	2.27 [ns]	0.82 [ns]	1.50 [ns]
Fertilizing (F)	3.42 *	0.46 [ns]	0.97 [ns]	0.65 [ns]	0.77 [ns]	1.00 [ns]	1.14 [ns]	0.78 [ns]	1.63 [ns]
S x F	0.32 [ns]	0.46 [ns]	0.97 [ns]	0.65 [ns]	1.04 [ns]	1.00 [ns]	0.62 [ns]	1.19 [ns]	0.97 [ns]
C.V. (%)	55.47	8.06	1.85	7.30	23.32	19.60	17.31	15.31	15.70
Average overall	5.96	9.38	3.89	0.30	1.54	1.04	23.50	2.88	26.38

Treatment	V #	m	S-SO$_4$ #	B	Cu	Fe	Mn #	Zn #	
	(%)	(%)	(mg dm^{-3})			(mg dm^{-3})			
Sprouts per strain									
1	9.83a	81.83a	45.25a	0.61	0.72	9.25a	4.35	0.33a	
2	11.92a	77.67a	16.75b	0.46	0.73	10.17a	1.59	0.27a	
S.M.D. (5%)	3.61	6.76	13.80	0.13	0.08	2.02	1.90	0.17	
Fertilizing (%)[+]									
0	10.17	80.83	35.33	0.53	0.68	9.67	1.52	0.23	
50	10.00	81.67	25.67	0.45	0.67	7.83	2.37	0.40	
100	11.67	78.00	27.33	0.52	0.73	10.17	3.50	0.15	
200	11.67	78.50	35.67	0.64	0.82	11.17	4.50	0.42	
F Test									
Sprout (S)	1.53 [ns]	1.74 [ns]	19.62 **	6.15 *	0.19 [ns]	0.94 [ns]	9.72 **	0.73 [ns]	
Fertilizing (F)	0.30 [ns]	0.32 [ns]	0.66 [ns]	1.78 [ns]	3.09 [ns]	2.19 [ns]	2.17 [ns]	2.78 [ns]	
S x F	0.81 [ns]	0.60 [ns]	0.54 [ns]	2.42 [ns]	4.86 *	1.46 [ns]	1.52 [ns]	0.64 [ns]	
C.V. (%)	16.71	9.69	22.39	27.78	12.95	23.80	26.60	10.98	
Average overall	10.88	79.75	31.00	0.53	0.72	9.71	2.97	0.30	

B: determined in hot water; Cu, Fe, Mn, and Zn: determined in DTPA. The chemical analysis was performed at the UNESP/FEIS Soil Fertility Laboratory. [+] The percentage refers to 200 kg ha^{-1} of formula 06-30-06 + 1.0% Ca + 3.0% S + 1.0% Mg + 1.5% Cu + 1.0% Zn. The means followed by the same letters in the column are not significantly different in Tukey's test at $p < 0.05$. * Significant at $p < 0.05$. ** Significant at $p < 0.01$. [ns] = not significant. [(1)] $Y = 0.5667 + 0.0616x$ ($R^2 = 0.98$ *); # Data corrected by the equation $(x + 0.5)^{0.5}$.

In terms of the effects of the interaction between fertilizer dosage and the number of sprouts per strain for the content of Mg and B and the CEC of the soil at a depth of 0.40–1.00 m (Table 5), there was an adjustment to the linear function that was increasing. For the Mg content in the soil and the CEC, there was no difference in the number of sprouts per strain. However, the B content in the soil at the highest fertilizer dosage was significantly higher in the one sprout per strain condition.

For the Cu content in the soil at a depth of 0.40–1.00 m, there was an interaction between the number of sprouts and the fertilizer dose. In the control, the highest Cu content in the soil was noted for the two sprouts per strain condition. In the management of the one sprout per strain condition, the increase in the fertilizer rates linearly increased the Cu content in the soil. This situation was not verified with the two sprouts per strain condition due to the higher absorption of this micronutrient.

The interaction between the number of sprouts per strain and the fertilizer dosage on the Mn content in the soil at a depth of 0.40–1.00 m differed for the one or two sprouts conditions at the two highest rates (100 and 200% of the recommended fertilization). The highest soil content was obtained for one sprout and the linear function for one sprout per strain was also adjusted.

Table 5. The interaction of the number of sprouts per strain and the fertilization rates for B, Cu, CEC, Mg, and Mn in the soil at a depth of 0.40–1.00 m, 44 months after choosing the *Eucalyptus* sprouts.

	B (mg dm^{-3})		Cu (mg dm^{-3})		CEC (mmol$_c$ dm^{-3})		Mg (mmol$_c$ dm^{-3})		Mn (mg dm^{-3})	
					Sprouts per Strain					
Fertilizing (%) [+]	1 [(1)]	2	1 [(2)]	2	1 [(3)]	2	1 [(4)]	2	1 [(5)]	2
0	0.59a	0.47a	0.57b	0.80a	25.30a	27.23a	1.00a	1.00a	1.50a	1.53a
50	0.40a	0.49a	0.63a	0.70a	25.00a	22.37a	1.00a	1.00a	3.43a	1.30a
100	0.61a	0.42a	0.80a	0.67a	27.33a	25.87a	1.00a	1.00a	5.57a	1.43b
200	0.83a	0.46b	0.87a	0.77a	32.03a	25.90a	1.33a	1.00a	6.90a	2.10b

[+] The percentage refers to 200 kg ha^{-1} of formula 06-30-06 + 1.0% Ca + 3.0% S + 1.0% Mg + 1.5% Cu + 1.0% Zn. The means followed by the same letters in the column are not significantly different in Tukey's test at $p < 0.05$. [(1)] Y = 0.4740 + 0.0015x (R^2 = 0.57 *); [(2)] Y = 0.5800 + 0.0016x (R^2 = 0.91 **); [(3)] Y = 24.2400 + 0.0363x (R^2 = 0.91 *); [(4)] Y = 0.9333 + 0.0017x (R^2 = 0.77 *); [(5)] Y = 2.0067 + 0.0268x (R^2 = 0.93 **).

3.2. Concentrations of N and P in the Leaves and Litter

Table 6 lists the results of the N and P concentrations in the leaves and the *Eucalyptus* litter, respectively. For the sprouts per strain, differences were only found in the N concentrations in the leaflet, and the highest content was obtained when there were two sprouts per strain (Table 6). This finding may be due to the higher concentration effect on the leaves with only one sprout per strain. It is important to keep in mind that N is part of the proteins and chlorophyll. However, the N and P concentrations of the *Eucalyptus* leaf at 44 months after choosing the sprouts were not influenced by the increase in fertilization (Table 6).

Table 6. The concentrations of N and P in the leaves and in the litter of the *Eucalyptus* at 44 months after choosing the sprouts, according to the number of sprouts per strain and the fertilization rates.

	N Leaves	P Leaves	N Litter	P Litter
	(g kg^{-1} de D.M.)			
Sprout per strain				
1	18.76a	1.93a	7.32b	0.44a
2	19.67a	1.98a	9.47a	0.51a
S.M.D. (5%)	0.93	0.13	0.67	0.09
Fertilizing (%) [+]				
0	19.45	1.90	8.16	0.44
50	18.58	2.00	8.00	0.44
100	19.70	1.94	8.80	0.52
200	19.12	1.98	8.62	0.50
F Test				
Sprout (S)	4.51 [ns]	0.79 [ns]	47.59 **	3.44 [ns]
Fertilizing (F)	1.27 [ns]	0.48 [ns]	1.47 [ns]	1.04 [ns]
S x F	1.52 [ns]	1.14 [ns]	0.62 [ns]	0.67 [ns]
C.V. (%)	5.51	7.63	9.11	20.77
Average overall	19.21	1.95	8.40	0.48

[+] The percentage refers to 200 kg ha^{-1} of formula 06-30-06 + 1.0% Ca + 3.0% S + 1.0% Mg + 1.5% Cu + 1.0% Zn. The means followed by the same letters in the column are not significantly different in Tukey's test at $p < 0.05$. ** Significant at $p < 0.01$. [ns] = not significant.

3.3. Leaf Chlorophyll Index, Plant Height, Diameter at Breast Height, Total Volume of Wood with Bark, and the Average Annual Increment

The results of the LCI, H, DBH, VW, and AAI of the *Eucalyptus* after 44 months according to the number of sprouts and fertilization rates are listed in Table 7. The LCI of the *Eucalyptus* was not influenced by the fertilization rate. Much like the effect of the number of sprouts per strain, there was a difference between the treatments and the strains, with the one sprout condition yielding higher LCI.

Table 7. The leaf chlorophyll index (LCI), plant height (H), diameter at breast height (DBH), total volume of wood with bark (VW), and the average annual increment (AAI) of *Eucalyptus* at 44 months after choosing the sprouts, according to the number of sprouts per strain and the fertilization rates.

	LCI	H (m)	DBH (cm)	VW (m^3 ha^{-1})	AAI (m^3 ha^{-1} ano^{-1})
Sprout per strain					
1	62.42a	25.28a	17.53a	411.07a	111.28a
2	55.44b	22.21b	13.26b	408.39a	113.32a
S.M.D. (5%)	4.46	0.58			17.47
Fertilizing (%) [+]					
0	61.07	23.95	14.66	370.52	110.20
50	58.80	23.50	15.66	414.50	112.94
100	56.39	23.58	15.66	429.60	110.45
200	59.47	23.94	15.60	424.30	115.61
F Test					
Sprout (S)	11.27 **	128.78 **	145.06 **	0.019 ns	0.06 ns
Fertilizing (F)	0.88 ns	0.75 ns	1.46 ns	1.88 ns	0.10 ns
S x F	2.53 ns	1.35 ns	1.70 ns	2.17 ns	1.30 ns
C.V. (%)	8.64	2.79	5.95	11.73	17.77
Average overall	58.93	23.74	15.39	409.73	112.3

[+] The percentage refers to 200 kg ha^{-1} of formula 06-30-06 + 1.0% Ca + 3.0% S + 1.0% Mg + 1.5% Cu + 1.0% Zn. The means followed by the same letters in the column are not significantly different in Tukey's test at $p < 0.05$. ** Significant at $p < 0.01$. ns = not significant.

There were larger H and DBH values for the one sprout per strain condition. Regarding the VW, there was no significant difference as a function of the number of sprouts per strain.

Surprisingly, the H, DBH, and VW, 44 months after the sprouts were chosen were not influenced by the amount of mineral fertilizer (Table 7), even in sandy soils with a low fertility content. However, it should be noted that there was an increase of approximately 14% in the VW when the fertilizer dosage (100%) used in the region was applied (200 kg ha^{-1} of formula 06-30-06 + 1.0% Ca + 3.0% S + 1.0% Mg + 1.5% Cu + 1.0% Zn) compared with the control without fertilization.

The AAI of *Eucalyptus* 44 months after the sprouts were chosen averaged 112.3 m^3 ha^{-1} year^{-1}. The AAI did not exhibit a significant difference for either the one or two sprouts per strain conditions. For the management of the two sprouts condition, the AAI was only about 2% larger compared to the one sprout per strain condition.

There was no adjustment for the increase in the dosage of the mineral fertilizer for AAI. However, for the 200% recommended fertilization dosage, we observed only a 5% increase compared to the control (without fertilizer).

3.4. Pearson Correlation

The correlations between the chemical attributes of the soil in the 0–0.20-m layer and the other biometric assessments or N and P concentrations in the *Eucalyptus* leaves and litter, as well as the correlation coefficients, are listed in Tables 8 and 9.

Negative correlations were observed between the number of sprouts per strain and the variables K (−0.53 **), Mn (−0.55 **), H (−0.89 **), DBH (−0.90 **), and LCI (−0.52 **). For fertilizer rates, a significant correlation was observed only with m% (0.41 *).

The P content in soil correlated significantly only with B (0.64 **), Cu (0.96 **), and Zn (0.55 **), and inversely with the pH (−0.40 *). The pH, in addition to the content of P, was inversely correlated with the levels of H + Al (−0.62 **), Al (−0.86 **), m% (−0.83 **), Fe (−0.52 **), and Zn (−0.47 *), and positively with OM (0.43 *) and V% (0.60 **). Meanwhile, the organic matter correlated significantly with Ca (0.55 **), Mg (0.48 *), SB (0.52 **), CEC (0.51 **), V% (0.56 **), and Mn (0.43 *) with relatively low coefficients and it correlated inversely with m% (−0.56 **).

The potassium content exhibited a significant correlation with the Fe content (0.42 *), the Mn content (0.51 *), and DBH (0.47 *). The Ca and Mg contents were significantly correlated with SB (0.99 ** and 0.99 **), CEC (0.92 ** and 0.97 **), V% (0.89 ** and 0.82 **), and inversely correlated with m% (−0.70 ** and −0.60 **), respectively.

The cation exchange capacity, as well as the V% and m%, did not correlate with any of the biometric evaluations. Fe was significantly correlated with B (0.46 *), Cu (0.55 **), and Zn (0.43 *) whilst B was correlated with Cu (0.60 **). However, the m% correlated negatively with the P concentration in *Eucalyptus* litter (−0.447 *).

The Mn content exhibited a significant correlation with H (0.61 **) and DBH (0.51 *). The biometric evaluations correlated significantly with one another: H with DBH (0.91 **) and LCI (0.65 **). In turn, the LCI exhibited a significant correlation with DBH (0.60 **) and VW with the AAI (1.00 **).

The N leaf concentration correlated positively with the N in the *Eucalyptus* litter (0.489 *) and with the concentration of P in the leaves (0.448 *). However, there was a negative correlation between the concentration of N in the leaves and the DBH (−0.439 *).

As for the N and P concentrations in the *Eucalyptus* litter, a positive correlation was observed between N and the number of sprouts per strain (0.771 **), and a negative correlation was noted between the concentration of N and the DBH (−0.652 **), H (−0.624 **), and LCI (−0.577 **). For the concentration of P, there was a positive correlation with the N in the *Eucalyptus* litter (0.632 **).

Table 8. The correlation coefficients of Pearson's (r) with significant values ($p < 0.01$ ** and $p < 0.05$ *) highlighted among the chemical attributes of soil, the biometric characteristics and the P and N concentrations in the leaves and in the litter of the *Eucalyptus* cultivated in the Brazilian Cerrado according to the number of sprouts per strain and mineral fertilization at a depth of 0.20 m.

	Sprout per Strain	Fertilization Rates	P resin	O.M.	pH	K	Ca	Mg	H + Al	Al	SB	S-SO4	CEC	V%	m%
Sprout per strain	1.000														
Fertilization rates	0.000	1.000													
P resin	-0.154	0.351	1.000												
O.M.	-0.178	-0.321	-0.153	1.000											
pH	-0.050	-0.265	-0.405 *	-0.070	1.000										
K	-0.535 **	0.011	-0.042	0.071	-0.070	1.000									
Ca	0.172	-0.364	-0.059	0.552 **	0.270	0.071	1.000								
Mg	0.210	-0.404	-0.142	0.480 **	0.175	0.071	0.957 **	1.000							
H + Al	-0.010	0.015	-0.236	0.028	-0.623 **	0.262	-0.121	0.025	1.000						
Al	0.011	0.227	0.324	0.520 **	0.218	0.085	-0.312	-0.215	0.544 **	1.000					
SB	0.188	-0.390	-0.106	-0.164	-0.120	0.116	0.987 **	0.991 **	-0.037	-0.259	1.000				
S-SO4	0.165	0.164	-0.136	0.512 *	0.033	0.356	-0.143	-0.112	0.242	0.158	-0.122	1.000			
CEC	0.179	-0.374	-0.171	0.566 **	0.600 **	0.158	0.921 **	0.967 **	0.251	-0.095	0.958 **	-0.049	1.000		
V%	0.117	-0.402	-0.109	0.049	-0.827 **	0.087	0.894 **	0.822 **	-0.443 **	-0.547 **	0.863 **	-0.152	0.709 **	1.000	
m%	0.002	0.414 *	0.258	0.042	-0.604 **	-0.067	-0.696 **	-0.603 **	0.542 **	0.823 **	-0.651 **	0.210	-0.475 *	-0.902 **	1.000
B	-0.142	0.197	0.637 **	0.430 *	-0.525 **	-0.097	-0.086	-0.114	0.278	0.466 *	-0.104	-0.216	-0.020	-0.320	0.458 *
Cu	-0.185	0.386	0.961 **	-0.320	0.308	-0.030	-0.081	-0.170	-0.312	0.343	-0.132	-0.176	-0.218	-0.085	0.246
Fe	-0.169	-0.013	0.597 **	0.279	-0.468 *	0.423 *	0.176	0.172	0.170	0.364	0.181	0.064	0.224	0.030	0.143
Mn	-0.551 **	0.063	-0.238	0.181	0.037	0.515 **	0.157	0.154	0.147	-0.183	0.164	-0.003	0.201	0.257	-0.317
Zn	0.102	0.366	0.555 **	0.090	0.020	-0.054	-0.197	-0.170	0.305	0.294	-0.184	0.151	-0.091	-0.359	0.422 *
H	-0.891 **	0.021	0.182	0.315	-0.013	0.387	-0.043	-0.070	-0.043	-0.002	-0.053	-0.153	-0.064	0.023	-0.086
DBH	-0.904 **	0.057	0.220	0.090	0.020	0.475 *	-0.187	-0.250	-0.002	0.028	-0.218	-0.140	-0.211	-0.121	0.002
VW	0.058	0.097	0.140	0.006	-0.165	-0.100	-0.020	-0.056	-0.043	0.098	-0.042	0.022	-0.053	0.033	-0.022
LCI	-0.525 **	-0.078	0.102	-0.100	0.088	-0.100	0.061	0.090	0.110	-0.128	0.079	-0.269	0.108	-0.025	-0.074
AAI	0.058	0.097	0.140	0.090	-0.136	0.084	-0.020	-0.056	-0.043	0.097	-0.042	0.022	-0.053	0.033	-0.022
N Leaves	0.401	-0.005	-0.167	0.090	-0.313	-0.256	0.271	0.319	0.065	0.142	0.297	-0.007	0.307	0.162	0.008
P Leaves	0.194	0.134	-0.303	-0.100		0.052	0.254	0.236	0.004	-0.216	0.247	-0.014	0.240	0.189	-0.243
N Litter	0.771 **	0.161	-0.086	-0.378		-0.386	-0.137	-0.110	-0.077	0.253	-0.128	0.257	-0.146	-0.108	0.213
P Litter	0.377	0.266	-0.127	-0.360		-0.359	-0.277	-0.258	0.212	0.392	-0.274	0.108	-0.205	-0.368	0.447 *

Forests **2018**, *9*, 290

Table 9. The correlation coefficients of Pearson's (r) with significant values ($p < 0.01$ ** and $p < 0.05$ *) highlighted among the chemical attributes of soil, the biometric characteristics and the P and N concentrations in the leaves and in the litter of the *Eucalyptus* cultivated in the Brazilian Cerrado according to the number of sprouts per strain and mineral fertilization at a depth of 0.20 m.

	B	Cu	Fe	Mn	Zn	H	DBH	VW	LCI	AAI	N Leaves	P Leaves	N Leaflet	P Leaflet
B	1.000													
Cu	0.599 **	1.000												
Fe	0.456 *	0.551 **	1.000											
Mn	−0.241	−0.146	−0.106	1.000										
Zn	0.404	0.368	0.428 *	−0.303	1.000									
H	0.136	0.245	0.220	0.610 **	−0.124	1.000								
DBH	0.213	0.239	0.218	0.515 *	−0.006	0.910 **	1.000							
VW	0.121	0.144	0.118	0.077	0.143	0.271	0.359	1.000						
LCI	0.324	0.093	0.233	0.222	0.024	0.647 **	0.598 **	0.314	1.000					
AAI	0.121	0.144	0.118	0.077	0.143	0.271	0.359	1.000 **	0.314	1.000				
N Leaves	−0.112	−0.088	−0.158	−0.044	−0.360	−0.251	−0.439 *	−0.065	−0.290	−0.065	1.000			
P Leaves	−0.190	−0.313	−0.217	0.049	−0.240	−0.211	−0.132	0.012	−0.115	0.011	0.448 *	1.000		
N Litter	−0.199	−0.021	−0.137	−0.363	−0.045	−0.624 **	−0.652 **	0.199	−0.577 **	0.199	0.489 *	0.106	1.000	
P Litter	0.065	−0.066	−0.190	−0.195	0.038	−0.223	−0.208	0.372	−0.103	0.372	0.295	−0.026	0.632 **	1.000

4. Discussion

4.1. Soil Chemical Analysis

We observed a linear increase in contents of P and Cu in the soil and an increase in the m% with the increase in the fertilizer rates, which can be explained by the uptake of the basic cationic nutrients and the relative acidification, which can be attributed to the mineralization of the deposited OM in the forest area alone. Dick et al. [19], in studies of five-year-old *Eucalyptus dunnii* Maiden, planted seedlings in an area of low natural fertility in the Pampas Gauchos of Brazil and reported a similar situation. These authors verified an increase in the content of P at the end of the *Eucalyptus* production cycle. According to Bazanii et al. [20], the average contents of P $_{resin}$ in the soil for *Eucalyptus* varies from 5 to 7 mg dm^{-3}. In our study, in the superficial layer (0–0.20 m), the average levels of P $_{resin}$ were roughly 7.2 mg dm^{-3}, which is within the average range.

The pH values of the soil under the *Eucalyptus* remained acidic and did not vary as a function of the number of sprouts per strain or the fertilization rate. Santana et al. [21] evaluated the effect of acacia and *Eucalyptus* forest sites on the chemical properties of soil and concluded that the pH values of soil are influenced more by liming and fertilization practices than by forest vegetation in *Eucalyptus* plantations where no liming occurred, in which significant changes in the pH content and the Ca^{2+} and Mg^{2+} levels were observed.

We verified that there were higher contents of K and Mn, when we selected the one sprout per strain condition. This finding indicates that the nutrient uptake of the soil is differentiated according to the number of shoots per *Eucalyptus* strain. We observed that an increase in the fertilizer dosage decreased the Mg content, the SB, CEC, and the V%, indicating that there was a higher uptake of Mg. This situation likely corresponded to maintaining the nutritional balance compared with the larger uptakes of N and K, which were supplied by the mineral fertilizer.

In terms of the relation between the fertilization dosage and the number of sprouts per strain for the B content in the soil at a depth of 0–0.20 m (Table 2), we noted an increasing linear function for the one sprout condition. Therefore, we again observed a lower uptake of this micronutrient in the management of the one sprout condition compared to the two sprouts per *Eucalyptus* strain condition since there was no adjustment for such an evaluation in the latter treatment. The application of B into soil is essential for the development of the *Eucalyptus* species, especially when these plants are managed in sandy soils in the second production cycle. The lack of this micronutrient can lead to a decrease in the growth rate of the height and the DBH [5] and, consequently, reduce the VW produced.

The results presented in Table 3, in relation to the contents of the MO, K, S-SO$_4$, B, and Mn at a soil depth of 0.20–0.40 m, confirm that the nutrient absorption differed for plants with different numbers of sprouts. This finding highlights the importance of the subsurface layer for *Eucalyptus* nutrition which presents a deep and vigorous root system. According to Kolm [22], roughly 60% of *Eucalyptus* roots are found in the 0–0.30 m layer. However, it is in the first 0.10 m that the largest number of fine roots, responsible for the absorption of the cycled nutrients from litter deposition and mineralization, are found.

The contents of P, S-SO$_4$, B, Cu, Mn, Zn, H + Al, and CEC at a depth of 0.20–0.40 m and the contents of P, Cu, and Mn at a depth of 0.40–1.00 m increased linearly in the soil with the increasing fertilizer dose. We inferred that the sandy texture of this soil enabled the leaching of these elements to the deeper layers. The increase in the potential acidity is also an indication of the increase in the uptake of cationic nutrients. In soils with forest litter, the dynamics of P can be effected via the leaching of this nutrient to layers below the surface layer. The high rates of the decomposition of litter lead to the production of organic acids with a low molecular weight, which is subject to translocation to the lower layers and can lead to the removal of the P zone via increased uptake by roots [23].

The content of the OM content decreases with the depth in all soils; the lowest OM content of the soil was found in the layer at 0.40–1.00 m. This finding is to be expected, demonstrating the importance of the superficial layer for nutrient cycling. Menegale et al. [24], evaluating the effect of different types of wood harvesting and fertilizer application in a settlement of *Eucalyptus grandis* Hill Ex Maiden, reported that maintaining the forest harvesting residues in the soil contributes to the increasing C and N contents in the soil and is important for the supply of nutrients to the plants. Among the treatments evaluated by the authors, this process is more important in sandy soils, highly productive forests, and successive harvest cycles. In the second production cycle, the treatment in which the forest remains were maintained and there was fertilization and liming, there were higher levels of plant biomass as well.

At a depth of 0.40–1.00 m (Table 5), the B content in the soil at the highest fertilizer dosage was significantly higher for the one sprout per strain condition. The result was similar for the Mn content in the soil at this depth. The highest soil level was obtained in the course of the one sprout per strain condition. These results indicate that the higher absorption of these micronutrients occurred when the two sprouts per strain condition was selected.

The correlations between all the variables related to the chemical attributes of the soil in the 0–0.20 m layer and the biometric variables were evaluated (Tables 8 and 9). In general, the strong positive correlations between the chemical attributes of the soil were observed.

The pH of the soil is negatively correlated with the contents of P, H + Al, Al, m%, Fe, and Zn and positively correlated with the OM and V%. Brunello [25], studying forest in an Oxisol located in the Amazonas mesoregion, found results similar to those of this investigation: the Al exchangeable soil content was correlated negatively with pH of the soil (-0.617 *), and there was a positive correlation between the pH of the soil and the exchangeable bases contents. This researcher verified that an increase in the pH of the soil when there was an incorporation of the bases in the soil increased the nutrients available to the plants and reduced the toxicity of the elements such as Al and Mn.

The CEC, as well as the V% and m%, did not correlate with any of the biometric evaluations of the *Eucalyptus*. This result can be potentially explained because these evaluations did not change with the same magnitude as the changes in the values of CEC, V%, and m%. That is, they did not present significant correlations. Nutrient cycling in deeper soil layers and the influence of the organic and inorganic reserves in the root system may have influenced the initial growth of the sprouts.

4.2. Concentrations of N and P in the Leaves and in the Litter

According to the appropriate leaf concentrations for the majority of planted *Eucalyptus* species in Brazil cited by Dell et al. [26], the N concentrations were slightly below the range considered to be adequate (18–30 g kg^{-1}); the P concentrations were slightly above the suitable range (1.0–1.3 g kg^{-1}), even for acidic soil with a low P content, which is commonly found in the Brazilian Cerrado. This may be due to the supply of P fertilizer, but what drew our attention was that even in plants that did not receive phosphate fertilization in the second production cycle of the *Eucalyptus*, the contents of this nutrient were above that which was considered adequate, which reinforces the hypothesis of the use of root reserves.

However, the leaf concentrations of N and P in the *Eucalyptus*, 44 months after choosing the sprouts, were not influenced by the fertilizer dose (Table 6). This finding is probably due to the already-established root system of the first reproduction cycle of the *Eucalyptus* and, mainly, because of the nutrient cycling contributing to the supply of these nutrients to the plants.

The leaf concentrations of the N and P in the *Eucalyptus*, 44 months after the sprouts were chosen were not influenced by the fertilizer dose (Table 6), which confirms that the already-established root system of the first *Eucalyptus* cycle and, to a large degree, the nutrient cycling, contributed to the supply of the plant's nutrients.

The concentrations of P and N in both the leaves and the litter exhibited a positive correlation with one another. Santos et al. [27] explained that there was a stoichiometric relationship between the N and the P in leaves and that a high N content in the soil implied a higher demand for P by the *Eucalyptus*. This explanation confirms a synergistic effect among these nutrients for *Eucalyptus* forests.

4.3. Leaf Chlorophyll Index, Plant Height, Diameter at Breast Height, Total Volume of Wood with Bark, and the Average Annual Increment

The LCI was not influenced by the increase in the fertilization, indicating that N was not in short supply for chlorophyll formation and that the importance of the root system was already established for the second cycle of *Eucalyptus* growth. The decreased LCI for the two sprouts per strain can be explained by the dilution effect of N because the accumulation of dry matter in the two sprouts condition is higher than of the one sprout per strain condition. However, the LCI was positively correlated with H and DBH, indicating a larger light uptake when the taller *Eucalyptus* trees were planted and, consequently, s larger stem diameter growth. Reis and Reis [12] found that each sprout left in the *Eucalyptus* strains behaved like an isolated plant and that there was greater pressure on the resources of the environment with an increased number of sprouts per strain, which competed with one another for growth resources.

The *Eucalyptus* with one sprout per strain exhibited higher H and DBH compared with the *Eucalyptus* with two sprouts per strain (Table 7). This finding indicates that the two sprouts per strain condition divides the consumption of nutrients, water, and light, increasing the competition of the sprouts in the same strain and resulting in a reduced plant height. There was a strong negative correlation between the number of sprouts per strain, the H, and the number of sprouts and DBH, confirming that the H and DBH of the sprouts in the same strain decreased as the number of sprouts per strain increased. According to Couto et al. [28], the growth in the stem diameter is influenced by the area available for the shoots. Therefore, the conduction of the strains with two sprouts results in smaller diameters. Depending on the intended purpose of the growth, it may be better to leave only one sprout per strain.

The K and Mn contents in the soil and the H, DBH, and LCI were negatively influenced by the number of sprouts per strain; the larger the number of sprouts per strain, the lower the K and Mn contents in the soil and the lower the plant height, DBH, and LCI. These findings can be explained by the significant requirement of these cationic nutrients associated with the higher uptake by *Eucalyptus*. A moderate Mg deficiency was observed in *Eucalyptus*, which explains this reduction in the LCI and plant growth. Gazola et al. [29], under the same soil and climatic conditions, found that increasing the K dosage resulted in an increase in the K concentrations in the leaves and a decrease in the Ca and Mg concentrations in the *Eucalyptus*.

The VW did not exhibit a significant difference in terms of the number of sprouts per strain. This result is explained by the total sum of the two sprouts, despite them being smaller in the H and DBH. Rezende et al. [30] found that larger numbers of sprouts were correlated with the increased growth in height and VW after both the first and the second cut. On the other hand, Simões and Coto [31], evaluating the effect of the sprout number and the mineral fertilization on the growth of *Eucalyptus saligna* sprouts in the second cycle, verified that the final volume of the cut and stacked wood positively correlated with the number of sprouts and the amount of mineral fertilization.

The H, DBH, and VW were not influenced by the mineral fertilization; the primary explanation for these results may be the nutrient cycling from the litter coming from both the first *Eucalyptus* production cycle and from the second production cycle. In addition, another explanation may be the consumption of the root reserves for the growth of the *Eucalyptus* sprouts. However, Silva [32], studying a five-year-old *Eucalyptus* forest managed by the coppice system with one or two sprouts per strain, verified that fertilization interferes with these variables when the soil and the residues from the previous cycle are not enough to fulfill the plant's needs. This author cited that there was a positive effect of fertilization on the DBH, H, and VW in this situation, finding that the trees that were fertilized exhibited a diameter 22.5% larger than the trees that received treatments without fertilization. Adequate fertilization carried out in the first cycle of *Eucalyptus* in the present research yielded a litter and roots with higher concentrations of nutrients, culminating in an increased nutrient cycling for *Eucalyptus* trees in the second cycle.

Under the same soil and climatic conditions, in the first cycle of *Eucalyptus*, Gazola et al. [29] verified that the maximum yield of *Eucalyptus* was obtained at 21 months of age with the application of 71 kg ha^{-1} of N, 100 kg ha^{-1} of P$_2$O$_5$, and 125 kg ha^{-1} of K$_2$O. Celestrino [33] observed positive results for plant height, DBH, and VW with a dose of 1 kg ha^{-1} of B applied to the *Eucalyptus*.

The AAI was not influenced by the number of sprouts per strain or the mineral fertilizer dose. However, for 200% of the recommended fertilization dose, there was an increase of 5% in the AAI compared with the control (without fertilizer). It is worth noting the high values of the AAI obtained for the cultivation of *Eucalyptus* even in the sandy textured soil with low fertility (on average 112.3 m^3 ha^{-1} year^{-1}). Melo et al. (2016) obtained 48-month-old AAI values of 54 m^3 ha^{-1} year^{-1} in the first cycle of the *Eucalyptus* forest; Faria et al. [34] also verified an AAI of 64.3 m^3 ha^{-1} year^{-1} in the first cycle of *Eucalyptus* production (clone I-144) at 57 months of age.

Significant correlations between the chemical attributes of the soil and the biometric variables were rare (Tables 8 and 9), which reinforced the hypothesis of the use of root reserves. Therefore, nourishing the *Eucalyptus* well in the first production cycle can be a more sustainable way to obtain satisfactory wood productivities in the second production cycle.

5. Conclusions

The chemical attributes of the soil at the evaluated depths varied according to the number of sprouts per strain. Forty-four months after choosing the sprouts, in sandy soils with low fertility, there was a larger difference between the numbers of sprouts per strain in the 0.20–0.40 m layer compared with the others soil layers. There were also higher contents of OM, K, S, B, and Mn when one sprout per strain was selected.

The mineral fertilization affects the chemical attributes of the soils at depths of 0–1.00 m and largely provides higher P and B contents, which may increase the potential acidity.

The one *Eucalyptus* sprout condition yielded higher LCI, DBH, and H, being more interesting for the commercialization of the *Eucalyptus* stem; with the two sprouts per strain condition, the N leaflet concentration was higher. However, the N and P leaf concentrations, as well as the total VW with bark, were similar for both the one or two sprouts per strain conditions, regardless of the mineral fertilizer dose.

The LCI, H, DBH, and total VW with the bark of the *Eucalyptus* were not influenced by an increase in the fertilizer dose, even in sandy soil with low fertility. The AAI of the *Eucalyptus* forest was high and similar for both the one or two sprouts per strain conditions. It was not influenced by increases in the mineral fertilization.

The adequate fertilization during the first cycle of *Eucalyptus* growth can supply almost the entire nutritional demand of the forest during the second production cycle. Comparing the control (without fertilization) with the dose of fertilizer commonly used in the region, approximately 86% of the total VW was obtained as a function of the supply of soil and the nutrient reserves of roots, and 14% of this productivity was due to the fertilization minerals.

Author Contributions: M.C.M.T.F. and S.B. conceived and designed the study; N.M.I.G., R.d.N.G., A.C.d.S. and T.d.S.C. prepared the data sets; M.C.M.T.F., N.M.I.G. and S.N.d.S.A. performed the statistical analyses; M.C.M.T.F. and N.M.I.G. interpreted the results; M.C.M.T.F., S.N.d.S.A. and N.M.I.G. wrote the paper with contributions from T.A.R.N. and S.B.

Funding: This research was funded by the São Paulo Research Foundation (FAPESP) grant number (2016/11613-1) and the Foundation for the Development of São Paulo State University (UNESP) grant number (0271/001/14).

Acknowledgments: Authors would like to thank the Cargill Agricola S/A, the São Paulo State University (UNESP) and the São Paulo Research Foundation (FAPESP) for supporting our research.

Conflicts of Interest: The authors declare no conflict of interest.

References

1. IBÁ. INDÚSTRIA BRASILEIRA DE ÁRVORES. Report IBÁ-2017. Indicators of Performance of the National Sector of Planted Trees for the Year 2016. Available online: http://iba.org/images/shared/Biblioteca/IBA_RelatorioAnual2017.pdf (accessed on 8 September 2017).
2. Foelkel, C. Minerals and Nutrients of Eucalyptus Trees: Environmental, Physiological, Silvicultural and Industrial Aspects about the Inorganic Elements Present in the Trees. In *Eucalyptus Online Book Newsletter*, 2nd ed.; 2005. Available online: http://atividaderural.com.br/artigos/538773282706a.pdf (accessed on 22 April 2016).
3. Faria, G.E.; Barros, N.F.; Novais, R.F.; Lima, J.C.; Teixeira, J.L. Production and nutritional status of *Eucalyptus grandis* stands in second rotation in response to potassium fertilization. *Rev. Árvore* **2002**, *26*, 577–584. [CrossRef]
4. Rocha, J.H.T.; Goncalves, J.L.M.; Gava, J.L.; Godinho, T.O.; Melo, E.S.A.C.; Bazani, J.H.; Hubner, A.; Arthur Junior, J.C.; Wichert, M.P. Forest residue maintenance increased the wood productivity of a Eucalyptus plantation over two short rotations. *For. Ecol. Manag.* **2016**, *379*, 1–10. [CrossRef]
5. Silveira, R.L.V.A.; Takahashi, E.N.; Sgarbi, F.; Camargo, M.A.F.; Moreira, A. Growth and nutritional status of *Eucalyptus citriodora* sprouts under boron rates in nutrient solution. *Sci. For.* **2000**, *57*, 53–67.
6. Silveira, R.L.V.A.; Gava, J.L. Nutrition and phosphate fertilization in Eucalyptus. In Proceedings of the Lecture presented at the Symposium on Phosphorus in Brazilian Agriculture, 2003, POTAFOS realization, São Pedro, State of São Paulo, Brazil, 14–16 May 2003.
7. Sgarbi, F. *Eucalyptus* sp. Productivity as a Function of Nutritional Status and Soil Fertility in Different Regions of the São Paulo State. Master's Thesis, Superior School of Agriculture "Luiz de Queiroz", University of São Paulo, Piracicaba, Brazil, 2002.
8. Costa, M.G.; Gama-Rodrigues, A.C.; Gonçalves, J.L.M.; Gama-Rodrigues, E.F.; Sales, M.V.S.; Aleixo, S. Labile and Non-Labile Fractions of Phosphorus and Its Transformations in Soil under Eucalyptus Plantations, Brazil. *Forest* **2016**, *7*, 15. [CrossRef]
9. Melo, E.A.S.C.; Gonçalves, J.L.M.; Rocha, J.H.T.; Hakamada, E.R.; Bazani, J.S.; Wenzel, A.V.A.; Arthur Junior, J.C.; Borges, J.S.; Malheiros, R.; Lemos, C.C.Z.; et al. Responses of Clonal Eucalypt Plantations to N, P and K Fertilizer Application in Different Edaphoclimatic Conditions. *Forest* **2016**, *7*, 2. [CrossRef]
10. Barros, N.F.; Novais, R.F. Eucalypt nutrition and fertilizer regimes in Brazil. In *Nutrition of the Eucalypts*; Attiwill, P.M., Adams, M.A., Eds.; CSIRO Publishing: Clayton, Australia, 1996; pp. 335–356.
11. Guedes, M.C.; Poggiani, F. Variation of leaf nutrient contents in eucalyptus fertilized with biosolids. *Sci. For.* **2003**, *63*, 188–201.
12. Reis, G.G.; Reis, M.G.F. Physiology of eucalyptus sprouting with emphasis on its water relations. *Ser. Tecnica IPEF* **1997**, *11*, 9–22.
13. Cacau, F.V.; Reis, G.G.; Reis, M.G.F.; Leite, H.G.; Alves, F.F.; Souza, F.C. Deceptive of young eucalyptus plants and management of sprouts in an agroforestry system. *Pesq. Agropecu. Bras.* **2018**, *43*, 1457–1465. [CrossRef]
14. Embrapa; Empresa Brasileira de Pesquisa Agropecuária—EMBRAPA. *Centro Nacional de Pesquisa de Solos. Brazilian System of Soil Classification*, 3rd ed.; EMBRAPA: Brasília, Brazil, 2013; 353p.
15. Van Raij, B.; Andrade, J.C.; Cantarella, H.; Quaggio, J.A. *Chemical Analysis for Fertility Evaluation of Tropical Soils*; IAC: Campinas, Brazil, 2001; 285p.
16. "Koppen Climate Classification Climatology". Encyclopedia Britannica. Available online: https://howlingpixel.com/wiki/K%C3%B6ppen_climate_classification (accessed on 4 August 2017).
17. Malavolta, E.; Vitti, G.C.; Oliveira, S.A. *Evaluation of the Nutritional Status of Plants: Principles and Applications*, 2nd ed.; Brazilian Association for the Research of Potash and Phosphate: Piracicaba, Brazil, 1997; 319p.
18. Campos, J.C.C.; Leite, H.G. *Forest Measurement Questions and Answers*; UFV: Viçosa, Brazil, 2002; 407p.
19. Dick, G.; Schumacher, M.V.; Momolli, D.R. Characterization of soil fertility in a settlement of *Eucalyptus dunnii* Maiden in the Pampa biome of Rio Grande do Sul. *Ecol. Nutrição Florest.* **2016**, *4*, 68–77. [CrossRef]
20. Bazanii, J.H.; Gonçalves, J.L.M.; Rocha, J.H.T.; Melo, E.S.A.C.; Prieto, M. Phosphate nutrition in *Eucalyptus* plantation. *Int. Plant Nutr. Inst.* **2014**, 148.
21. Santana, G.S.; Knicker, H.; González-Vila, F.J.; González-Pérez, J.A.; Dick, D.P. The impact of exotic forest plantations on the chemical composition of soil organic matter in Southern Brazil as assessed by Py–GC/MS and lipid extracts study. *Geoderma Reg.* **2015**, *4*, 11–19. [CrossRef]

22. Kolm, L. Nutrient Cycling and Microclimate Variations in Eucalyptus grandis Hill ex Maiden Plantations Managed through Progressive Thinning. Master's Thesis, Superior School of Agriculture "Luiz de Queiroz", University of São Paulo, Piracicaba, Brazil, 2001.

23. Chaer, G.M.; Tótola, M.R. Impact of organic waste management during the reform of eucalyptus plantations on soil quality indicators. *Rev. Bras. de Ciência do Solo* **2007**, *31*, 1381–1396. [CrossRef]

24. Menegale, M.L.C.; Rocha, J.H.T.; Harrison, R.; Goncalves, J.L.M.; Almeida, R.F.; Piccolo, M.C.; Hubner, A.; Arthur Junior, J.C.; Ferraz, A.V.; James, J.N.; et al. Effect of Timber Harvest Intensities and Fertilizer Application on Stocks of Soil C, N, P, and S. *Forests* **2016**, *7*, 319. [CrossRef]

25. Brunello, A.T. Availability and efficiency in the use of nutrients in secondary forests under different levels of change in the region of Santarém and Belterra, eastern Amazonia. Master's Thesis, Intituto Nacional de Pesquisa da Amazônia (INPA), Manaus, Brazil, 2016.

26. Dell, B.; Malajczuk, D.; Xu, D.; Grove, T.S. *Nutrient Disorders in Plantation Eucalypts*; ACIAR: Camberra, Australia, 2001; 188p.

27. Santos, F.M.; Chaer, G.M.; Diniz, A.R.; Balieiro, F.C. Nutrient cycling over five years of mixed-species plantations of Eucalyptus and Acacia on a sandy tropical soil. *For. Ecol. Manag.* **2017**, *384*, 110–121. [CrossRef]

28. Couto, H.T.Z.; Mello, H.A.; Simões, J.W.; Vencovsky, R. *Eucalyptus saligna* Smith sprout management. *IPEF* **1973**, *7*, 115–123. Available online: http://ipef.br/publicacoes/scientia/nr07/cap05.pdf (accessed on 4 February 2017).

29. Gazola, R.N.; Buzetti, S.; Teixeira Filho, M.C.M.; Dinalli, R.; Moraes, M.L.T.; Celestrino, T.S.; Silva, P.H.M.; Dupas, E. Rates of N, P and K in the cultivation of eucalyptus in soil originally under Cerrado vegetation. *Semina Ciências Agrárias* **2015**, *36*, 1895–1912. [CrossRef]

30. Rezende, G.C.; Suiter Filho, W.S.; Mendes, C.J. Regeneration of the forest masses of the Agricultural and Forestry Company Santa Bárbara. *Boletim Técnico SIF* **1980**, *1*, 1–24.

31. Simões, J.W.; Coto, N.A.S. Effect of sprout number and mineral fertilization on growth of *Eucalyptus saligna* Smith on second rotation. *IPEF* **1985**, *31*, 23–32.

32. Silva, N.F. Productivity, Demand and Nutritional Efficiency of Eucalyptus Clones in High Stem and Coppice Regime. Master's Thesis, Federal University of Viçosa, Viçosa, Brazil, 2013.

33. Celestrino, T.S.; Buzetti, S.; Teixeira Filho, M.C.M.; Gazola, R.N.; Dinalli, R.P.; Silva, P.H.M.; Carvalho, A.C.; Sarto, G.D. Sources and application methods of boron in Eucalyptus crop. *Semina Ciênc. Agrár.* **2015**, *36*, 3579–3594. [CrossRef]

34. Faria, G.E.; Barros, N.F.; Cunha, V.L.P.; Martins, I.S.; Martins, R.C.C. Evaluation of productivity, content and efficiency of nutrient utilization in genotypes of *Eucalyptus* spp. in the Jequitinhonha valley, MG. *Ciênc. Florest.* **2008**, *18*, 363–373. [CrossRef]

MDPI

St. Alban-Anlage 66

4052 Basel

Switzerland

Tel. +41 61 683 77 34

Fax +41 61 302 89 18

www.mdpi.com

Forests Editorial Office

E-mail: forests@mdpi.com

www.mdpi.com/journal/forests

www.ingramcontent.com/pod-product-compliance
Lightning Source LLC
Chambersburg PA
CBHW051730210326
41597CB00032B/5674